THE UNIVERSAL FRAME

THE UNIVERSAL FRAME

HISTORICAL ESSAYS IN ASTRONOMY
NATURAL PHILOSOPHY AND
SCIENTIFIC METHOD

J.D. NORTH

THE HAMBLEDON PRESS
LONDON AND RONCEVERTE

Published by The Hambledon Press, 1989

102 Gloucester Avenue, London NW1 8HX (U.K.)

309 Greenbrier Avenue, Ronceverte WV 24970 (U.S.A.)

ISBN 0 907628 95 8

British Library Cataloguing in Publication Data

The Universal frame: historical essays in astronomy,
    natural philosophy and scientific method.
    1. Science – History
    509   Q125

Library of Congress Cataloging-in-Publication Data

North, John David.
    The universal frame: historical essays in astronomy,
    natural philosophy, and scientific method/J.D. North.
    Includes bibliographical references and index.
    1. Astronomy – History – 17th century. I. Title.
    QB29. N67 1989
    520'. 9'032 – dc19   88-38307 CIP

Printed and bound in Great Britain by W.B.C. Ltd,
Bristol and Maesteg.

CONTENTS

ACKNOWLEDGEMENTS

The articles in this volume first appeared in the following places. They are reprinted by kind permission of the original publishers.

1   *Mathematics and its Applications. Studies Presented to Marshall Claggett*, ed. by E. Grant and J. Murdoch (Cambridge, 1987), pp. 173-88.

2   *Studia Copernicana*, ed. by P. Czartoryski *et al.* (Ossolineum, Orbis, 1975), pp. 169-84.

3   *Times Literary Supplement* (1985), p. 1266.

4   *Gregorian Reform of the Calendar. Proceedings of the Vatican Conference to Commemorate its 400th Anniversary, 1582-1982*, ed. by G.V. Coyne, M.A. Hoskin and O. Pedersen (Vatican City, Pontificia Academia Scientiarum and Specola Vaticana, 1983), pp. 75-113.

5   *The Light of Nature*, ed. by J.D. North and J.J. Roche (Dordrecht, Martinus Nijhoff, 1985), pp. 145-74.

6   *Thomas Harriot, Renaissance Scientist*, ed. by J.W. Shirley (Oxford, Clarendon Press, 1974), pp. 129-65.

7   *Physis*, 11 (1969), pp. 418-57.

8   *Old and New Questions in Physics, Cosmology, Philosophy and Theoretical Physics*, ed. by A. van der Merwe (New York, Plenum, 1983), pp. 689-717.

9   *Johann Maurits van Nassau-Siegen, 1604-1679. A Humanist Prince in Europe and Brazil*, ed. E. van den Boogaart *et al.* (The Hague, The Johan Maurits van Nassau Stichting, 1979), pp. 394-423.

10   *Nature Mathematized*, ed. by W. Shea (Dordrecht,
      Reidel, 1983), pp. 113-48. Copyright © 1983 by
      D. Reidel Publishing Company, Dordrecht, Holland.

11   *Times Literary Supplement* (1976), p. 173.

12   *Times Literary Supplement* (1986), pp. 131-2.

13   *On Scientific Discovery*, ed. by M.D. Grmek *et al.* (Dordrecht,
      Reidel, 1980), pp. 115-40. Copyright © 1980 by D. Reidel
      Publishing Company, Dordrecht, Holland.

14   *Times Literary Supplement* (1972), pp. 1301-2.

15   *The Study of Time*, ed. by J.T. Fraser *et al.* (New York,
      Heidelberg, Berlin, Springer Verlag, 1972), pp. 1-32.

16   *Times Literary Supplement* (1975), p. 6.

17   *Probabilistic Thinking, Thermodynamics, and the Interaction
      of the History and Philosophy of Science*, ed. by J. Hintikka
      et al. (Dordrecht, Reidel, 1980), pp. 271-81. Copyright
      © 1980 by D. Reidel Publishing Company, Dordrecht,
      Holland.

18   *Archives Internationales d'Histoire des Sciences*, 117 (1986),
      pp. 362-73.

PREFACE

The writings in this collection of essays concern themes in the history and methodology of the exact sciences from a period stretching from the sixteenth century to the twentieth.* In the early chapters there is much retrospection to earlier times: coordinate geometry had medieval as well as ancient antecedents; astronomy after Copernicus was conservative to a fault; the Gregorian calendar was an answer to Grosseteste's prayers; Aristotle and his medieval commentators, if they did not utter the last word on infinity, were the minds behind most seventeenth-century ideas on the subject; and so on. The mood of retrospection does not, I hope, prevail, for I would hate to be thought afraid to face the charge of having Whiggish tendencies. Copernicus was one of those on whose shoulders Newton stood: the Gregorian reform was a triumph of medieval rationality, not to say of European bureaucracy; and a renewal of concern with the logic of the infinite and the continuum was of prime importance to the future of deep mathematical understanding. While not the most single-minded of Whig historians, I confess willingly to being a fellow-traveller.

If there was an air of excitement surrounding cosmology and astronomy in the seventeenth century, then it certainly had other than medieval sources. Much of it came in the wake of a relatively simple invention. The Dutch telescope – an instrument that would typically have fetched only a fraction of the price of the splendid equatorium discussed in chapter 7 – brought about radical changes in seventeenth-century mentality. To a man like Harriot, it offered new worlds to be investigated by proven astronomical techniques. To Galileo it promised new cosmological truths. To Markgraf it was just one of many instruments that would allow him to establish a new astronomical dominion in Brazil. To Rømer and Bradley it brought new discoveries, but now with a bearing on fundamental physics, and not on astronomy alone; and this was a sign of things to come. These are among the themes treated in chapters 6, 8, and 9.

*A collection covering an earlier period has been published as *Stars, Minds and Fate* (The Hambledon Press, 1989).

Most of the remaining chapters relate to the physical sciences, and to associated problems of historiography, from the editing of texts to the question of whether we should be concerned with – or disregard – the scientific quality of the arguments we chronicle. There are short pieces on Newton, the cultural ambit of Halley's comet, styles in scientific survival (Faraday, Huxley, and Lockyer), and the growth of laboratory science in Cambridge and Oxford. Chapter 13 is an analysis of several historical forms of analogical argument in science (from Newton, Thomson, and Maxwell in particular, and underscoring the naivety of J.S. Mill's account) and here, as again in chapter 15 on some historical aspects of the notion of time-dilation, I have tried to focus on the problematic dividing line between conjecture and proof. Chapter 17, like the first in the collection, sketches ways in which historians of science may impose non-historical categories on their material. More precisely, it is a portrayal of a type of historian who, sharing a strong incentive to explain, does so only at a certain cost. There is an eternal triangle with historians, philosophers, and their subject-matter as its vertices. It behoves us to reduce as far as possible the separation of the first pair, but as for the third vertex, we must face up to the fact that, far from being invariant, it is as inconstant as the fashions pursued by the other two.

LIST OF ILLUSTRATIONS

1

COORDINATES AND CATEGORIES:
THE GRAPHICAL REPRESENTATION OF FUNCTIONS
IN MEDIEVAL ASTRONOMY

"Dixit Ptolomeus . . ." There is something to be said for this medieval
way with intellectual history, where the main concern is simply to
repeat what has been said, rather than provide a wider context for it.
Historical interpretations are always dependent to some extent on current orthodoxy, and the greatest danger occurs when we try to look
across a significant historical divide, perhaps even one so great that
later scientific groups have not been aware of it, and have revived an
earlier style of thought without realizing it. In such instances, we are
prone to use the wrong categories to describe what we find. What we
cannot categorize we cannot compare. The tradition to be considered
here – from the late Middle Ages – is a case in point. It is a case of
"coordinate geometry" from the domain of astronomical computation
that cannot, without distortion, be assimilated either with Greek analytical geometry or with later "Cartesian" forms. It will be shown to
have much more in common with what was later called "nomography"
– itself a byway in mathematics, and one that presents its own problems
of historical analysis. Those who wrote on nomography seem to have
been quite oblivious of their mathematical forebears.

There have been many attempts to show the existence of "Cartesian
geometry" before Descartes. That this has been so is a simple consequence of the important status that has always been assigned to his
work. It has more to do with historians' yearning for continuity than
with the excitement to be had from furthering a priority dispute. Of
course, the analysis of the Cartesian revolution in geometry has not
gone unqualified. Descartes has been seen by some as marking a stage
in the evolution of deeper trends in intellectual history, these having
supposedly been revealed in the light of later developments in mathematics; and the question again arises as to just how far it is right to
allow later trends to dictate historical categories. How far, in other

words, may historians be allowed to stray from the "Dixit Ptolomeus" mentality?

Some idea of this problem, as it relates to our present subject, can be had as soon as we have drafted a rough definition of analytical (or better, "algebraic") geometry in general terms. It may be described as the correlation of what had earlier been regarded as "geometric" concepts and objects – no matter what was meant by the appellation "geometric" – with certain algebraic concepts and objects, which are then proved, by algebraic methods, to have such and such properties. These properties will now therefore be describable in geometric as well as in algebraic terms. As it is often more briefly put, the essence of the analytical method in geometry is the study of geometrical loci by means of their equations. This is a narrower definition, and the primary concern of those who use it is with *geometry*. Modern mathematicians have tended to lose this perspective, but the two-way traffic between algebra and geometry is not new. Fermat, for example, in studying the intersections of the circle and the parabola as a problem equivalent to that of finding the solutions to a quartic equation in a single variable, seems to have been as interested in the one problem as the other, and one might take many similar examples from him, or Descartes, or even – with a few qualifications as to the meaning of "algebra" – from Greek precursors such as Menaechmus or Apollonius.

Of course, analytical geometry has taken many new forms since the seventeenth century, and some of the more recent revisions of mathematical attitudes have left historians unsure of how best to describe events. With the advent of a better appreciation of the axiomatic methods that geometry had bred from the time of the Greeks, geometers were able first to shed the metrical element. The effect of this move was to make some historical writers feel that projective geometry was what really mattered, and that they should be casting history into categories dictated by the projective approach. So much the worse for the fashionable historian, then, when mathematicians unashamedly reintroduced axioms for metrical bases.

There have been many such changes in conceptual signposting, by which mathematicians have left historians over the last century wondering where fashion was leading. There have been numerous new analytic constructions of geometry, for example, developed from linear algebra and the theory of vector spaces over a field, that have suggested new ways of relating the axiomatic structures with the algebraic. Thus, as a now conventional basis for the geometry of vector spaces, there is the theorem that the Pappus planes are simply the affine-coordinate planes defined by a two-dimensional vector space over a field (a skew field).[1] This insight immediately lends itself to a new way of looking at the history of Cartesian geometry, and at what Descartes was "really" doing; but then, in much the way I have suggested already, the temptation is checked somewhat by the realization that there are geometers who favor other approaches. The systems of axioms for the Euclidean plane that lead to metrical theorems, or that are themselves overtly metrical, are somehow closer not only to the world of common experience but to the way history has always – at least, to the superficial gaze – seemed to have developed. The feeling is reinforced that perhaps after all the greatest sin is anachronism, and the best way of avoiding it is only to look backward in time. The ideal is at best dull and at worst unattainable. There is something to be learned from the modern tendency to look at algebra and geometry in an evenhanded way, when considering algebraic geometry, and we should not allow ourselves to slide back into the old ways of introducing the history of the subject through the Roman *agrimensores*, Oresme's "latitudes of forms," and the like.

Speaking generally, there have been three influential attitudes to the historiography of early analytical geometry. The first has emphasized the heavy dependence of seventeenth-century mathematics on Fermat and Descartes, to the exclusion of anything medieval. It has been held that medieval contenders for the title to the invention of their methods were helpless, ignorant as they were of the use of free variables. The second approach finds coordinate geometry in innumerable places – for example, in the Egyptian practice of dividing the land into rectangular districts, in Hipparchus' use of stellar coordinates, and in early medieval graphs of planetary movement in latitude. Such examples and more were produced more than a century ago by Sigmund Günther, who was in turn answered by H. G. Zeuthen.[2] If priorities are in dispute, Zeuthen's position was surely the most satisfactory. In broad terms, this was that analytical geometry was an invention of the Greeks.[3] Fermat and Descartes had, of course, a far better understanding of the geometrical equivalents of the six basic algebraic operations, and obviously a great deal of latitude is required in our understanding of what the word "algebra" means in a Greek context.[4]

The neat modern division between algebra and geometry, the two branches of mathematics that are to be put into correspondence with each other, is historically misleading, inasmuch as it obscures the way in which specific correspondences were first usually established. In the first place, there is more to the subject than simply setting up a correspondence between elements. As a minimum requirement, a transference of rules amounting to an algebra of geometrical line segments is called for, so that addition, subtraction, multiplication, division, raising to a power, and extracting a root can all be given geometrical equivalents – if possible with homogeneous results. Descartes was particularly clear on the order in which he should proceed, but in the great majority of cases, mathematicians dipped into Euclid without any regard for conceptual tidiness, and the seventeenth-century situation was that much of the algebra was tacitly borrowed along with such calculational rules of standard Euclidean geometry as Pythagoras' theorem. The fact that so much of the "algebraic" work had already been done by "Euclid," means that it has not always been easy to see the nature of the correspondence between the *structural* elements of the two subjects. There is some truth in the view often expressed, that much algebraic geometry amounted to a trivial rewriting of old geometry with a new terminology. It is this intermingling of elements, in fact, that has obscured a fundamental difference between most of what was done then and in the Middle Ages.

An important stage in the evolution of algebraic geometry is reached – and was reached, as Zeuthen showed, by Menaechmus – when the new algebraic relationship is looked upon as standing proxy for the geometrical object – say a curve – so that the connections between it and other geometrical objects may be deduced by the manipulation of their algebraic equivalents. A third stage comes when the algebraic relationship is considered to be, as it were, that in which is encapsulated *all* that can be said about the curve, once it is properly interpreted. When Fermat and Descartes showed, for example – in their different ways – that the equation to a curve determines the direction of the normal at any point, there was a tacit assumption of this principle.

For something to merit a historical description as "algebraic geometry" it is clearly not enough that it involve a formal correspondence between algebra and geometry. There is also the question of motive, and of the wish to further the one subject by the help of the other. This does not mean that assistance had to be offered in only one direction. Descartes' third and crucial book of his *Géométrie* contains a novel algebraic theory of equations, and is therefore quite in keeping with his ambition, as expressed in a letter to Isaac Beeckman in 1619, to formulate an algebra by which all questions may be resolved as regards any sort of quantity, continuous or discrete. It is in keeping, too, with Descartes's general concern to *construct* points satisfying a certain relationship, and later to construct the roots of equations of various degrees. This is not the Descartes so many have imagined, creating a royal road to geometry that anyone could tread, but, on the contrary, a Descartes who seems almost to have meant geometry as a substructure for his algebra. His characteristic philosophical priorities and his great confidence in geometry as a basis for truth were certainly not those of the modern mathematician. At all events, the graphical representations of preexisting algorithms that we shall be considering were meant to assist not in the establishment of general mathematical truths but in numerical computation. One might say of these cases that "algebra was being assisted by geometry," but certainly not in any strong epistemological sense.

The algebraic or quasi-algebraic relationship whose graphs we shall consider have as their main purpose the correlation of sets of numbers – such as the temperature of an object and its weight, or the coordinates of a planet and the time. These coordinates do not have to be empirical, of course, nor do they have to be positional. If they represent physical magnitudes, then we require only that their functional relationship is known. As for positional quantity, as soon as one mentions position there is a potential confusion, for we wish to go a stage farther and introduce a correspondence between our algebraic relationship and a geometrical locus, or other geometrical entity. In short, we wish to consider the notion of a geometrical curve (or whatever) corresponding to an algebraic relation that is of central interest and is regarded as having been antecedently established. This "graph of such and such a function," in short, is to be thought of not as coming before the function but as something derived from it. The sequence is the reverse of that with which we began our catalogue of approaches to algebraic geometry. The medieval episode to be considered stands quite outside the history of algebraic geometry in the "Fermat–Descartes" sense.

This last type of correspondence between algebra and geometry, which has so often been treated as a precursor of what is more usually meant by "algebraic geometry," fails to live up to this claim twice over. As explained, it lacks the necessary generality, and it is pointed, so to speak, in the wrong direction. In the conventional phrase, it does not "provide a knowledge of geometrical loci through their equations," but rather a knowledge of equations, and the correlation of the variables they contain, through geometrical loci. In most cases, the word "knowledge" is misleading, although graphing might certainly help us with visualization, empirical understanding, or calculation. It was precisely with these things in view, though, that the medieval astronomer developed what – on account of the direction in which it is aimed – might reasonably be called his "geometrical algebra." The confusion between this and "algebraic geometry" in the usual sense stems to a great extent from the fact that the graphed quantity has so often been a positional coordinate.

Consider such typical astronomical tables as are found in single- or double-entry form. (Many tables ostensibly in the latter form are only trivially describable as double-entry, such as when days are entered down the left column and months across the top.) The chances are that such a table – at least if it is not Babylonian – was arrived at on the basis of a geometrical algorithm. Consider the simplest of all single-entry tables, those for mean motion. These give a steadily increasing

angle as a function of the time, and to anyone with the barest knowledge
of rectangular graphing techniques they beg to be turned into straight-
line graphs (broken, of course, at 360-degree intervals). At a more
advanced level of conceptualization, the resulting planetary positions
recorded, after much calculation, in an almanac, would have been very
usefully graphed, whether in rectangular or polar coordinates. At the
very least, interpolation problems would have been better understood
and more easily executed. These steps, however, were not taken, and
while there are medieval survivals that look very much like our
"squared paper," not to mention polar coordinate grids, we must resist
the temptation to suppose that they were always viewed in the way
we find most natural, namely, as the bearers of some previously evaluated function, the contents, so to speak, of a table of values. This
way of looking at things – of which we shall find evidence only in the
later Middle Ages – is quite alien to an earlier period, as may be very
well illustrated by the most notorious of medieval graphical representations, those belonging to what might be called "Plinian" astronomy,
which had a revival between roughly the ninth and the twelfth
centuries.[5]

The diagrams in question simply show the variation of planetary
latitude (the ordinate) with time, or rather with zodiacal position – and
this in a very loose sense, as will be explained later. That the theory
behind the representation was ill understood may be guessed from the
very notion of capturing within a single circuit of the zodiac all potential
latitude variations, in a situation where the nodes of the planetary orbit
are not stationary. What matters more to us is that the genesis of the
graphs that are now best known, and that are seemingly based on rectangular coordinates, was not at all along the route that we, with our
expectations based on the modern graphical tradition, are inclined to
suppose. The route becomes plain as soon as we consider the diagrams
that had gone before. These were *circular* in form. They comprised a
background of thirteen evenly spaced concentric circles, with each
planetary path an eccentric circle, the eccentricity being settled by the
maximum and minimum potential latitudes attained by the planet in
question. At this stage, although the diagrams might be grandiosely
described as being in polar coordinate form, all that has really been
done is to put a three-dimensional picture on the two-dimensional page.
(To speak of a projection would be too strong. There is no sense of
perspective here.) At the later, rectangular, stage, the band of latitudinal degrees is straightened out, and, of course, inevitably broken in
the process. What is more, at this later stage something very odd happened: The planetary paths, while having the roughly sinusoidal forms
that are to be expected, and while having the right amplitudes in latitude, no longer follow a single cycle, as the original diagrams suggest

they ought to have done. Instead, they show a completed number of longitudinal waves that are usually inversely proportional to the latitudinal amplitude, the moon defining the basic unit.[6] It would be impossible to deny that these diagrams carry information, but it is in a form so garbled, and so far removed from the geometrical algebra we are seeking, that there is little point in discussing it further here.

There is something to be learned, even so, from the Plinian latitude graphs, namely, that at an early historical stage a coordinate mesh is more likely to have been superimposed on the pre-existing figure – a picture of reality, so to speak – than to have been the route to that figure. This is the case in a yet older and much more important cluster of traditions, those centered on the astrolabe.

The astrolabe may be characterized in many different ways, but for our purposes there is much to be said for describing it as an analogue computer. It provides the user with numbers – times, altitudes, azimuths, longitudes, and the like – as functions of other numbers – altitudes, ascendant ecliptic longitudes, and whatever. Its two main parts are the fretwork rete that represents the sphere of the fixed stars, and the plate of local coordinates over which it turns. Their relative disposition is analogous to that of the two main sets of circles as conceptualized in elementary spherical astronomy.[7] As is well known, they are produced on the instrument by stereographic projection, and this precise geometrical way of turning one figure into another is a far cry from the crudely impressionistic transformation found in the former example.

The astrolabe illustrates several of the different elements that ought to be distinguished in any account of the evolution of geometric algebra. These include the notions of (1) coordinate assignment; (2) secondorder graphical representation; (3) coordinate transformation; (4) the graphing of quasi-algebraic functions, that is, over and above any geometrical substructure that might have led to them; (5) contour lines; and (6) the use of these graphs as substitutes for the algorithms on which they are based.

The first of these conceptual elements, *coordinate assignment*, is a minimal requirement, of course. In the broadest sense, coordinates are a means to solving the problem of saying where things are, so as to be able to locate them on another occasion, or have others do so. The discovery of the historical use of simple coordinate systems does not seem to me to be an occasion for great excitement. It is a matter of great importance that in astronomy, where the most significant developments in the use of coordinates took place, coordinates were most commonly angular, and therefore often cyclical. The most common ways of assigning coordinates to points of the celestial sphere were through ecliptic latitude and longitude, right ascension and declination, or mediation (the degree of the ecliptic that culminates with the object) and declination.

By *second-order graphical representation* nothing more is meant here than a mapping, a turning of one figure or space into another, not necessarily as a self-consciously mathematical action. It is what was done by the "Plinian" astronomers who turned the zodiacal belt on the sphere into a belt of thirteen concentric circles on the plane surface of their books. It is what a painter does; and if painters had needed to wait for mathematical rules of perspective, which arrived on the scene only in the late Middle Ages, their art might never have begun. This is not to say that perspective painting is without interest: Those who practiced it were, after all, generally doing a better job than the "Plinian" astronomers. It leaves something to be desired, even so, from a mathematical point of view. When a painter turns three dimensions into two, there is a loss of information along the way. Each point in the painting may correspond to any or all points of a visual ray in the three-dimensional original. Information in this case is tacitly restored by the topological sense of the onlooker, who knows that eyes belong with heads, feet belong with the ground, and so on. In the astronomical case, where only directions and not radial distances are held to be significant, the two-dimensional representation used on an astrolabe loses nothing of importance. We shall later come across ways in which three independent variables (and more) were represented in a plane without loss, in the context of astronomy.

In effecting the sort of graphical representation found on an astrolabe, and doing so in a consciously correct geometrical way, one is practicing a sort of second-order geometry, somewhat analogous to what is done in the geometrical algebra and algebraic geometry of which we have already spoken. It is a stage on the road to the former, in fact, as will be shown later, from a number of medieval examples.

Coordinate transformation is one of the principal functions of an astrolabe. A typical example is where the rete is positioned for a particular moment in the day, allowing us to deduce the altitude and azimuth of a star whose latitude and longitude (or mediation and declination, perhaps) were needed to position it on the rete in the first place. Astrolabe techniques provided a basis for certain instruments used for conversion between one system of celestial coordinates and another. Such instruments were used, for instance, to turn one type of star table into another – and, of course, they could not be very precise, by comparison with the usual long trigonometrical methods of calculation. Oddly enough, the reason for effecting the conversion, in many cases, was to have coordinates (mediation and declination) that were more convenient for placing stars on an astrolabe. Some of the more sophisticated techniques of coordinate transformation were developed in connection with the *universal* astrolabe projection, usually known in the West as the *saphea*.[8] There were many other forms of coordinate transformation used by astronomers, however. Angle-measuring devices (such as the triquetrum, Jacob's staff, even most of the numerous forms of sundial) that use a linear scale to yield an angular coordinate are clearly being used to effect a transformation of coordinates, and although this might seem to be a pretentious way of describing what is done in these cases, familiarity with them had at the very least a psychological importance for the evolution of stage (4) in our progression. The message was simply that measures are not always what they seem, that a distance is not always the measure of a distance, and that even when it is so, the one distance need not be directly proportional to the other.

The *graphing of quasi-algebraic functions* is something best approached in stages, for it is easily confused with second-order graphical representation. Consider the case of the lines on an astrolabe that show the unequal hours. It is not important to know here how they are used, but only to realize that they are arcs of circles that the point marking the sun (on the rete) crosses in the course of the daily rotation, its crossing marking the traditional divisions of day and night. (A twelfth part of the day, a "seasonal" or "unequal" hour of the day, will only be equal to a twelfth part of the night at the equinoxes.) It is how these arcs are obtained that matters to us. For every position of the sun on the ecliptic, its movement from setting to rising – to take the case of the night hours – around the lower part of the astrolabe plate describes an arc, which we may suppose to be circular. (Since the sun's longitude changes by a trifle, in the course of the night, there is a slight approximation involved.) A point a twelfth part of the way along that arc will measure the position of the sun at the end of the first unequal hour of the night. The totality of all such points, for all possible polar distances of the sun, will be the form of the hour-line we are seeking. Each point on the hour-line is such that an algorithm was known for calculating it. It would have been possible in principle to have drawn that hour-line on the basis of the known algorithm, in which case we should have had what I have called the graphing of a quasi-algebraic function. In fact it was drawn otherwise, that is, as an arc of a circle passing through three points of symmetry that are easily constructed.

There are many other cases in medieval astronomy that should be classified in much the same way as the hour-lines example, although they are often instances where the construction is an exact one. The lines on an astrolabe by which the astrological houses are computed using what I have elsewhere called the "fixed boundaries methods," lines that are often confounded with the unequal-hour lines, are of the same general sort.[9] They are *contour lines*, moreover, in a way that should be obvious: the hour-lines, for example, are labeled from 1 to 12 with the hour variable, if we may call it that. In a sense, many other lines on an astrolabe are analogous to them, and of most the same could be said: There are algorithms, and there are tables embodying those algorithms, to which they correspond, *but there is no immediate genetic relationship between algorithm and graph.*

The astrolabe was used for problems in spherical astronomy, and was not as such connected with the problem of computing planetary positions. Here there was a large class of instruments available, under the generic name "equatorium." Despite their being described by this one name – the name stems from their function of evaluating the "equations" to be added to the mean positions of the planets to give their true positions – they may be classified in many different ways, and one of the most enlightening of classifications concerns the degree of simulation of the fundamental planetary (usually Ptolemaic) model. Whether or not the Ptolemaic planetary models are themselves meant to simulate reality is beside the point here: We shall suppose that they are agreed as the means to the end of calculating planetary positions. With each geometrical model goes a whole series of trigonometrical procedures. The geometrical model, with its deferent circles, epicycles, apselines, equants, and the rest, may be simulated precisely, in principle, by graduated circles of metal or parchment or whatever, rods, pivots, and so forth, the whole arrangement being set for a particular moment of time in accordance with the mean position-angles that the tables of mean motions give for that moment. Most equatoria fall into this general class.

There is another sort, however, that follows more closely the sequence of algorithms gone through in an ordinary Ptolemaic calculation of planetary position. Perhaps the most intricate of these equatoria was the so-called Albion of Richard of Wallingford, dating from around 1326, but his was not by any means an isolated example.[10] Speaking very generally, there is an element of simulation, but only in the component computations. Thus, if an adjustment term, an "equation," depends on a sine factor, one might find a procedure for using certain scales of the instrument to perform a construction that amounts to the evaluation of that sine, just as one might do it with a ruler and protractor on paper; and so for more complex algorithms. There is simulation here, but not of models in their entirety, and certainly not of anything the astronomer might have been tempted to describe as reality.

Once this sort of procedure had been established, once the spell had been broken that held astronomers to a procedure simply picturing the Ptolemaic picture, the way was clear to a new phase in experiment with more powerful graphical techniques. It is certainly not possible to say that the inventors of equatoria in this second style were alone responsible for this new phase, but it does seem to be true that the most significant combination of graphical (subsidiary) instruments came along with Richard of Wallingford's Albion, which was in an unmistakably "trigonometrical" form. I shall here give attention to just two of the graphical devices incorporated in it, one for conjunctions and oppositions of the sun and moon, and the other for eclipses of the sun and moon.[11]

The use of both instruments is extremely involved, and it is impossible here to give more than the barest sketch of the general principles behind them. At first sight, the instrument for conjunctions (which also gives lunar velocity) is a series of arcs arranged spirally, each with a certain number attached, exactly as on the unequal-hour lines of an astrolabe. The idea for these contour lines undoubtedly came from that ubiquitous instrument. The moon's argument settles the radial distance of a bead on a thread through the center of the instrument. At this distance, we may imagine a scale whose graduations are where the imaginary scale crosses the numbered contours. The thread is then set to a certain longitude difference (that between the sun and the moon at mean syzygy, in fact) on a peripheral scale, and the bead will then indicate a certain "equated (corrected) time" on the imaginary scale graduated by the contours. In fact, as with all contours, as every map user knows, the method of interpolation is intuitively obvious, but carrying out the task is only easy when the contours are closely packed.

The difference in principle between this instrument and the hour-lines on an astrolabe, with which it is functionally related, is one of intuitive appeal. The hour-lines mark out quite simply the divisions over which the sun, in its daily course, may be considered to pass. The syzygy instrument lends itself to no such interpretation. It simulates nothing that can be readily pictured as happening in the heavens. It is a series of (polar) graphs that together allow one variable to be evaluated as a function of two others, and that have their justification in a fairly complicated algorithm due to Ptolemy. The *form* of those graphs is still not very complex. They are still arcs of circles through three points. There is good reason for suggesting, however, that here we have a case of what I called earlier the graphing of quasi-algebraic functions.

The eclipse instruments that form a part of the Albion are no less complex in their entirety, but briefly they may be said to contain a series of eccentric (semicircular) arcs that are polar graphs of different functions of certain dimensions (lunar, solar, and earth's shadow at the lunar distance) that matter to anyone calculating an eclipse. The radial scale is a linear scale of those (angular) dimensions, and the angular argument is the lunar argument during eclipse. The loci in question are not quite what they should be according to Ptolemaic eclipse

theory, but the important thing is that Richard of Wallingford was trying
to make them so. He was working within the limits set by circular arcs.
(And lest anyone be tempted to weave a story of the hidebound medieval astronomer, constrained by a Greek philosophical obsession
with the perfection of the circle, let it be said that no one used ovals
more intelligently than Richard of Wallingford, whether on his Albion
or in his astronomical clock.)

The Albion was an instrument of great complexity, with its many
subsidiary parts, and over sixty scales – its very name meant "all by
one." It is hardly surprising that later astronomers tended to abstract
its subsidiary instruments from it. The only set of surviving fragments
of an Albion made in metal is now in Rome.[12] Richard of Wallingford's
treatise on the instrument was extremely influential, however, and
some of those writers whom it influenced were themselves influential,
notably in central Europe. There are at least a dozen editions or treatises that derive immediately from it, not to mention several minor
tractates that make use of its parts, and these last usually concentrate
on the instrument for conjunctions and the double eclipse instrument.
The first new edition was done by Simon Tunsted, before 1369. The
Viennese astronomer John of Gmunden produced what was perhaps
the most widely copied, around the year 1430, not altering the instrument significantly but adding explanations where the original was too
cryptic. He added a thousand words or so in connection with the instruments singled out for attention here. Regiomontanus (1436–76) produced a rather careless version. John Schöner published an equatorium
treatise in 1522 that included Richard of Wallingford's eclipse instruments. The most striking printed work to make use of the treatise
Albion, however, was the lavishly illustrated *Astronomicum Caesareum* of Peter Apian, published at Ingolstadt in 1540.

The writers named here not only were intrigued by the instruments
mentioned but show distinct signs of having tried to extend the underlying principles to the solution of other problems. Apian was not
alone in this, nor was he the first, but his book was a rich source of
new ideas, and will serve to illustrate another stage in the evolution of
the graphing of functional dependences.

In two *enunciata* (20 and 21), he made use of the principle that Richard of Wallingford had used in his conjunctions instrument. The
problems are somewhat similar: They are to determine small changes
in the aspects of the moon with the planets and the planets with other
planets. As before, one variable is evaluated as a function of two others,
the dependent variable being given by spiraling contour-lines, the independent by polar coordinates. *Enunciatum 26* concerns an instrument of the same general character, to relate the time to the hourly
changes in lunar and solar argument at syzygy, and this begins a re-

markable series of instruments that develop the theme. All are of the same general form, that is, they express a variable as a function of two others, and all are used in the same way as those already explained. On them, however, the loci which on the Albion and immediately derivative instruments are circular arcs are now curves of a much more intricate form. They are contour lines such as might have charted the form of a sand dune whipped up by a tornado. They are contour lines of constant time difference, that might have been found a "Ptolemaic" algorithm, although that would have been extremely complex, for reasons that will shortly emerge. In fact they were *plotted from previously calculated tables*. The data in the worked examples included in the text are there far too precise to have been found from the instruments. On the other hand, there is no doubt that when he was designing his latitude instruments, he had in front of him the latitude instruments designed by Sebastian Münster, who had not proceeded in this way. Münster's work was published in his *Organum Uranicum* (Basle, 1536).[13] The change of style marks an important step in the evolution of graphical methods.

We need not enter into the details of Münster's solution to the problem of latitudes. It followed the same general principle as that adopted by Richard of Wallingford in his conjunctions instrument, and there are clear links between the two. The radial polar coordinate (set by a seed pearl on a radial thread) was to be set, now, with the help of an ancillary circular scale of degrees, a sort of epicycle. This first variable (the radius vector) was thus effectively itself a function of another (the "true argument" set in the epicycle). A peripheral scale gave the variable known in astronomy as the "true center," while, as a function of these two, the latitude was given by the contour on which the seed pearl fell – or, rather, the imaginary scale graduated by the contours. The important thing to notice is that the contours were still *arcs of circles*. What Apian did to Münster's design was highly significant: he created a *linear* radial scale of the (true) epicyclic argument. As explained, there had not previously been a radial scale at all, since the length of the radius vector had been found as a function of an epicyclic argument, but the effective radial scale was nonlinear. By simplifying the instrument in this respect, Apian had made it necessary to produce noncircular contours. Their form was complicated by yet another fact: he set the 360 degrees of the peripheral scale, that of the true center, into about 340 degrees of the instrument, leaving space in the other 20 degrees for his radial scale. His were therefore not polar coordinates as we know them, but they were none the worse for that. Apian had realized that what others might have seen as the distortions of his coordinate scales did not matter in the slightest, if he had at his disposal tabulated information to plot on the plane of his instrument. It was not what we usually mean by "empirical" information, but the same principle would have applied, had it been so.

Apian's latitude instruments had a more influential career than their publication in the sumptuous "Imperial Astronomy" alone would have guaranteed them. They were adapted, for example, by Jacques Bassantin in his work *Astronomique discours* of 1557, having already been given a more public form on the face of the extraordinary astronomical clock by Philip Imser of Tübingen, dated 1555.[14] It is hard to believe that many of those who contemplated with the intricately engraved discs of contour lines on Imser's clock knew how to interpret them.

Apian's *Astronomicum* has the eclipse instruments of the Albion, in a heavily disguised form, but here there is nothing essentially new.[15] In some versions of the original eclipse instruments, it should be added, the semicircles are backed by a square mesh, giving the appearance of our modern squared paper. Its function here is quite different, however, for it is simply to assist the user in dropping perpendiculars to the scales, as the canons require. As so often in the history of "coordinate geometry," appearances are deceptive.

It is not suggested that the examples given here are unique, or the first of their kind. There is a very different type of instrument for planetary latitudes, for instance, as drawn by Johann Werner in his *Organa* of 1521, which may be generously interpreted as yielding the product of two numbers m and x as the ordinate to the straight line $y = mx$.[16] The technique of "geometric multiplication" was, of course, as old as Euclid, but the graphical form of Werner's figures, with their rectangular mesh, certainly enriched the "nomographic" tradition being discussed here. I have said nothing of the contour lines on sundials – for example, on cylinder dials – or of influences from outside Europe. The tradition of engraving the so-called sinecal quadrants on Indian astrolabes, for example, although seemingly late, would repay investigation. Doing so would not, however, affect significantly the thesis being put forward, that by the early sixteenth century the path to a "geometric algebra" was all but complete, if we interpret the word "algebra" liberally enough. The algorithms mentioned have been essentially trigonometrical, and the great notational advantages of the sixteenth and seventeenth centuries were still in the future, but the astronomers we have been discussing were undoubtedly conscious that they were using the graphs as substitutes for the underlying algorithms, and that the graphs were very *immediate* representations of the working-out of those algorithms. This might even be argued in the case of Richard of Wallingford, although he seems to have made no use of the representation of tabulated data.

The works of Richard of Wallingford, Apian, and their intermediaries, did not, in any direct way, concern the graphical representation of empirical data. One does not find this conspicuously in contour-form before the eighteenth century. It seems that the first important example might have been Philippe Buache's use of contours in his charting of submarine channels.[17] The older tradition is much more reminiscent of the graphical scales introduced by Louis Pouchet in 1790, as substitutes for the double-entry tables used in metric conversion. Pouchet's work in turn seems to have given rise to an entire subject, and one that was thought to be new – nomography. (The name is due to Maurice d'Ocagne, and dates only from 1891.) It is a subject that seems to have owed nothing to Fermat, Descartes, and their immediate followers. It made use of graphical procedures representing three and more coordinates in a plane. As it happens, Apian – or whoever was his source of inspiration – had already shown the way to introducing any number of variables into the plane, by the successive use of different "instruments" on the same page: c is a function of a and b; e is a function of c and d; g of e and f; and so forth. In short, the technique exploited by Richard of Wallingford lent itself to enormous generality. Nomography in its later manifestation came into its own in a technological milieu, where much rapid calculation was needed in any one of a class of essentially similar problems. This is very much the situation in which medieval astronomers had found themselves. But what is more to the point: Medieval astronomy had yielded techniques very similar to those found in the eighteenth and nineteenth centuries, and had done so in a way that many historians, aware of the chronology of developments in algebraic geometry, would have judged impossible a priori. Such a mistaken judgment can only arise from a poor historical categorization of "coordinate geometry." To this there are clearly more sides than one.

Notes

1 The phrase "Pappus planes" is a conventional expression having only a tenuous link with Pappus, and I use it simply because it is now a fairly standard way of referring to planes defined by the affine form of the famous theorem of Pappus and Pascal concerning collineations in connection with a hexagon. They are sometimes called "Pappus–Desargues planes," since in an affine plane Desargues's theorem follows from the limited form of that of Pappus and Pascal.

2 S. Günther, "Die Anfänge und Entwicklungsstadien des Coordinatenprinciples," *Abhandlungen der naturforschenden Gesellschaft zu Nürnberg* 6 (1877):3–50; H. G. Zeuthen, "Note sur l'usage des coordonneés dans l'antiquité, et sur l'invention de cet instrument," *Oversigt over Danske Videnskabernes Selskabs Forhandlinger* 1 (1888):127–44. The

literature spawned by these two classic papers is very considerable but not very relevant here.

3 For Zeuthen's most important statement of this, see his *Die Lehre von den Kegelschnitten im Altertum* (Copenhagen, 1886).

4 Descartes has an especially clearheaded statement of the equivalence, at the beginning of the first book of the *Géométrie*. For a convenient edition, with the French edition of 1637 facing an English translation (from the French and Latin), see D. E. Smith and M. L. Latham, *The Geometry of René Descartes* (repr. New York: Dover, 1954).

5 For more information about the background to these and other "Plinian" illustrations, and reproductions of many of them, see B. S. Eastwood, "Plinian Astronomy in the Middle Ages and Renaissance." *Science in the Early Roman Empire: Pliny the Elder, His Sources and His Influences,* eds. Roger French and Frank Greenaway (Beckenham, England: Croom Helm, 1986), pp. 197–251.

6 This far from obvious interpretation was offered by B. S. Eastwood (Ibid.).

7 For an elementary introduction to the instrument, see my "The Astrolabe," *Scientific American* (January 1974):96–106. *Stars, Minds and Fate,* ch. 14.

8 The universal astrolabe is so described because it is valid for all geographical latitudes. A conventional astrolabe needs a separate plate for every latitude for which it will be used.

For four tracts concerning instruments for conversion, all of them possibly associated with Richard of Wallingford (ca. 1292–1336), see my *Richard of Wallingford: An Edition of his Writings, with Introductions, English Translation and Commentary,* 3 vols (Oxford: Clarendon Press), vol. 2, 302–8. For an introduction to the *saphea,* and its various Western forms – it was incorporated in many Renaissance instruments, for instance – see vol. 1, 331–3; vol. 2, 187–92; and vol. 3, 36.

9 See my *Horoscopes and History* (London: The Warburg Institute, forthcoming). The "standard method" may actually be operated with the hour-lines themselves.

10 The instrument was the subject of a treatise edited in the edition cited in note 8. For a general study of equatoria, see E. Poulle, *Equatoires et horlogerie planétaire du XIIIe au XVIe siècle,* 2 vols. (Geneva: Droz, 1980).

11 See the edition of *Albion* and the commentary on it (as cited in note 8; above), sections III. 18–19 (for the first instrument's use) and III.20–34 (for the use of the second).

12 See plates XXIII and XXIV of my edition (note 8 above). There is another part, not illustrated there. For amplification of the remainder of this paragraph, see especially vol. 2, 127–36, 270–86, and appropriate places in the text, commentary, and illustrations (vol. 3).

13 There are many manuscript versions of Münster's work on equatoria. For a general survey, see Poulle, *Equatoires et horlogerie planétaire* [note 10 above], especially chapter B1. For the latitude instrument mentioned below, see Poulle, figure 123 and plate L.

14 See E. Poulle, *Equatoires et horlogerie planétaire,* plates LII–LIII.

15 For a survey of Apian's work, see *Richard of Wallingford* (note 8, above), vol. 2, 278–285.

16 The remarks about nomography in this paragraph are largely based on a paper read by Dr. H. A. Evesham to the British Society for the History of Mathematics in September 1983.

17 These are illustrated from Bodleian Library MS Digby 132 in Poulle, *Equatoires et horlogerie planétaire* (see note 10, above), plates XLVII–XLIX.

2

THE RELUCTANT REVOLUTIONARIES:
ASTRONOMY AFTER COPERNICUS

There is, I believe, a Polish saying to the effect that only one thing is difficult to predict, and that is the past. The saying is obviously that of an historian, conscious of changing fashions in historiography, rather than of an astrologer, for whom a little freedom in the interpretation of the past is no bad thing. In the history of astrology, however, fashion is as significant a determinant of the truth as in any other form of history, and even now the prevailing fashion is to be censorious. It is as though the achievements of a Copernicus would be diminished if he were ever seen to have encouraged, even unwittingly, doctrines as intellectually reprehensible as those of astrology. There are many sorts of astrological belief, and many possible degrees of belief, and it is far from easy to establish where exactly Copernicus stood. But this is not my aim. I want only to examine his influence on the climate of astrological opinion, and it will be readily acknowledged by all who know his extant writings that they were not, by dint of categorical astrological statement, calculated to influence posterity one way or the other[1]. But Copernicus lived among scholars and clerics for many of whom the study of astronomy was to be justified first and foremost in terms of its potential application to the art of prognostication, whether of human fortunes, or of such natural phenomena as the weather. It was inevitable that the Copernican scheme, once it was shown to be astronomically practicable, should be adapted to astrological needs.

Why this was so can be easily appreciated. That the art of prognostication was difficult and uncertain, no one denied. But might not the

---

[1] Carlo Giacon, in his *Copernicus, la filosofia e la teologia*, "Civilta cattolica", 94 (1943), pp. 281 – 90 and 367 – 74, is one of the few modern historians to have maintained that Copernicus was an astrologer, but he rests his case on evidence relating to Rheticus. Most discussions of the life of Copernicus avoid the subject entirely. See, however, p. 24 below.

uncertainty be a consequence of astronomical imprecision? This was a distinct possibility which, it was felt, must be investigated. There were a few writers deeply conscious of the need to confirm, and where necessary modify, astrological hypothesis, in the light of subsequent experience, but this they could only do if they first had confidence in their astronomy. For most, however, it was enough of a recommendation that an astrological doctrine be sanctified by tradition. This is not to say that astrology was without its critics, and these, broadly speaking, we may divide into three camps — which were not, however, mutually exclusive.

First there were those who, like the Fathers of the Church, saw in astrology the enticements of the devil, and who classed it with witchcraft, necromancy, and other spiritually suspect forms of divination. Such writers thought it enough to preach against astrology as they preached against the devil, that is to say, on religious grounds but without too close a familiarity. Petrarch, Pico della Mirandola, and John Wycliffe, were among those who considered this a legitimate and effective form of attack.

Second, there were those many theologians who abhorred what they took to be the fatalism of much Islamic thought, who feared Averroistic doctrines denying the freedom of the will, and who were determined to preserve Augustine's doctrine of predestination, or something like it. And third, there were those who thought that astrology would succumb to an informed, mathematical, and rational attack. Why, they would ask at the simplest level, do twins often have vastly different characters, when they were born under closely similar stellar influences? Why, they would ask after they had learned of the slow drift of the divisions of the zodiac with respect to the constellations, do you, the astrologer, not constantly change the rules concerning the influence of the signs? Why, asked that excessively rational questioner, Nicole Oresme, does the astrologer suppose that he can meaningfully base his art on regularities of planetary aspect, when, by the incommensurability of the planetary motions, such regularities cannot occur, except trivially? Awkward questions of this sort probably influenced far fewer people than the attack on the spiritual front, but even taken together, it must be admitted that the different forms of assault seem to have had very little effect on the tide of astrological belief in the Middle Ages[2].

---

[2] There is a considerable literature attacking and defending astrology, but John Chamber, *A Treatise against Judicial Astrologie*, London, 1601, covers most of the traditional points in a restrained manner, while Sir Christopher Heydon, *A Defence of Judicial Astrologie*, Cambridge, 1603, gives a very detailed answer, the latter book being more than four times as long as the former. To avoid misunderstanding, Chamber published, in the same year as his *Treatise*, an *Astronomiae Encomium*, a per-

There was, nevertheless, a gentle decline in the popularity of astrology at higher academic levels in the 16th and 17th centuries[3], and not unnaturally this has often been taken as an indication of the progressive scientific temper of the age. It is very difficult to make such an idea precise, and to do so is certainly outside the scope of my paper. For what the comment is worth, however, I will point out that from a study I have made of the interests of scholars in medieval Oxford, it seems to me that while there was a fairly constant concern with the affairs of astronomy, yet in logic, in natural philosophy, and in the life sciences, there was a drastic decline after the mid-14th century, with a trough a little before 1500. That there was a rapid recovery of the empirical sciences after that date is of itself therefore no explanation of the corresponding decline in astrological enthusiasm within the university, unless we can say that the characters of the sciences were radically changed, as between the late Middle Ages and the 16th and 17th centuries. In fact the first conspicuous change in the character of the sciences was somewhat superficial, dependent as it was on changing literary tastes, and in particular on the preference shown for new Greek texts, as opposed to treatises based on Islamic writings.

Copernicus shows by his frequent reference to classical witers that he shared the new literary tastes, as one might imagine a man educated in Italy was almost bound to do. But not all his contemporaries shared this advantage, and one, in particular, reveals through his writings that study in Zürich, Wittenberg, Nürnberg, and Tübingen did not predispose an astronomer to tastes of quite the same sort. The impression, given by even a casual acquaintance with the pattern of 16th-century publishing, that more astrology was printed in the German-speaking towns than in Italy, is borne out — at least for meteorological astrology in studies made by G. Hellman[4]. In the first decade of the century, there was little to choose between the two regions; but by the end, the Italian output had almost come to a halt, whereas in Germany, Austria, and

oration delivered 27 years earlier to the University of Oxford. In this he praises Melanchthon, Ramus, Stringelius, Peucer, and Reinhold, but does not mention Copernicus.

[3] This decline is not to be studied statistically on the basis of the publication of almanacs and ephemerides (for which see following paragraph). Although these were necessarily the work of men with some scholastic training, the work was essentially derivative, with the *Tetrabiblos* of Ptolemy and the *Prutenic tables* as typical basic texts.

[4] *Wetterprognosen und Wetterberichte* (Neudrucke von Schriften und Karten über Meteorologie und Erdmagnetismus, No. 12), Berlin, 1899; *Versuch einer Geschichte der Wettervorhersage im XVI Jahrhundert*, Abhandlungen der preussischen Akademie, phys.-math. Kl., 1924, No. 1. There is some distortion in the ratios given here, since Italy was subject to a relevant papal bull of Sixtus V in 1586.

Switzerland the rate of publication had more than trebled. This was the Kulturgebiet from whence came Rheticus.

The fortunes of Georg Joachim (Rheticus) were, needless to say, intimately linked with those of Copernicus. Rheticus was not only the author of *Narratio prima* (1540 and 1541) in which the new astronomy was first given to the world in published form, but also the man who supervised the printing of the first sections of *De revolutionibus*. He was not yet thirty when his host and tutor died, and we cannot argue that the generation gap alone can explain the gap in their attitudes to astrology, without also supposing that Renaissance enlightenment was suffering a minor reverse. In fact, as his biographer shows, Rheticus was throughout his life immersed in an atmosphere wherein astrological practice was the norm[5]. In the *Narratio prima* there are two or three passages in which old astrological ideas are adapted, surely for the first time, to the new astronomy. The first of these concerns the dependence of the changing fortunes of the kingdoms of the world on the changing eccentricity of the Sun[6]. In the older astrology — as, for example, in Alkabucius — it was often the position of the planet on the epicycle, or the position of Mercury's deferent center on its auxiliary circle, which turned the Ptolemaic heavens into a Wheel of Fortune. Now it is the position of the mean Sun (i.e., the center of the earth's orbit) on its small auxiliary circle which decides the rise and fall of princes, and which will decide the time of the second coming of Christ[7]. On a later page, Rheticus shows himself to have been irritated by Pico della Mirandola's *Disputationes adversus astrologos*, which he maintains could not have been written had Pico known so accurate a system as that of Copernicus[8]. And towards the end of the *Narratio*, in an astrologically embroidered section given over to praise of Prussia, which by tortuous reasoning he likens to Rhodes, he again returns to the problem of accuracy[9]. He tells us that, until he was persuaded to do otherwise by Tiedemann Giese, Copernicus had decided to avoid philosophical controversy by simply composing tables and issuing them with rules, but no proofs. Common mathematicians would have their calculus, while true scholars would be able to arrive at the fundamental principles from which the tables were derived.

---

[5] Karl Heinz Burmeister, *Georg Joachim Rhetikus, 1514—1574*, vol. 1, Wiesbaden, 1967, pp. 6—9, 166—72, etc.

[6] See p. 121—3 of Edward Rosen's excellent translation, in *Three Copernican Treatises*, second edition, 1959.

[7] Rosen, ibid., n. 57 refers to a comparable Rheticus passage in the preface to Werner's *De triangulis sphaericis*. He notes that, in a letter to Tycho Brahe, Christopher Rothmann censures Rheticus for the whole idea.

[8] Ibid., pp. 126—7.

[9] Ibid., pp. 188—96.

The common mathematicians undoubtedly included the astrologers. (In certain medieval contexts, the word *mathematici* is actually synonymous with *astrologi*; but this is by the way.)

Thus even before the publication of the *De revolutionibus*, and with his knowledge, if not his approval, Copernicus was presented to the world in such a way as to dispose astrologers favourably towards him. An impatient audience, apprised of the virtues of the Copernican scheme, was pressing for details. Gemma Frisius wrote from Louvain to John Dantiscus, bishop of Ermland, showing that he anticipated, with the publication of Copernicus' book, an end to the astronomical errors and uncertainties that beset the astrologer[10]. It was of no consequence to Gemma Frisius whether the hypotheses were true or false. Copernicus clearly understood his public. Two years later Gemma was still writing impatiently[11]. In the end his satisfaction with the new astronomy seems to have been muted. He did, nevertheless, imbue his astrologer son Cornelius with great respect for Copernicus[12].

It was a constant plaint by contemplative astrologers, for decades to come, that dubious or certain though astrological principles may be, without good tables accurate prediction is impossible. They showed a marked tendency, therefore, to turn to the *Prutenic Tables*, rather than to Copernicus' own writings, thus again showing that Copernicus' instincts were right. The accuracy of tables was often respected, as it had been throughout the Middle Ages, even when it was wholly illusory; but as Melanchthon remarked, there was no comparing such works as the *Tabulae resolutae* of Schöner with the wooden doves and automata of the time, which gave as much planetary information as the masses required[13].

Schöner, be it noted, was the one to whom the *Narratio prima* of Rheticus was addressed. In 1545 he printed three books of judgements of nativities, introduced by that inveterate writer of introductions, Melanchthon, and containing worked examples of an autobiographical sort[14]. He naturally found the Copernican system to be not unfavourable to

---

[10] Maximilian C u r t z e, *Fünf ungedruckte Briefe von Gemma Frisius*, "Archiv der Mathematik und Physik", vol. lvi, 1874, pp. 313—25, esp. the fourth letter of 20 July 1541, pp. 318—20.

[11] Ibid., p. 322 (letter of 7 April 1543, also to John Dantiscus).

[12] See p. 182 below.

[13] *Tabulae astronomicae* ... Nuremberg, 1536, f.v.verso: "Quis non miretur tam brevi libell i tot annorum descriptionem includi posse ? Mirentur alii ligneas, columbas, aut alia opera automata. Hae tabulae multo magis sunt dignae admiratione, quae omnium siderum positus ostendunt, nec unius tantum anni, sed multorum seculorum".

[14] *De iudiciis nativitatum, libri tres*, Nuremberg, 1545.

astrology. Had there been any widely accepted theory of the mechanism of planetary influence on terrestrial affairs — had the centre of the universe been likened to a focus, for example — then moving the earth out of its central position might have worried astrologers more. In fact as a class they appear to have been better able by far to cope with a heliocentric world that with the new star of 1572, for which the standard astrologies gave no precedent. The matter was later well put by Sir Christopher Heydon, in his defence of Astrology[15]. Whether the Earth be at the center, or in the Sun's place between Mars and Venus, or whether Tycho's hypothesis be correct,

> the Astrologer careth not. For so that by any of these Hypotheses, he may come to the true place and motion of the Starres, this varietie of opinions, whether such things be indeede, and in what order they be, is no impeachment to the principles of the Arte.

Melanchthon reiterated Rheticus' idea that the Sun's eccentric motions were connected with terrestrial changes, and this became a leading topic for astrological debate, being the subject of attack by Julius Caesar Scaliger and Jean Bodin. Bodin ascribed the *Narratio prima* with its astrological content to Copernicus himself, and in this he was followed by a number of writers[16]. Are we to suppose that Copernicus concurred, when Rheticus mingled astrology with his master's astronomy? Thorndike thinks that Copernicus probably agreed with the new twist which had been given to what, as I have already pointed out, was essentially an old doctrine about the rise and fall of earthly powers[17]. In support, Thorndike notes that Copernicus owned a copy of Albohazer Haly[18], and that he added notes to his copy of Ptolemy's *Tetrabiblos*. Each must judge for himself whether this is a convincing mode of argument. Copernicus was taught astronomy by men who, for the most part, must have been astrologically inclined (John of Głogów, who taught him at Cracow, had written on the consequences of the conjunction of Saturn and Jupiter of 1504). Copernicus does not appear to have been so hostile to the

---

[15] *Op.cit.*, p. 371. The same idea occurs again on p. 386.

[16] Jean Bodin, *Les Six Livres de la Republique*, Paris, 1576, pp. 441 — 2: "Quant a ce que dit Copernicus, que les changements et ruines des monarchies, sont causees du mouvement de l'eccentrique: cela ne merite point qu'on en face ny mise, ny recepte: car il suppose deux choses absurdes: l'une que les influences viennent de la terre, et non pas du ciel: l'autre que la terre souffre les mouvements, tous Astrologues ont tousiours donne aux cieux hormis Eudoxe". A rough and ready account of the Copernican Theory of the earth's triple motion follows; see Thorndike, for other evidence, *A History of Magic and Experimental Science*, vol. v, Chicago, 1941, pp. 418 — 9.

[17] *Ibid.*, p. 419.

[18] *Liber completus in iudiciis astrorum*, Venice, 1485.

subject as to have ever tried to persuade a later generation to abandon its ways, but his silence might equally well indicate his disdain. There is no doubt that Rheticus was a firm believer for most or all of his days, and he is even known to have begun, at the suggestion of friends, a separate astrological commentary on Copernicus' work, occasioned by the conjunction of Saturn and Jupiter on 25 August 1563[19]. The commentary was presumably never completed.

Whatever the attitude of Copernicus himself, the man to whom he dedicated *De revolutionibus*, Pope Paul III, is known to have been exceedingly superstitious. He was the recipient of numerous astrological predictions made for him by such men as Thomas Griphus of Spoleto, Paris Ceresarius of Mantua, and Alfonso Pisano. In the very year of Copernicus' dedication, Paul III commissioned Luca Gaurico to select the most favourable moment for laying the corner-stone of the building about the church of St. Peter[20]. An assistant, Vincentius Campanatius of Bologna, observed the sky with an astrolabe to judge the most propitious moment, whereupon a cardinal, clad in white stole and red tiara, set up the symbolic block of marble. Lest there be room for doubt as to Paul's state of mind, Vincentius Francisucius Abstemius gave an oration that throws much light on the pontiff's devotion to astrology, which he is said to have loved and cherished, and on which he devoted no small part of his time[21].

It is impossible to summarize in a few sentences the highly complex attitudes of the different churches to astrology, but the doctrine of a simple association of judicial astrology with popery was initially no more than a product of crude protestant propaganda. The Counter-Reformation Church took a firm stand against the occult arts after the Decrees of the Council of Trent and the Papal Bulls of 1586 and 1631, even though we rarely hear of astrologers falling foul of the Inquisitions. On the whole, necromancy, chiromancy, nigromancy, geomancy, onomancy, and like forms of magic, were reckoned to be much more dangerous to their practitioners' souls[22]. The inquisitors were enjoined to proceed, not against the application of astrology to husbandry and medicine, but only against

---

[19] "Hoc tempore in manus sumpsi opus Copernicii, et cogito illud illustrare nostris commentariis". Letter to Thaddaeus Hayck, in Prag. Printed first in 1584. See Burmeister, *op.cit.*, vol.iii, 1968, letter 46.

[20] Thorndike, *op.cit.*, vol. v, p. 259. Gaurico had twice predicted that the papacy would come to Paul, when he was yet Alessandro Farnese. Gaurico was duly rewarded by being knighted, and made a table companion, and then Bishop of Giffoni.

[21] *Ibid.*, p. 265.

[22] See, for an especially good account of the whole subject, D. P. Walker, *Spiritual and Demonic Magic from Ficino to Campanella*, 1958, pp. 205—20.

those who predicted events in such a way as to suggest a restriction on the freedom of the human will. Agrippa, Porta and Cardan, all of whose works were prohibited, were thought guilty of propagating spiritual error, but not in their astrological writings. Cardan was actually imprisoned and kept under house arrest for a total period of 163 days, and yet, as though to emphasise the uncertainty of the Church, he retired with a pension from Pius V[23]. The Venetian nobleman Francisco Barozzi was an astrologer who was sentenced by the Inquisition, and one who positively sided against Copernicus, but the condemnation was consequent upon the discovery of books of magic in his possession, and did not relate to any erroneous astrological doctrine[24]. Astrology as such, whether or not it was tainted with Copernicanism, does not appear to have been, in practice, a very hazardous pursuit within the orbit of Rome.

I have previously mentioned some slight evidence for supposing that astrology was more of a Protestant than a Catholic weakness, but again this must be qualified in numerous ways. John Calvin, Theodor Beza, Heinrich Bullinger and Peter Martyr were amongst those who preached actively against astrology, and the Protestants in England under Edward VI followed their example. The Puritan George Gylby translated Calvin's treatise into English in 1561, and of the many others in England who wrote in the same vein in the century ending with the Civil War of the 1640s, only a small minority were without affiliations to the Calvinist or Puritan schools of religious thought[25]. The arguments advanced by the several opponents of astrology were, however, less intellectual than spiritual, and few of them were prepared to rule out the possibility that prognostication would become a precise and powerful discipline. They feared rather that in order to be successful, divination required an alliance with the Devil. "Let us not be ashamed to be ignorant in a matter in which ignorance is learning", Calvin had written, adding that we should willingly abstain from the search after knowledge, to which it is both foolish as well as perilous and even fatal to aspire[26]. It is amusing to observe that the most implacable opponents of astrology, with its often tacit fatalism, were precisely those who themselves preached predestination, albeit predestination of a different sort. But among the

[23] *De vita propria liber* (translated by Jean Stoner as *The Book of My Life*, London, 1931), end of cap. 4. For reference to several biographies more reliable than Cardan's own autobiography, see Jean Stoner's introduction.

[24] Thorndike, *op.cit.*, vol. vi, pp. 154 – 5.

[25] Keith Thomas, *Religion and the Decline of Magic*, London, 1971, p. 367. Thomas gives many examples. He names as the two prominent exceptions the crypto-Catholic Henry Howard, earl of Northampton, and the future Straffordian, John Melton.

[26] Quoted by Thomas, *ibid.*, p. 370 (Calvin, *Institutes*, III, xxi. 2).

leaders of the English Church as a whole, for example, particularly among the Laudians (including Archbishop Laud himself) and Arminians, as well as among the Independents and radical sects of the Civil War period, and even among those who set themselves up to turn the attention of the universities to more rational pursuits, there were supporters of astrol-ogy far outnumbering those who actively opposed it[27]. It seems, there-fore, that religious orthodoxy, however defined, can give at best a rough and ready index of astrological belief. It is also noteworthy that in all the religious controversy which centered on this subject, the name of Copernicus is inconspicuous.

Religious considerations apart, an enormous variety of attitudes to astrology was manifest after Copernicus. Certainly the more competent a man was in mathematical astronomy, the more probable it is that he supported astrology; but this is rather like saying that the more a man knows of musical harmony, the more likely he is to play a musical in-strument. Exceptions to the rule are not hard to find, but even they are often more apparent than real. Alessandro Piccolomini, for instance, argued that astrology was at fault, but only because it was not sufficiently well developed. The English Puritan theologian William Fulke, who took a very similar view, chose to attack individual astrologers rather than the art as a whole[28], and later in life even went so far as to publish an astrological table game[29]. Many of those who preached against astrology, however, like Savonarola, were sadly misinformed as to the details of the subject they were attacking; others, like the medical man Jacques Fontaine[30], believed in only selected astrological doctrines; while others thought that astrology had reached an impasse, which would be crossed only granted a fuller knowledge of the demonic agencies that were sup-posedly intermediaries between stars and men. This last idea was spread through northern Europe by the books of such writers as Paracelsus and Cornelius Agrippa, who were in turn carrying the neo-Platonic and Hermeticist message of the Florentines, Ficino and Pico Della Miran-dola[31]. Their work stimulated an intellectual concern with the demimonde of the occult, far surpassing anything done for astrology by Coperni-canism. Hermeticism, on the other hand, proved to be far more suscep-

---

[27] Thomas, *ibid.*, pp. 358—85.

[28] *Antiprognosticon*, 1560. The book makes no mention of Copernicus.

[29] "*OYPANOMAXIA, hoc est, astrologorum ludus, ad bonarum artium, et astro-logiae in primis studiosorum relaxationem comparatus, nunc primum illustratus, ac in lucem aeditus*, London, 1572.

[30] *Discours de la puissance du ciel sur les corps inferieurs, etc.*, Paris, 1581. Fon-taine believed in general, but not specific, astrological prediction, and believed in the potency of demonic influence.

[31] Frances A. Yates, *Giordano Bruno and the Hermetic Tradition*, 1964, passim.

tible to scientific fashion than was astrology, and its underlying animistic philosophy led to its eventual neglect in the 17th century. Astrology, which has for most of its long history been a dogma with a code of practice requiring no clearly expressed philosophy to explain claimed correspondences, muddled through to a point well beyond the middle of the 17th century, its intellectual pretensions as great as ever.

These pretensions were first widely seen to be hollow when they could be contrasted with the more palpably scientific achievements of the later part of the century. Before this happened, and notwithstanding the revolution in astronomy, tradition was of supreme importance to the astrologer, although in some quarters, humanistic influences led to a mild reaction against Arab strains of thought. (It is significant that Cardan's long astrological commentary was on Ptolemy's *Tetrabiblos*.) Others were indiscriminate in their praise of any early authority. Oger Ferrier, who was writing in Copernicus' lifetime, but who was still popular a century after *De revolutionibus*, made this characteristic affirmation of trust in tradition:

> Leaving aside the Animodar of Ptolomie, and the meetings of Schoner, and all others uncertaine waies (although they have their Authors) to verifie the houres of Nativities, I will presently follow the method of Hermes, approved by long experience and confirmed by Ptolomie in his Centiloque, and by Abraham Avenesrus, by Alphonse, Leopold, Haly, and other most expert Astrologians[32].

This passage reveals an obsession with an essentially geometrical problem, which to the average astrologer must have seemed a much more fundamental matter than the heliocentric doctrine. It was an obsession perpetuated by Kepler, whose English interpreter John Booker, in 1652, contrasted "the Rationall way Alcabitian or Arabian" with the "Radicall scheame"[33]. The new method proposed by Regiomontanus for casting the houses was, if we are to judge from the work of astrolabists, not unpopular; and yet |that the different methods gave radically different results seems neither to have worried astrologers unduly, nor to have induced them to make tests to decide on the most effective method. It was a rather ludicrous argument which Jacques Peletier offered, to the effect that astrology proves its merits by surviving so fundamental

---

[32] Quoted from *A Learned Astronomical Discourse of the Judgement of Nativities*, tr. Thomas Kelway, 1642, p. 3. The *Ashmole copy* (Bodleian Library, Ashm. 539) is heavily annotated, much of the annotation being in shorthand. The first edition was printed at Lyons in 1550, and was reprinted in 1582. The author was sometime physician to Catherine de Medici.

[33] *MS Ashmole 348*, f. 36r: "The most excellent and true way of Direction of all the Significators in a Nativity invented by the Learned and accurate Mathematician John Kepler as in his Sportula Genethliacis missa, in *Tabulis Rudolphinis*, pag. 124, etc., in 92 et 3". A date 1652 occurs at f. 48r.

a difference of opinion! Peletier, as one might have expected, was not a thorough-going traditionalist, and he seems even to acknowledge the possibility of innovation in Copernican terms, although he apparently never fulfilled a vague promise to expatiate futher on the Copernican scheme[34].

Astrology was a deeply ingrained habit of the western mind, a habit satisfying a need more religious than rational. Astrology purported to be systematic, and perhaps for this reason it has been claimed as an intellectual exercise. But this it was in a very limited way; it is an odd sort of intellectuality which dismisses or ignores all normal criteria of test, logical and empirical, whenever they appear hostile to the guiding tradition; whose intellectual origins are not even known. This is where astrology parts company with Copernicus. After Copernicus, no less than before him, astronomers fell back on tradition and authority. Copernicus himself did so. But most who did so were aware, however dimly, of the logical character and status of the argument they were accepting. When the same men took up their astrological pens, they entered a world where there was always a ready-made excuse for a wrong prediction. In the last resort, God himself could be invoked as the unknown influence, whose judgement and will alone disposed of all contingency[35].

The pretentiousness of the astrologers was of course closely related to the relative poverty of alternatives. Astrologers never tired of boasting that there was no human event which did not have its counterpart in the heavens, from the prediction of which an adept could prejudge all human affairs. There was in those days no psychologist or sociologist ready to offer a battle of wits. As the fashion for scientific discourse in the Baconian manner took hold of the seventeenth century, however, attempts were made to bring astrology into line with the rest of science. A Baconian who tried to do justice to Copernicus was the English schoolmaster and clergyman Joshua Childrey. In his first short tract of 1652, Childrey was obliged to confess that Bacon himself, like the ancients, was wrong to argue that the Earth is stationary[36]. Galileo, Philolaus, and "other

---

[34] *Commentarii tres* ... III. Basle, 1563, p. 52, "De constitutione horoscopi, hoc loco veterum opinionem sequor, huic argumento minime contrariam; alias Copernici sententiam pulchris illustraturus argumentis, Deo juvante".

[35] Tycho Brahe was wise enough to be uncertain, and religious enough to make God his unknown. Round the sides of a royal horoscope, for example, he could write Potest — fata augere — Deus — tollere fata. See J. L. E. Dreyer, *Tycho Brahe*, 1890, p. 153 (c.f. p. 152).

[36] *Indago astrologica, or, A Brief and modest enquiry into some principal points of astrology, as it was delivered by the fathers of it, and is now generally received by the sons of it.* London, 1652, preface. The tract is only 16 pp. long. For a specimen of Bacon's mature opinions, see *Works*, ed. Spedding, Ellis and Heath, (1857—74), vol. vi, p. 127.

accurate observers", had demonstrated the movement. Even so, he complained, planetary aspects were still calculated according to the system
of Ptolemy. They should be calculated not *quoad nos* but *quoad naturam*,
taking into account the orbital positions with respect to the Sun[37]. He
never questioned the significance of the standard interpretations of the
aspects, however, which he argues were vouchsafed to the ancients (empirically, of course) at a time when aspects as judged from the Earth happened to coincide with true aspects[38].

Childrey followed this publication with an ephemeris in which he
further argued for the Baconian approach, introduced planetary latitudes into his calculations, and tried his hand at weather prediction.
One result of his "experimental astrology" was to confirm the old idea
that there is a 35-year cycle in the weather[39]. This was Copernicanism,
but of a sort which Copernicus would have found difficulty in recognizing.

Childrey's astrology, whether or not it was in principle scientifically
laudable, led nowhere; but it was at least preferable to the apologetics
of those who, from the earliest times to the present day, have maintained
that the stars incline but do not compel, and that an astrologer's fee
is given for making, not a definite prediction or judgment, but only a probable conjecture by natural causes of what may possibly happen if the
influence of those celestial bodies be not restrained[40]. There is something
ironical about a profession whose adherents will make such apologies
as this, and who will yet give planetary longitudes to minutes, seconds,
thirds, and even smaller sexagesimal fractions of a degree. This, however,
was the astrologer's dilemma: he was a man at least dimly aware of the
rich intellectual content of the most intricate and quantitatively precise
science then known, who yet aspired to predict, and by conceptually
similar methods, what we still find unpredictable. Tycho Brahe in his
early years was, on the surface at least, an orthodox astrologer, but
he could see the folly of trying to work a system when the chief astronomical tables available were divergent. It was widely felt that this

---

[37] I have no wish to suggest that there were no precedents for Childrey's astrology. Thomas Bretnor, for one, a dyed-in-the-wool Copernican Almanac maker
of the early years of the century, preferred to speak of earth in Libra rather than
of Sun in Aries, and so on. See his *A Newe Almanacke and Prognosticacion for ... 1616*,
London, 1615, sig C 3r. There must have been many others who wished to make such
terminological adjustments.

[38] See sections 2 and 20. In section 17, it is noted that the geocentric conjunction of Jupiter and Saturn of May 1652 really occurred on approximately 10 April.

[39] For further remarks on Childrey's meteorological work, and his *Britannia
Baconica* (1660), see my article in D. S. B., vol. iii, p. 248.

[40] Nathaniel Sparke (quoted by Keith Thomas, *op.cit.*, p. 335), *Ashmole* 356,
f. 4v.

must be remedied; and here Copernican astronomy had its greatest influence on astrology, an influence which must not be gauged by astrological success, but by the fervour of those who calculated tables from partly astrological motives.

The discrepancies to which Tycho drew attention will give some idea of the problem. He noted on one occasion that the *Alphonsine tables* differed from the *Prutenic* by nineteen hours as to the time of the vernal equinox of 1588, and that it was therefore not surprising to find two different astrologers arguing for different rulers for the year[41]. Tycho had previously noted that the *Alphonsine tables* were a whole month in error over the conjunction of Saturn and Jupiter in 1563, while the *Prutenic* could hardly find the day correctly, let alone the minutes and seconds[42].

It was upon the *Prutenic tables*, nevertheless, that the best astrologers based their ephemerides and individual predictions after 1551, the year in which they were first published by Erasmus Reinhold at Tübingen. John Stadius used them to produce perhaps the most widely used ephemeris of the period 1554–70 in 1556, and in the preface to Philip II he affirmed the possibility of weather prediction and the prediction of the fortunes of principalities by their use[43]. (The capture of Francis I was predicted to the day and the hour by a Franciscan friar of Mechelen, and so forth.) In the 1570 edition, the preface by Stadius, besides praising Copernicus and Reinhold, includes much new astrological material, among which is the notorious Cardan horoscope for Edward VI, the boy king of England. (Details of a disease which, it was said, would afflict Edward at the age of 55 could not be confirmed, since he died shortly after Cardan's prediction, at the tender age of 16.) Gemma Frisius wrote a letter that also prefaces the 1570 edition, weighing the merits of the Ptolemaic and Copernican theories. The popularity of these ephemerides is beyond doubt. John Dee found an unexpected bonus in their generous margins, in which he wrote his diary[44].

Stadius' predilection for astrology is to be seen in his *Tabulae Bergenses* (Cologne, 1560), where he reminds us of Rheticus' astrological ideas. The changing solar apogee will at length, Stadius believes, produce

---

[41] Dreyer, *op.cit.*, p. 155. Tycho was being consulted by Duke Ulrich of Mecklenburg-Güstrow, who had, by one man, been told that the year was governed by two beneficent planets, by another that it was governed by two maleficent!

[42] *Ibid.*, pp. 52–3.

[43] *Ephemerides novae et exactae ... ab anno 1554 ad annum 1570*, Cologne, 1556. It is possible that the ephemerides were at first published annually. A later edition of 1570 covered the period 1554 to 1600.

[44] The diary is now in the Bodleian Library, as *Ashmole* 487 (continued in *Ashmole* 488).

the final conflagration of the world, in keeping with the prophecy of Elias.

Cornelius Gemma (the son of Gemma Frisius), like Stadius, found fault with the *Alphonsine tables*, and greatly admired the work of Copernicus, while still being dissatisfied over the *Prutenic* figures for Mars and Mercury. (The error for Mercury could exceed eleven degrees.) Unlike Stadius he began to question the doctrines of *Tetrabiblos*, which often failed to give accurate weather prediction[45]. He accordingly planned to reform astrology, perhaps seeing himself as a worthy astrological counterpart to the astronomer he so admired. Many more took over the *Prutenic tables* without a murmur. John Feild's derived ephemeris for 1557 was provided with a preface by the author and another by John Dee, both heaping praise on Copernicus, Rheticus, and Reinhold[46]. Feild, like Stadius, was driven by astrological motives. Not until the end of his work did he know of the earlier ephemeris, and in fact he felt constrained to criticize its accuracy[47]. The precise nature of Dee's Copernicanism is still somewhat obscure. On the whole, he seems to have regarded its computational advantages as outstripping its philosophical desirability. But as Meric Casaubon was to remark, Dee was a Cabalistical man, up to the ears, while his feeling for astronomy as an intellectual pursuit independent of astrological applications is, to say the least, uncertain.

Others followed Stadius' example. Francesco Giuntini's *Speculum astrologiae* (Leyden, 1573), a compilation of earlier astronomy, included tabulae resolutae for planetary positions in keeping with Copernican data[48]. Ephemerides with a like astrological purpose, and based on the *Prutenic tables*, were issued by many other writers, including Thomas Finck, Michael Maestlin (who first took the Stadius ephemerides from 1577 to 1590, but not without noting the faults inherent in the *Prutenic tables*), Edward Gresham, Thomas Bretnor, David Dost (Origanus), and Andrea Argoli[49]. Argoli computed a second series of ephemerides

---

[45] *Ephemerides meteorologicae anni MDLXI*, 1661.

[46] *Ephemeris anni 1557 currentis iuxta Copernici et Reinholdi canones ... supputata et examinata ad meridianum Londoniensem*, etc., London, 1556, Sig A i-iv.

[47] *Ibid.*, See *Lectori*, on the penultimate printed page of the book.

[48] *Op.cit.*, (*Ashmole* 487), p. 17. In the Stadius volume containing his diary the words *Revolutionum non tam efficax doctrina*, do not refer to Copernicus, as has been supposed. Dee's *Propaedeumata aphoristica* (London, 1558) is at no point Copernican, so far as I can see. Junctinus is the more familiar Latin version of Giuntini's name.

[49] Many compilers of almanacs used their prefaces to argue in the Tychonic/Copernican/Ptolemaic controversy, and it is a recurrent theme that astrology is not much affected one way or another. English almanac makers of the 17th century worth mentioning in this connection are Arthur Hopton, Thomas Bretnor, Nicholas Culpeper, Thomas Streete, Vincent Wing, and John Gadbury.

(covering 1630—1700), calculated from his own tables, which were in turn based on Tycho's observations.

Although astrologers continued to use the *Prutenic tables* far into the 17th century, many were conscious of the tables' imperfections. Tycho Brahe's deathbed request to Kepler concerning the publication of the *Rudolphine tables* is well known. After his death, Tycho's heirs dedicated to Rudolph a work they had seen through the press, on the nova of 1572. They praised the Emperor's forebears (Alfonso of Aragon and Castile, Albert of Austria, Frederick and Charles V — astronomers all), but commented on how lame and defective were all the costly tables of Alfonso. They acknowledged that the defects "divers learned men have endeavoured to supply, and especially the most famous Copernicus, who was yet much hindered by wanting fit instruments"[50]. Tycho, of course, had gone far towards supplying the true alternative, but had died before his work was complete. This prompted James VI of Scotland (I of England) to commend Tycho in an elegy; the translator (V. V. S.) to write a verse underlining the astrological import of Tycho's observation of the new star; and a reader of Elias Ashmole's copy of the work, now in the Bodleian Library, to write these words under the Preface:

"The loss wee received by Tycho Brahe his death no toung can express: oh that hee had finished the tables of the celestiall motions, ere deathe had taken him"[51].

News apparently travelled slowly. The *Rudolphine tables* had been published by Kepler in 1627 (at Ulm), and on their title page there was no hiding the fact that they had been conceived and effected by that Phoenix of astronomers, TYCHO.

I realize that in singling out those astrologers who owed something to Copernicus, I have given a distorted image of 16th and 17th century astrological thought as a whole. One cannot capture the thought of a nation by reference only to what is spoken by its political leaders. I have said nothing of the charlatans who exploited the credulity of educated no less than uneducated men, and virtually nothing of those who, like Dee, were up to the ears in natural astral magic, and whose celestial hieroglyphics were closer to a linguistic and psychical parlour game than to astronomy. I have said nothing of astrological medicine, with which Copernicus must have been very familiar, but for the decline of which

---

[50] Quoted from an English translation, *Learned Tico Brahe, his Astronomicall Coniectur of the new and much Admired Star which Appeared in the year 1572*, London, 1632. Astrology was very much a royal pastime, of course. As Shakespeare wrote "When beggars die, there are no comets seen" (*Julius Caesar*, Act II, Scene 2, 11 30—1).

[51] Bodleian Library, *Ashmole* 161 (2).

he can hardly take any credit[52]. Copernicus created an instrument with which the Peripatetics could be chastised, but which left astrology unscathed. Radicals of all persuasions, with Galileo as their archetype, made of it what they could, and the name of Copernicus was on the lips of many who knew nothing of his work, beyond the fact |that it offered a means of overthrowing Aristotelian cosmology. If astrology was a casualty, it was by accident rather than design. The second (in point of time) of the three greatest astronomers of the 16th century, Tycho Brahe, was conservative to a fault. The third, Johann Kepler, was temperamentally incapable of using any Copernican argument without first examining it in the minutest mathematical detail. Not one of the three astronomers offered direct refutations which could possibly be held responsible for the eventual eclipse of astrology. Two of the three, at least, were themselves deeply implicated in the subject at one time or another. Each in his own way, nevertheless, helped to bring about precisely this eclipse of astrology, by providing in its place an intellectually more rewarding cosmology. And it was Copernicus who took the first step in this direction.

---

[52] The decline was probably a consequence of simple ignorance on the part of medical practitioners, vis-à-vis astronomy, together with an improvement in medical knowledge *per se*. Note, however, that the almanac-maker and astrologer Richard Forster was twice made President of the Royal College of Physicians in the early 17th-century, and others of the Fellows were practising astrologers, even into the second half of the century.

3

BETWEEN BRAHE AND BAER

The *Apologia pro Tychone contra Ursum* was written around the end of the year 1600 by the best theoretical astronomer of his age, in defence of the best observing astronomer – who happened to be his employer. Its author, Johannes Kepler, was not yet thirty. Tycho Brahe, Danish nobleman and by now a satellite of the court of Rudolph II in Bohemia, was in his fifties, and had less than a year to live. The *Apologia* was an embarrassment to the younger man, but he was under a strong moral obligation to issue it. In 1595, as a young man with his way to make in an inhospitable part of the world, he had written to Nicholas Reymers Baer (the "Ursus" of the title) in praise of his mathematical talents – a safe enough gesture to make towards the Imperial Mathematician. Baer included the letter in a work of 1597 that bitterly attacked Tycho, who was not amused. When Tycho died, the unpublished manuscript copy of Kepler's defence passed into that limbo to which only historians have access.

It was first published by C. Frisch in 1858 for his collected edition, and has been discussed sporadically since then, but never with the care now accorded to it by Nicholas Jardine. Nearly half of his book is taken up with a new edition and translation of the original – the former done from the autograph manuscript, now in the USSR, although parts of this have been lost since it was seen by Frisch. Variant readings from Frisch's edition are something that historians of the history of science will have to provide for themselves. The letter is one thing, the spirit another, and this its editor has captured well. His title gives away his belief "that if any one work can be taken to mark the birth of history and philosophy of science as a distinctive mode of reflection on the status of natural science it is Kepler's *Apologia*".

A number of different views have been taken of Kepler's posture in this *opusculum*, even so. Karl von Prantl(1875) held in esteem its solid inductive method – so different, he thought, from the wild metaphysics and theology of Kepler's earlier writings. Not that this was a new

JARDINE, N. [1984], *The Birth of History and Philosophy of Science: Kepler's "A Defence of Tycho against Ursus"*, Cambridge: Cambridge University Press, pp. 301.

perspective on Kepler: of his *Astronomia nova* Robert Small had written in
1804 that "it exhibited, even prior to the publication of Bacon's *Novum
Organum*, a more perfect example, than perhaps ever was given, of
legitimate connection between theory and experiment; of experiments
suggested by theory, and of theory submitted without prejudice to the
test and decision of experiments". Such were the prevailing attitudes to
scientific respectability in the last century. Jardine, like a number of other
modern interpreters, is able to find common ground between the
Apologia and the earlier *Mysterium Cosmographicum*. Both works include a
plea for the founding of astronomical hypothesis in physics and
metaphysics. The central theme of the *Apologia* concerns the
establishment of criteria for the resolution of theoretical dispute, and the
occasion for its composition was a sceptical attack on a position towards
which – in the last analysis – he was essentially sympathetic.

It is ironical that it should all have begun with a dispute with Tycho
over priorities, for Tycho's own view of his supposedly plagiarized
world-system was not cast in the same sceptical mould as Baer's, by any
means. As a taste of Tychonic argument: the Earth cannot have the triple
motion Copernican astronomers advocate, since (a good Aristotelian
argument this) a single body can only have a single natural motion.
Tycho was a creature of the cold northern nights; it was easier to be a
sceptic in the warm logic schoolroom. Although its history is venerable,
the notion that astronomical hypotheses carry with them no implications
as to the reality of the things they seem to describe had been given
excellent publicity in the sixteenth century through the anonymous
preface to the *De revolutionibus*. (This had been penned by Osiander – as
Kepler was the first to prove – and was not true to Copernicus's
sentiments.) The Tychonic system is closely related to the Copernican,
although for Tycho, the Earth was at the centre of the planetary system.
It was as though he had simply taken out the drawing pin from the
middle of Copernicus's diagram, where it was near the Sun, and inserted
it through the Earth. To us it might seem rather obvious, but in fact
Tycho was stung by Kepler's perfectly justifiable remarks made on a
number of occasions, that the transformation was a simple one, and that
it had been affected by others, probably independently. Tycho would
have had them all stigmatized as plagiarists – and indeed one of the more
colourful stories of supposed scientific plagiarism centres on Baer's visit
to Uraniborg, Tycho's Danish observatory. A student warned Tycho of
Baer's suspicious behaviour. (He was, after all, a farmer's boy from
Ditmarsch.) A certain Andreas searched Baer in his sleep, and removed
from his pockets papers, the loss of which began a mental disturbance
that led to his eventual loss of patronage. Things are much better in
Denmark now.

There is more than colour in this dispute, for from it we may learn
something of what the disputants thought was the nature of the stolen

intellectual property. They complained of the appropriation of constructions, inventions, formulae and associated tables, but not of fully articulated world-systems. It was here that Kepler showed a deeper insight than either of them, in insisting that Tycho's claims hinged on what Jardine calls "his articulation of a detailed system of astronomical hypotheses answerable to careful observation . . . something approaching the modern notion of a theory . . .". A particularly telling sentence in the *Apologia* includes the words "when we speak in the plural of astronomical hypotheses . . . we designate thereby a certain totality of the views of some famous practitioner".

In addition to the philosophical content of its opening section, Kepler's defence was steered into a historical direction by Baer's having maintained that Tychonic hypotheses had been anticipated by Apollonius, as well as by Copernicus. If his essay seems at times historically naive, that is in part to be explained by a lack of texts – and of this he was very conscious. His interpretation of those he possessed is generally well balanced, even when he is constrained by the need to refute Baer point by point. It cannot have been easy, with Tycho at his elbow.

Jardine does not acknowledge the existence of a "historiographical category even approximately corresponding to history of science in the sixteenth century", but accepts the genre of what he calls "prefatory histories" – such as are embedded in dedicatory letters, prefaces and so on. No doubt he is right to see in them at least a part of the tradition of the *Apologia*, but it would be a mistake to overlook the preoccupation astronomers have always had with history. History, for astronomers, was a potential source of empirical data, and was valued accordingly. It was a part of the astronomer's experience that did not usually reveal itself in his writing, although there was a long tradition of listing genealogies of doctrines (Adam, Abraham, Hermes, Timocharis, Hipparchus and so on) which has left its traces in the present work. Then there was the Renaissance ambition, if not to recover the great achievement of antiquity, at least to provide one's subject with dignity by association. Thus Kepler – as Bruce Eastwood has shown – tried hard to establish a pedigree for the geoheliocentric notion going back to Plato. He also made use of a third historical genre fostered by astronomers – that which relates human affairs, in particular religious affairs, to the conjunctions of the planets. Although assembled with his customary verve, these elements in Kepler's writing are not in character very startling. Jardine is hardly stretching a point, though, when he insists that "at a deeper level . . . [Kepler] is concerned to validate both his own view of astronomy as a discipline that has yielded ever greater insight into the constitution of the world and his own account of the means whereby such progress can come about through integration of mathematical astronomy into natural philosophy".

This brings us to the philosophical treatment accorded to the arguments offered by the sceptical (as well as virulent) Baer, on the status of astronomical hypotheses. Both he and Kepler indulged in somewhat fanciful histories of the meaning of the word, but Kepler took the biscuit for rhetorical skill in his use of the device of *coloratio*, insinuating, as Jardine has it, "the impropriety of a sceptical attitude to astronomical hypotheses, intimating that in the transfer of the term from geometry *via* logic to astronomy the term preserved its original connotations of certainty and evidence". But then we come to the nub of their disagreement. For Ursus, it was a well-worn truth that accurate prediction does not guarantee the validity of any astronomical hypotheses on which it is based, since true conclusions may follow from false premises. This was a simple logical point, and one often made in the same context in the Middle Ages. Astronomy has long been rich in observationally more or less equivalent models, whose truth was thus seen as either an open question, or one to be settled in some other way.

To this appealing argument Kepler answered essentially that (to take an example) the astronomical system of Ptolemy merely predicts phenomena, whereas the equivalent Copernican astronomy reveals in addition their *cause*. Copernicus did not deny the ancient hypotheses, but incorporated them in his system, which was to be preferred (so he seems to be saying) by virtue of its "most beautiful regularity". But then he seems to wish to speak of at least some of the kinematically equivalent hypotheses as false – as capable of yielding accurate predictions only by chance, and in the short term. Let them be used long enough, and we shall find them out. Jardine is right to remark the difficulties of the relevant passages from the *Apologia*, but it has to be said that Kepler's argumentation, while rhetorically strong, is essentially dependent on unsubstantiated hints as to aesthetic qualities, simplicity, harmonic elegance and the sort of arguments from physics that he had offered in the *Mysterium*, and was currently deriving from Gilbert's *De magnete*. It is easy to agree with him that such things matter; and yet as soon as one's opponent has been persuaded to accept the additional criteria for the comparison of theories, the same old problem as started the dispute might very well reappear – and where then does Kepler hope to turn?

If Kepler's position is a difficult one to hold – and I suspect that he has more supporters than critics – this is not to say that there was much to be said for the opposite camp in the sixteenth century. The tendency for philosophers today is to regard the opposition as monolithic, as "fictionalist", to use the current term of mild opprobrium. This is to blur a great many shades of opinion. Jardine does a useful service in drawing some of the necessary historical distinctions. There were, for instance, those who denied the possibility of knowledge of the heavens – not necessarily a universal scepticism, for even in modern times astronomy has been regarded as unusual in that we cannot substantially alter its subject-matter. There were those who took Ptolemy to be acceptably accurate but doubtfully true. There were the theologians like Osiander, who wanted to play safe, for doctrinal reasons. There were the astronomers who wanted to avoid debate with the natural philosophers. Most scholars of the time would indeed have thought such debate improper, for reasons Aristotle had made plain, when he split the world into terrestial and celestial realms. There were neo–Platonic sceptics, even those who argued for an astronomy without hypotheses as the Babylonians were supposed to have had, deriving their precepts (hypotheses?) from tables of observations. This was the view of the influential Ramus. Throughout, however, one thing is noteworthy: no one was proposing what is sometimes misleadingly called the instrumentalist view, namely that truth and falsity are not predicable of astronomical theories, or indeed of scientific theories of any other sort. Pierre Duhem has misled philosophers on the fringe of history for far too long into thinking that what Jardine calls the "pragmatic compromise" of Kepler's opponents was either a general epistemological stance or an endorsement of sceptical epistemology. What they shared was a reluctance to assimilate natural philosophy to astronomy. Kepler's greatness was not that of a philosopher (in the modern sense), but of an architect of such assimilation.

4

THE WESTERN CALENDAR: *INTOLERABILIS, HORRIBILIS ET DERISIBILIS*; FOUR CENTURIES OF DISCONTENT

> [*Kalendarium*] *est intolerabilis omni sapienti, et horribilis omni astronomo, et derisibilis ab omni computista.*
>
> Roger Bacon, *Op. ter.*, cap. 1 xvii

When it comes to calendars, we are most of us realists.[1] Although a calendar is essentially a means of referring a particular event to a day or a year unambiguously, so that other events may be correlated with it, the calendar is a system of names that have taken on a life of their own. It is a system of associations — of religious sentiment with season, for example, and of season with Name. "Rough winds do shake the darling buds of May", said Shakespeare at some date after the Gregorian reform, but well over a century before his own country had adopted it. He would have been ill served by his fellow countrymen had the seasons been allowed to slip through the calendar to a point where May became a winter month. There you have an obvious problem that most can see; but the Church's calendar is less obvious, for it is not only bound to the seasons; it is bound to three sorts of natural phenomena, or at least idealized natural phenomena, namely those marking the limits of year, month, and day. We, with our cut and dried view of time-keeping, we who gather together to celebrate events four hundred tropical years after they have occurred, are all too easily inclined to overlook the real reason for all the fuss in the middle ages about calendar reform. It was not primarily a matter of losing ten or eleven days of one's life (which seems to be the only consequence the average student of history is aware of), or of getting the agricultural cycles back to where they had once been in the calendar. (Had early Jewish history been known, where Passover was made to coincide with the earing of the corn, medieval astronomers would perhaps have been consumed with less scientific self-righteousness.) It cannot be said too often that the problem was more than that of getting the right value for the length of the tropical year, and inventing a rule to give a high level approximation to it. The real key to the movement for reform was the desire to celebrate Easter at the "correct" time. This was of concern, as Bacon said, to the computist and the wise, as well as to the astronomer; but of course it was of central importance to *every Christian*, for whom the death and resurrection of Christ were the most important events in human history.

1. *The Early History of the Christian Calendar and Its Astronomical Frame*

The "correct" time of Easter. It is hard to look into the history of Easter celebration without growing despondent at the excesses of misplaced scientific zeal to which it testifies. I have to say this at the outset, because I shall shortly be judging my characters by their scientific worth, that is, accepting their premises as to the way in which Easter should be computed, and assessing the extent to which they managed to satisfy those premises. As for Easter, the rule finally agreed was that it must be celebrated on the *Sunday next after (and not on) the 14th day of the Paschal moon, reckoned from the day of the new moon inclusive.* The Paschal moon is *the calendar moon whose 14th day falls on, or is the next following, the vernal equinox, taken as 21 March.*[2] At last we can see why the Gregorian reform mattered so much. The vernal equinox used to determine Easter was a conventional one, just as the calendar moon was conventional; both were clearly meant to be *good* conventions, and yet as the centuries went by, they both got steadily worse, judged by the phenomena. You might ask who cares, if they are truly conventional, whether Moon and equinox correspond with the phenomena. Not all theologians agreed as to the rightness of celebrating Easter when the positions of Sun and Moon are propitious, and some even thought that to insist on the "correct" (historical) disposition of Sun and Moon was to court astrological principles. This is not an aside to the subject of our meeting, for the scientific chronologists whose praises we are here to sing are the very ones who wanted an exact correspondence with the phenomena!

As for those Christians who wanted Easters to fit the historico-astronomical pattern, Dante is the supreme example.

Conventions as to Moon and equinox were inevitably imperfect, when measured against the phenomena — that, after all, is what the astronomical sciences are all about — but unfortunately the debate was often entangled with one over conventions having nothing to do with the observable. When scholars argued, for instance, for the use of the 84 year cycle in preference to the 19 year cycle, they would not have been inconsistent had they said, in effect, that they knew that it gave inferior results, but that there were virtues in the whole church sticking to the same deviation from the empirical norm. And yet the empirical test was *available*. In arguing, on the other hand, for or against an Easter falling as late as days xxi and xxii of the moon, only traditions could help. (The traditions included, of course, traditions of inclusive and exclusive counting!)

There are several possible ways of presenting the conventions of the first, empirical, kind, but there is nothing to be gained by losing ourselves in a sea of figures, so I will confine myself to two rough and ready assessments, one of the "slip" of the lunar cycle, the other the "drift" of the vernal equinox. (I use these words to avoid confusion). Suppose there are m days (actual) in a lunation. Those using a 19-year cycle (235 calendar months in 19 years) say in effect that the length of an average year is $235\,m/19$ days. They do not have to accept a Julian year, but if they do so their calendar year will be too long by $(365\frac{1}{4} - 235\,m/19)$ days, judging it, that is to say, by the length of the year implied by their lunar calendar *in conjunction with* a value for the lunation. Suppose now that *we* supply a reasonably good figure for this — say 29.530589 days, which is acceptable for our purposes (to a small fraction of a second) for the whole of our era. It is easily verified that if this and the "Metonic" ratio be accepted, the Julian year must be 0.00324 days too long. If we accept the Julian calendar as our norm, then we can say that the lunar calendar based on the "Metonic" cycle slips by one day in roughly 308 years. I shall here denote this important parameter by "M", and suppose that as a reasonable ideal, M = 308.5 during the historical period under review. Note that the value of M has nothing to do with the figure we should offer if asked independently for the length of the tropical year. It is a measure of the *joint* efficiency of two conventions, "Julian" and "Metonic" (although other conventions could be substituted for either of them), judged by the counting of days and the phases of the Moon.

To get an idea of the superiority of the 235/19 convention, as against the main historical alternatives, all taken with a Julian calendar, the old 99/8 convention (99 months making 8 years) leads to a slip of one day every 5 years, roughly speaking; the 136/11 convention slips by a day in roughly 7 years; and the 1039/84 convention slips by a day in roughly 66 years. One could make other comparisons, for example, between expectations, based on the calendar used, and those based on astronomical tables, such as the Toledan or Alfonsine. They would not be very different from mine. For the time being it should be enough to realise that a good observer of new moons (and lunar phases are not easy to observe in themselves) would have been very dissatisfied with any but the 235/19 convention within a century or so, at most, of the drafting of the moon tables he inherited; and that even tables drawn up on that convention (or its equivalents taking a 76-year cycle) would have shown their weaknesses within perhaps five or six centuries. Bede, in chapter 43 of his *De ratione temporum* (A.D. 725; a long work, since his brethren found the shorter *De temporibus* of 703 incomprehensible), noted that the full moon was ahead of its date as tabulated. Eclipses were often recorded as being ahead of the new moon of the tables. But more significantly, Robert Grosseteste gave a quantitative assessment of the accumulating error. In 304 years, he wrote, the moon is $1^d\ 6^m\ 40^s$ older than the calendar says.[3] Here he chose 304 as being four full cycles of 76 years. The minutes and seconds are minutes (sixtieths) of a *day*, and sixtieths of them — a point that some commentators have not appreciated. On this reckoning, the tables slip out of phase by one day in 273.6 years. Grosseteste's, like my 308.5, was a *calculated* figure, and had nothing to do with his own observations. In fact it rested on an approximation of $29^d\ 31^m\ 50^s$ (again minutes of a day) for the lunation, taken from Ptolemy but agreeing with *all* other serious writers, if we round off beyond the seconds, on this convention. (Ptolemy had 29; 31, 50, 04, 09, 20.)

The other conventions empirically testable are the tropical year length implicit in the Julian calendar itself, and the date assigned to the equinox. The length of the year may be taken as 365.2423154 days at the origin of our era, decreasing by about 6.13×10^{-6} days per century. Thus in the year 1000 the tropical year differed from the Julian by 0.00768 days, which means that on this account the whole calendar was drifting by one day in a little over 130.1 years. For the year 1500 the drift was one day in about 128.6 years. I will call the reciprocal of this number (J) the "Julian excess", so that the true length of the tropical year is $(365\frac{1}{4} - 1/J)$ days. The Gregorian reform effectively took $J = 400/3$, or 133.3. A better figure for 1582 would have been 128.5. (As a working rule, accurate to the nearest 0.1, the value of J for the year Y A.D. (between 1 and 1600) is given by 130.1 − Y/1000.) As we shall see, the general opinion prevailing in the sixteenth century was that J = 134, although there were rival opinions. The advantages of a system wherein we discount three bissextile (leap-year) days every four centuries are considerable, however.

The question then arises, where are we to find our value for J? One of the most striking things about the late middle ages is that very few of the figures quoted in this connection originated in the Christian world itself, vigorous though astronomical activity was. Even more significant — and I think there is a lesson to be learned here about the tenuous nature of medieval feeling for anything meriting the name of scientific progress — are the erratic and unprincipled changes in the values accepted for J. The reason is very simply the eclecticism of the time. Admittedly the Ptolemaic value of 300 is not much in evidence after the thirteenth century, but three important early sources, namely Al-Battānī, Thābit ibn Qurra, and Az-Zarqellu (Arzachel as I shall call him), with respectively J = 106, 130, and 136, all come to the surface regularly, and do so long after the Alfonsine 134, which enters the world of the church calendar around the mid-fourteenth century, a good seventy years after the original recension of the tables. (More precisely, when the value is implicit in "Alfonsine" work, rather than explicit, one finds J = 134.2) I shall discuss these unsystematic changes of heart as they occur. Finally, for future reference, I note that Clavius claimed that for the Gregorian reform the mean of Copernicus' maximum and minimum values for the tropical year was accepted, namely $365^d 5^h 49^m 16.4^s$, the mean of $365^d 5^h 55^m 37.7^s$ and $365^d 5^h 42^m 55.1^s$. The corresponding values for J are 134.2 (mean year), 329.4 (max.) and 84.3 (min.). The vital 134.2 is the Alfonsine value, and no juggling with figures is going to prove or disprove Clavius' contention, which must be left to the historian. Provisionally, we can say that the weight of Copernicus' authority was added to a long-discussed Alfonsine parameter.

The last empirical decision concerns the date of the vernal equinox and the rules for its adjustment — both necessary for a strictly "phenomenological" Easter calculation, if not for a workable civil calendar. If tropical coordinates are used in the theory of the Sun, then the "movement of the eighth sphere" (corresponding to the modern precession of the equinoxes, and in the middle ages often taken to be of the nature of a trepidation) is of no immediate consequence, and I shall have little occasion to mention it again. The drift of the equinoxes through the calendar by reason of the incommensurability of the Julian and tropical years is something of which I have spoken already; but I should add, for those who have not given the matter any thought, that the equinox's precise hour (and sometimes even day, when we are in the neighbourhood of the day division) [4] moves about in roughly six hour units from one year to the next, thanks to the odd quarter day in the tropical year. Most writers on the calendar are content to name the *date* of equinox (or solstice), "for the present time", showing at once their unscientific approach. As for the date, the general assumptions were that for the vernal equinox in the time of Julius Caesar, 25 March was correct; that at the time of the Council of Nicaea the date was 21 March, the canonical date; and that the current date was (from say 1200 onwards) of the order of ten days earlier. I will illustrate the dangers of these assumptions with an example. The Florentine monk John Lucidus, in a work on the emendation of the calendar written in 1525, and published in 1546, argued that since the equinox in Caesar's time was 25 March and since that in his own day it was 10 March, it had moved 15 days in 1590 years, making $J = 106$.[5] This comforted him, for it gave al-Battānī's figure! In fact the interval of 1590 years is not an integral number of leap year cycles; and in 64 B.C. ($- 63$, but an error for 45 B.C.), the first year after a leap year, the equinox was approximately 9 a.m on 23 March, at Greenwich. (In Florence roughly 45 minutes earlier.) In 1525, it was at nearly 5 a.m on 11 March (Greenwich). The difference would lead one to suppose that $J = 130.4$, substantially different from 106. Of course Lucidus was dependent on traditions he was simply incapable of verifying. He would have been only marginally safer had he turned to such astronomical data as those of *Almagest* for his information as to the date and time of an early equinox.

2. *Growing Dissatisfaction and Grosseteste's Proposal for Reform*

Complaints over the drift of the dates of equinoxes and solstices were as common in the early middle ages as those over the inadequacies of the ecclesiastical lunar calendars. Conrad of Strasburg, writing in the year 1200, commented on both phenomena, and set the retreat of the winter solstice at ten days, from the 25 March of antiquity.[6] The figure of ten days became canonical in some quarters for three centuries (although there was often confusion as regards the "base" date, since 21 March was that accepted by the Council of Nicaea) even though its progression was the very matter

under discussion. Conrad offered no programme of reform, but Grosseteste had done so perhaps before 1220, and Sacrobosco by perhaps 1232.
So many are the difficulties of dating their writings — for example on the
sphere — that the line of influence here is hard to establish with certainty,
but generally speaking Sacrobosco seems to follows Grosseteste rather
than lead him; and on the calendar he yet seems more superficial. As already explained, Grosseteste favoured a 76 year cycle (4 × 19), although
since this slips by a day in approximately three centuries (see above) he
recommended discarding one day in the lunar calendar exactly every three
centuries. He emphasized that the earliest Easter terminus must be 14
March rather than 21 March.[7] He stressed the importance of precise measurements of the length of the tropical year, and in his review of earlier data
(Hipparchus/Ptolemy, al-Battānī, Thābit, and others), as well as in his
admiration for the lunae-solar cycle of Arab time-reckoning, he set a fashion
that would last until the seventeenth century and beyond in Christian
intellectual circles. Sacrobosco, by contrast, looked back to more conservative sources, including Bede, who had said that the Julian year is too
long by a twelfth part of an hour (and thus $J = 12 \times 24 = 288$), and
Eusebius, who had said that in the year of Christ's birth there was a new
moon at 23 March. For my "M", the slip of the lunar calendar, he based
his calculation on Ptolemaic data — but with more approximations than
Grosseteste and made recommendations as to a shift in the lunar calendar
of four days (23 January to 19 January, 1232). He was prepared to go
against the Council of Nicaea as regards his final scheme, and in this he
helped to prepare the ground for more radical later reformers. As for the
ten day error in the date of equinox, he recommended tampering with leap
years — omitting the bissextile day for a period of forty years, thus shifting the calendar by the right amount. He never made it clear that he had
seen all the implications (especially of the interlocking of lunar and solar
calendars) of his scheme, but his writings complement those of Grosseteste,
and they provided patterns for future debate.[8]

Both by his criticism of Grosseteste and his borrowings, Campanus of
Novara made it clear that he had the Oxford chancellor and bishop of
Lincoln very much in his thoughts as he wrote on the calendar.[9] What
marks Campanus out from the great majority of contemporary computists
— take Vincent of Beauvais, for instance, as an almost exact contemporary,
still under Conrad's influence, but having somewhere picked up a value
of 120 for J — [10] is his astronomical expertise. It is not that Grosseteste
and Sacrobosco were incompetent, but that Campanus was a practised

calculator, *au courant* with much of the astronomy then creeping into Europe from the Islamic world; and here the theory of trepidation is a good example of something capable of separating sheep from goats. Like Grosseteste he was an admirer of the lunae-solar calendar (with its 30 year cycle of a 354 day year, into which 11 days are intercalated), as indeed was Roger Bacon, who in turn seems to have been an admirer of Campanus. (Bacon in 1267 named him as one of the four best contemporary mathematicians, and the English connection is an interesting one. Although it is not proven that he ever left Italy, he might have accompanied Cardinal Ottobono to England in 1265-8; and — whether or not *in absentia* — there is a good chance that he held the benefice of Felmersham in Bedfordshire from those years onwards).[11]   Perhaps his insistence that the cumulative errors in the Easter computus be corrected "by astronomical instruments and tables" carried some weight, but if so, there was no very obvious flurry of activity in the empirical determination of equinox and paschal full moon by his readers.

3. *Clement IV, Bacon, and His Time*

Much the same might be said of Roger Bacon, but with Bacon the whole subject takes on a new ecclesiastical dimension, with the sort of papal interest necessary — but unfortunately not sufficient — for an accepted programme of reform.   Bacon's entire career pivots around the papal mandate sent by Clement IV in 1266, and preceded (5 February, 1265) by a request by Clement (strictly speaking this was before he became pope, and was thus writing as Guy de Foulques) for a copy of his comprehensive philosophical writings about which rumours were then circulating. The *Opus maius*, no less than the *Opus minus*, a supplementary work with a suitably eulogistic introduction and dedication to Clement, not only reached papal eyes, but it came at papal behest; and the somewhat mysterious passage in the papal mandate that has been seen by some as an allusion to alchemical secrets might very well, it seems to me, be no more than a reference to an earlier suggestion (by Bacon to Foulques) concerning calendar reform.[12]   Note, incidentally, that a passage on the calendar occurs, word for word, in the *Opus tertium*[13] and in the *Opus maius*,[14] and that this repeats sentences taken from Grosseteste's *Computus*.[15]   Bacon himself acknowledged that other computists had found fault with the calendar before him; but one point is worth mentioning, namely his statement to the effect that $J = 130$, which he later changed to 125, both better than

Ptolemy. The first, admittedly, is found in other writers quoted from Thābit,[16] and need not have been original, whilst the second seems to rest on a calculation of the sort I gave earlier from the monk Lucidus, as liable to lead to appreciable error. Now it is that since the equinox in 140 A.D. (Ptolemy) was 22 March, and in 1267 13 March, the drift is to be reckoned as one day in 1127/9 (approx. 125) years. He was not to know that Ptolemy's measurement was 30^h in error.

Bacon's reform proposals, in brief, were thus the removal of a day every 125 years, together with either the adoption of the lunae-solar year of the eastern nations or a thoroughgoing astronomical approach to the fixing of Easter, even using Hebrew tables! If we are to divide our history into periods, Bacon would seem to mark the end of the first significant period of initiatives to reform. At more or less the time of his exchanges with Clement, the Alfonsine tables were being prepared, but they were not to penetrate northern Europe effectively for another seventy or eighty years,[17] and even after their arrival we shall find echoes of our first "heroic" age, the age of Grosseteste and Sacrobosco, Campanus and Bacon. After Bacon we find occasional evidence of equinox determinations — as for instance in the tract "to show the falsity of our calendar" by a brother "John of S." (not Sacrobosco) around 1273.[18] This is an important period, too, for the composition of computi: especially noteworthy from the point of view of his influence is William Durand (not Durand of St. Pourçain) and his *Rationale divinorum officiorum* of around 1280. This fundamental work in the history of western liturgy — and the first non-biblical work to appear in print, namely in 1459 — was in eight books, of which the last is a computus. Several computi appeared around the year 1292 (a year that can never be taken with confidence as a year of composition, even when it appears in titles, since it marks the beginning of a new lunar cycle, i.e. has golden number 1, whether on the 19 year or 76 year convention), including those of Peter of Dacia [19] and William of St. Cloud. William, who was writing for Mary of Brabant, Queen of France, helped to establish a fashion that cannot have failed to promote a sense of the fundamental importance of astronomy in the matter of the calendar. He added much highly technical material to his own — the solar altitude at noon, hours of new moon, diurnal and nocturnal arcs, and a table that allowed the conscientious calculator to make corrections to the basic table (covering four years only) of the entry of the Sun into the zodiacal signs, and *a fortiori* into Aries. Of course, this sort of conscientiousness led to the discovery of error and hence to frustration. Was it not Bacon who had drawn Clement's attention to the fact that in 1267 the Church was a week out, in its celebration of Easter? But then, frustration is the spur to reform.

4. *Changing Styles in Astronomical Tables. Firmin of Belleval, John of Murs, and Clement VI*

Apart from his *kalendarium regine*, William of St. Cloud compiled a planetary almanac — that is, an ephemeris, giving in his case planetary positions for twenty years commencing 1292, and based on the Toledan and Toulousan tables.[20] A whole generation later, such good astronomers as John of Murs and Geoffrey of Meaux were relying on the same well worn sources. John, in his work beginning "Autores calendarii nostri duo principaliter tractaverunt..." says categorically that the tables of Toulouse are the best available, and this in 1317; while in a calendar running from 1320 to 1340 (and therefore compiled around 1319 or 1320) Geoffrey said that he preferred the Toledan tables to the Alfonsine![21] This is the time of the Parisian reception of the later tables, and they were to make their presence felt in English circles shortly afterwards; and yet the Toledan tables were still in circulation in some quarters a century later. In his later work (1337?) on the art of the computus, John of Murs proposed certain reforms of a fairly conventional sort: he made complaints about the casual and inaccurate way computists had of expressing themselves; proposed that ten days be dropped either by suppression of the bissextile intercalation for forty years (Sacrobosco's proposal) or the shortening of a suitable number of months by a day each; and canvassed for the 76-year lunar cycle, which Grosseteste had done long before. The Church was not, however, in desperate need of anything more than the thirteenth century authorities, duly sifted, could provide, and John of Murs certainly had ecclesiastical interests at heart, as is clear from the advice he gave in response to Clement VI's request of 1344. (In the work of 1317, if indeed it was his, he had deplored the way in which the Jews had ridiculed Christians for their error in the date of Easter in 1291. This sort of complaint was not uncommon in the middle ages, and one must suppose that here we have another stimulus to reform, not exactly ecumenical, but anticipating some of the ecumenical motives of the fifteenth century.)

On 25 September, 1344 (rather appropriately, since the day was that of St. Firmin, Bishop of Amiens) Pope Clement VI wrote to Firmin of Belleval in the diocese of Amiens and to John of Murs, canon of the church of Mazières in the diocese of Bourges, inviting them to Avignon to consider and advise on the correction of the calendar.[22] Their expenses were to be met by the bishops of Amiens and Paris. The resulting treatise by the two scholars [23] begins by referring to the papal mandate, and names the cardinal of Rodez as the man who directed their work. The Easter calculation was uppermost in the pope's (and hence their own) reckoning; the solar calendar offered a simpler astronomical problem but a harder ecclesiastical one. If the equinox were to be moved through the calendar, what of the celebration of such fixed feasts as Christmas by the schismatic sects? What hopes for church unity, if different groups were to celebrate feasts on different days? Better to leave the solar calendar alone? Of course ideally they would have liked to shorten one year by the appropriate amount, even though it lead to tumult in the courts of princes, not to mention difficulties over civil contracts. In all their estimates of the true astronomical situation, be it noted, they were at last using the *Alfonsine* tables, with $J = 134$, for instance, and $M = 310.7$.[24] Solar reckoning apart, the lunar calendar should be adjusted, they said, by setting back the golden numbers (written down the side of calendars) by a day every 310 years. They expressed a preference for bringing the calendar Moon back to reality before instituting this scheme. (The calendar moon was then 4 days out.) To clarify matters, they prepared a calendar volvelle, of a general type well known to most students of the computus — and that meant most university students.

Firmin and John suggested that the reform take effect from the year 1349, the first after a leap year, and the first of golden number 1 after their report. This report was ready by 1345 ("Sanctissimo in Christo patri ac domino..."), but Clement was subject to many distractions in the four year interval. There was his struggle with the emperor Louis of Bavaria, ending in Louis's excommunication and death in October 1347; there were his protracted negotiations with the Greek emperor John Cantacuzenus, and with the Armenians, in the interests of church unity; and his resistance to the claims of the kings of England, Castile, and Aragon to a share in ecclesiastical jurisdiction, now in the interests of French unity; and then there was the terrible plague of 1347-8. By comparison, the calendar must have seemed a trifle. At all events, the joint report was not heeded. Clement died in 1352 and was buried at Auvergne. Ironically, his tomb was destroyed by Calvinists in 1562, during the final sessions of the Council of Trent.

If the Avignon reformers got nowhere, at least it can be said that papal patronage concentrated their minds, and gave rise to a moderately coherent proposal. One of the marks of those reforms that were proposed without strong ecclesiastical support is their tendency to eclecticism, and to a rather hopeless rehearsal of complaints. There are several plaintive fourteenth century texts surviving that for these reasons seem to me to be lacking in intrinsic interest. Here I might mention Robert Holcot, whose criticism, set down in Oxford around 1330, is very largely taken from Grosseteste, and John de Thermis, who in writing for Innocent VI in 1354 ("Easter in 1356 will be wrongly celebrated by five weeks") was perhaps even making use of so old an authority as Vincent of Beauvais.[25] There is the case of the English Franciscan John Somer, who is reputed to have written a work beginning "Corruptio calendarii horribilis est", but of this it is impossible to speak with any certainty, and those opening words do ring of the quotation from Roger Bacon with which I began this chapter. Such works were not, of course, the prerogative of the Western Church. The Greek church too had its would-be reformers, and there was also some interesting Jewish work on the calendar, but these are matters outside my scope. I will content myself with two or three references to influential writings. First, Nicephoras Gregoras' treatise on calendar reform:[26] the work was submitted to Andronicos II in 1324, and might at that time have seemed an irrelevancy to an emperor whose thoughts were chiefly occupied with making war on his grandson. Second, the solar and lunar tables of Levi ben Gerson: these are especially interesting because they represent an approach of two faiths. Levi lived in southern France, and in his tables used the Christian months and years, rather than the Jewish calendar and the rules derived from the Torah, doing so on the grounds of relative simplicity. It is unlikely that the claim was widely acknowledged by Levi's co-religionists, for as we have seen, and shall see again, there were many Christian supporters of the Jewish and Islamic calendars, while very shortly afterwards Immanuel Bonfils of Tarascon drew up a set of tables written in Hebrew for the Jewish calendar which not only became very popular in its original form, but was even translated into Byzantine Greek.[27]

5. *Pierre d'Ailly and the Councils of Rome and Constance*

The next important move within the Western Church came in the second decade of the fifteenth century, in the context of the Councils of Rome and Constance, and thus as one small — and to most contemporary observers no doubt trivial — aspect of deliberations aimed at countering the religious anarchy brought about by the Great Schism. Since I shall later introduce the proceedings of other Councils, it will be as well to point out here the important difference between those held under the threats implied by the Schism, when the question of the authority of the General Councils themselves (and their ability to limit papal authority) was one of the chief matters for dispute, and later and more effective reforming Councils, in particular that at Trent, where at last principles of discipline and dogma could be effectively settled, and where the greater threat came from without, rather than within, the Roman Church.

The first great fifteenth century Councils were the outcome of a meeting at Livorno between cardinals of both papal courts. Meeting in July 1408, they appointed the calendarically auspicious 25 March, 1409 for the meeting of a united General Council. This proved to be exceedingly well attended by both sides. Meeting at Pisa, it moved so quickly that by 5 June there was agreement of sorts on the dismissal of both popes — Benedict XIII and Gregory XII — as schismatics and heretics. Neither would withdraw, and thus the election of Peter Philargi, Archbishop of Milan (as Alexander V), now meant there were *three* claimants to the papal throne. Alexander did not live long. Within a year he was followed by John XXIII (Baldassarre Cossa), who was certainly not the pillar of virtue needed to end the Schism. There was an abortive Council of Rome in 1412, and then — under pressure from Sigismund, king of the Romans — another at Constance on 1 November, 1414. The three great objects of the meeting concerned faith, unification, and reformation. The calendar might seem to fall under all three heads, but it can hardly have attracted much attention, in view of the alternative excitements — the scandalous degradation and burning of Hus, the endless nationalistic struggles, especially those aimed at minimising the effects of the Italian vote, and the hounding of Pope John from the city as a prelude to his formal deposition. In fact within three years of the Council's first meeting a state of affairs had been reached where two claimants — John and Benedict — had been deposed, and the third, Gregory, had resigned. On 11 November, 1417 Oddo Colonna (Martin V) was elected pope, and the Schism was at an end. Not that the ecumenicity of the Council, or the legality of its actions, ceased to be matters for debate; but there is something to be said for judging schisms by the simple device of counting popes, and now there was only one.

This is a part of the background to the reforming polemics of the cardinal and scholar Pierre d'Ailly (c. 1350 - c. 1420). D'Ailly had a number of commentaries to his name — on the *De anima* and *Meteorologica* of

Aristotle, for instance, and on Sacrobosco's *De sphaera* — and he had been Chancellor of the University of Paris from 1389 to 1395. A critic of astrology, he was deeply imbued with it, and there is no doubt that a part of his concern with the calendar and chronology was a product of his belief that the Church was swayed by cosmic influences to which great conjunctions of Saturn and Jupiter were almost as important a guide as the scriptures. (He did, admittedly, modify the old doctrine: influences on the Christian and Jewish faiths was indirect, since they were more *resistant* to the celestial influence than were rival faiths.) [28] He was no great mind. It is hard to speak of plagiarism at a time when the notion of intellectual property was generally so vague, but even by the standards of his day he must be counted a plagiarist,[29] and in his plea for an improved calendar he borrowed heavily from his predecessors, especially Bacon. He was bishop of Cambrai when he was made cardinal in 1411, and at the Rome Council in the following year he tried to interest John XXIII in calendar reform. In his *De concordantia astronomice veritatis et narrationis hystorice* he wrote of discordant observations of the equinoxes (John of Saxony, Henry Bate, the Alfonsine tables, etc.), the Easter problem, and so on. He is best remembered, though, for his plea at the Council of Constance: calendrical calculation is more important to the Faith than financial calculation![30] As Kaltenbrunner remarked, the Cardinal's work is more important for the personality of its author than for its intrinsic merits.[31] He relied on Gerland, Grosseteste, Sacrobosco, and especially Bacon, from whom he copied whole sections — but I note that he was not so old-fashioned as to prefer Bacon's value for J to the 134 of the Alfonsine tables. A polemicist rather than an astronomer, he knew the importance of good titles: we may indeed use the *Arab* years and months, as Grosseteste said, to reckon lunations accurately, but we can do as well, and with greater authority, if we use the tables of the *Greeks and Hebrews*, "et ideo si quis considerat tabulas eorum ad occasum Jerusalem, plenam in his reperiat veritatem". Straight out of that other great polemicist, Bacon, of course![32]

John XXIII acknowledged the convincing character of Pierre's critique, and said that astronomers consulted had recommended these measures: each full moon on or next after the vernal equinox to be the spring full moon; Golden Numbers to be adjusted by four days; certain amendments to *termini*.[33] A decree of 1412 had no consequences, and although the subject was twice brought up at Constance (1415 and 1417), there was again no effective result. Pierre d'Ailly's initiatives might as well never have been taken.

6. *Nicholas of Cusa and the Council of Basel-Ferrara-Florence*

The next concerted attempt at calendar reform within a conciliar context was at Basel, although yet again more than one city was eventually connected with the Council in question. It had been decided at Constance that the pope convoke a Council at Pavia in 1423. Plagues, and other problems, led to its dissolution after transferring to Siena. Basel was chosen for a next meeting in 1431, but Martin V died before meetings began, and his successor, Eugenius IV, in part disenchanted by the status accorded to the papacy by many of the Basel delegates, and in part wishing to meet in an Italian town more accessible to Greek delegates with whose church it was hoped a union might be achieved, tried to dissolve the Basel assembly before the end of 1431. He was unsuccessful, and was obliged to accept a great many unpalatable decrees — especially unpalatable being the reassertion of the ruling at Constance that the decisions of a General Council take precedence over those of a pope. For the sake of completeness it should be mentioned that after Eugenius convened a Re-union Council at Ferrara in 1437, remnants of the Basel assembly claimed to depose him, and elect a new pope (Felix V), but the offspring of this new schismatic movement were never very influential, and finally submitted to Rome (Nicholas V) twelve years afterwards. As for the Ferrara assembly, which ultimately moved to Florence, again to escape plague, its affairs chiefly concerned union with the several Eastern churches. It is to Basel we should look for the proposal for calendar reform associated with the name of Nicholas of Cusa.

It was as a young man of about thirty that Nicholas of Cusa went to the Council of Basel in 1431, to plead the case of Ulrich von Manderscheid, claimant to the archbishopric of Trier. Already recognized for his accomplishments in canon law, not to say historical scholarship,[34] Nicholas quickly became one of the leading lights of the conciliar faction — arguing for the supremacy of an ecumenical Council in his *De concordantia catholica* of 1433, but taking the side of the papal faction on questions of church unity, so that, for instance, he was one of an embassy sent to Constantinople in 1437. His name occurs in John of Segovia's account of the Basel attempt at calendar reform.[35] On 18 June, 1434 a letter was read out to the assembly, drawing attention to the scandalous state of affairs, the derision of unbelievers, and so forth, and suggesting that a commission of expert astronomers be nominated to put matters right. The cardinal of Bologna was chosen to select scholars for the work, and in March 1437, in the refectory of the Minorites, their report was presented by Nicholas of Cusa. His talent for historical scholarship no doubt stood him in greater stead than his astronomical knowledge, which has been much exaggerated; but by now, as we have seen, the calendar had become virtually a historical problem. The proposals made were not very remarkable. Seven days should be dropped from June 1439, and a day should henceforth be inserted into the calendar every 304 years. The shortcomings of such a solution are more evident to us than they were to the assembly, for it seems that the only amendment proposed was by a certain Hermann, monk of Cloister N., who wanted the seven days to be dropped either from May or October.[36] The Commission was given the power to institute the reform, but prematurely, for this was the time of the secession (and attempted suspension) of Eugenius, and in the resulting turbulence within the Church there was no chance of the general acceptance of the reform for the time being. Had the attempt at reform been successful, we should no doubt have been gathering five years hence (or fifty-five) to celebrate the fact, and Nicholas of Cusa would inevitably have stolen the limelight, as is the wont of Famous Men. He did, after all, write a tract in 1436 in connection with the Basel Council, undoubtedly: *De reparatione calendarii*.[37] From it we can say that he knows his texts, his Ptolemy, Albategni, Sacrobosco, and the Alfonsine tables, and yet takes $J = 150$. Taking M, conventionally, as 304, he assesses the error in the Moon's position as $4^d 15^h$, apparently by dividing the year (1436) by 304 — giving the slip in the Golden Number correctly, by the lights of the parameter 304, but not the correct lunar position. The fact that the Basel commission placed so much emphasis on the preservation of Dominical letters (by dropping exactly a week from the calendar) suggests, though, that Hermann was the "expert" behind the scenes, and perhaps it is just as well that they were not successful, for there would almost certainly have been an outcry from better astronomers than they.[38]

Whether or not my next instance of a reform proposal has anything to do with the Basel Council, or whether its apparent date of 1434 is a mere coincidence, I cannot say. Its author was Richard Monk, an Oxford man of whom relatively little is known. If I pay more attention to his manuscript remains than they seem to merit, this is simply because he has not — unlike most of the writers I discuss here — been seriously discussed before. What I find attractive about his proposal is its refreshingly radical nature, although it must be said that it was as unrealistic as it was radical.

7. *Richard Monk, the "Year of the World", and a Muslim Comparison*

Richard Monk's calendrical tables are all that we have by him, and they are unfortunately accompanied by virtually no explanatory text. The tables are found only in MS Laud Misc. 594 in the Bodleian Library, where their author is described as "chaplain of England". In fact we know a little of his life from a lawsuit, recorded in the Calendar of Close Rolls,[39] where we find under 4 December 1439:

> Richard Monke of London, chaplain, to Thomas Gosse, mercer. Recognisance for £ 20 to be levied etc. of his lands and chattels and church goods in the City of London. Condition, that he shall abide and keep the award of John Stopyndoun clerk keeper of the chancery rolls concerning all debts, trespasses, debates, etc. between the parties to this date, *and certain opinions of certain articles of the science of astronomy*. [My emphasis]. Thomas Gosse to Richard Monke, [like] recognisance to be levied [etc.].[40]

It seems a strong likelihood that the point at issue was a calendrical one, although of this one cannot be certain. It is reasonably clear that Monk's views on biblical chronology were well known, and possibly controversial, for in MS Ashmole 369, which once belonged to John Dee, he has written "falsum" above a statement to the effect that there were 4909 years from the origin of the world to the birth of Christ, "secundum R.M. inchoando annum ab equinoctio vernali".[41] This must refer to Monk, for the date can indeed be extracted from his tables.

Perhaps the most interesting thing about Monk's tables is his use of the Egyptian year of 365 days, the year favoured by a few of the best astronomers of later centuries, but not the usual churchman's choice. His system was obviously heavily bound up with certain numerological tenets bearing on his chronology, and this was becoming increasingly the case in calendar work — as evidenced by Nicholas of Cusa, Paul of Middelburg, Scaliger, and Petavius.[42] Monk gave what he considered to be the "true year of the world" (on his reckoning), which he believed ran in seven cycles each of 924 years, and he correlated those years with the year of Christ, both "according to the Church" and "according to the truth". [A specimen entry from the seventh cycle: year of the world 6336 = year of Christ 1449 (true) = year of Christ 1427 (Church).][43] A key to his system is a 33 year cycle (and $924 = 24 \times 33$). His months have either 30 or 31 days (months 3, 5, 7, 10, and 12 have 31), and since day 21 of the very first of his months correlates with the conventional 1 April, he was clearly beginning his cycles at what he judged to be the current equinox. In fact in a table for the true place of the Sun in the ninth sphere, this was given for day 1 (presumably complete) as $0^s\ 0°\ 59'\ 08''$; and we note that the meridian in question was not Oxford, or Rome, or London, or Jerusalem, but "the place of the world between East and West", elsewhere called "the middle of the world" ("super situm medii mundi"), and presumably the place usually called Arin. He included much straightforward conversion material which it is unnecessary to consider here — such as dominical letters on the different ways of reckoning years, "primaciones" (Golden Number, beginning of the lunar cycle) according to the Church and from the origin of the world, and so on. He gave the entry of the Sun into Aries for the Oxford meridian, as a concession to his acquaintances, showing us that he took mid-world to be about 82° east of Oxford. But the most unusual thing about his system is its cycle of 33 Egyptian years.

It must first be said that Monk takes the Moon for granted, and concentrates on a calendar for the Sun. What his views on the Easter problem were, I cannot say. Secondly, I doubt whether his system is very profound. At the end of his tables are "tabule Solis vere atque perpetue" with a short canon that suggests no great sophistication. He there explains that a knowledge of the motion of the solar apogee is necessary if we are to find the true place of the Sun in the ninth sphere, as is that of the motion of the eighth sphere (equivalent to our precession of the equinox). In a final table he claims to offer a combined correction term for every degree of every sign, an extraordinary mish-mash, if I have analysed it correctly, of the "equatio diametri dimidii circuli" in the Thābit theory of trepidation (maximum $4°\ 18'\ 43''$) and the standard Ptolemaic theory for the solar equation. I point out this fact here since I think it would be a mistake to look too deeply for genius in his main calendrical tables, which I will now summarize.

The Egyptian year of 365 days Monk calls the "year of the world". The "zodiac of the sphere of the Sun" (0° to 360°) is the eighth sphere, in solar theory, and its degrees are correlated with two quantities: (i) the time interval between the moment at which the Sun reaches those individual degrees, and midnight; (ii) true motus at the nearest midnight. Another table gives the Sun's positions in the ninth sphere for each day of the 365-day years of his system. They are, of course, not the same on successive years, and this is where his 33 year cycle is needed. Monk assumed that 33 years of 365 days are exactly 8 days short of 33 tropical years. He accordingly gives us a table correlating the year of the 33 year cycle with what he calls "revolutions of the years of the world". A specimen entry may be taken to explain it: "year 30" found opposite "revolution 98° 1'" means that after 30 tropical years have passed, the earth has turned [7 times plus] 98° 1'.

No explanation of the way the calendar was to be implemented has survived, and some doubt attaches to Monk's intentions as to intercalation. A rigorous application of the principle of the 365-day year, the "year according to the truth", without intercalary days, would mean a continuing drift of the seasons through the calendar. The acceptance of a set of annual calendars, 33 in all, for a 33-year cycle, suggests that he had some sort of intercalation in mind. Perhaps he meant to run two calendars in harness, or perhaps the "true" years of the world were only for chronologists, for *conoscenti*. Without his intercalation plans we can say no more than that, at worst, his 33-year calendar would have made the seasons oscillate within the calendar with an amplitude of eight days. His contemporaries would surely have found this unpalatable. Easter, already a movable feast, would simply have moved differently (as far as its labelling was concerned, which is what we are talking about when we talk of its movable nature), but now a bigger problem arises with the placing of saints' days. This would surely have led Monk into far stormier waters than the "conventional" reformers were encountering, had his scheme ever come to the attention of the Church's counselors. As it was, however, it seems likely that at best it served as a topic for local academic discussion.

What Monk did is of interest for other reasons. His ideas bear comparison with those implicit in the Muslim calendar, and also (a separate issue) with some proposals by al-Khayyāmi (1048?-1131?). The Muslims had in a sense rejected in an even more radical fashion than Monk the agricultural year, that is, the solar year, by taking as a "year" 12 consecutive lunar months. This "year", drifting steadily through the seasons, implies that in about 33 tropical years an *extra* "lunar" year (i.e. 34 in all) will have been counted. This relationship could well explain part of the attraction of Monk's scheme, in his own eyes, although that is doubtful. The possibility of influence from al-Khayyāmi (the well known algebraist, polymath, and above all poet "Omar Khayyam") poses more interesting historical problems. He was at Isfahan for eighteen years with the best astronomers of the time under his guidance, where they compiled the "Malik-shāh astronomical tables", and where in 1079 he presented a plan for calendar reform, also writing a history of earlier reforms.[44] He followed a 33-year cycle for his Malikī era (or Jalālī era; in honour of Shams al-Mulūk, khaqan

of Bukhara in the first case or Jalāl al-Dīn Malik-shāh, the Seljuk sultan
at Isfahan in the second), selecting eight years out of the 33 as leap years
of 366 days. This is so reminiscent of the solution later offered by Monk
that one wonders whether there was some link — for example through
Tūsī? —, but if so I have not found it. I have calculated some of the data
for the entry of the Sun into Aries on *Alfonsine* principles, and found them
to fit quite closely with Monk's data (although about ten hours out from
reality; they are given for noon); but the interesting thing about the 33
year cycle is that it yields an appreciably *different* year length from the
Alfonsine ($J = 131.9$ rather than 134.2). The value is indeed an improvement; but one should be sparing of superlatives, since to accept a particular
solar cycle is always to accept some sort of compromise, and does not tell
us what precise year length its user accepted.

8. *From John of Gmunden and Regiomontanus to Paul of Middleburg and the Fifth Lateran Council*

As the fifteenth century went by, the calendar debate began to accrue
a corpus of subsidiary problems. An improved historical awareness led to
discussions of what was the "correct" date to which the equinox must be
restored — that for the time of Christ, of Caesar, of the origin of the world,
or some other? Adjustment of the lunar calendar was occasionally mixed
up with the problem of the precise date of the creation of the world — was
it spring or autumn, for example? What is the "correct" meridian for
Easter calculation? When the equinox nears the end of a day at the
meridian of Rome, it will be the next day in Jerusalem. Some of the best
astronomers were content to make long-lasting and usable calendars setting
forth both fact and convention — as did John of Gmunden, for example,
in Vienna. Judging by the number of extant manuscripts of his (and
similar) works, and the large number of sixteenth century editions, they
were more in demand than was a reform. Regiomontanus was another who,
rather than fulminate over risible errors in Easter, set to, and tabulated
solar and lunar positions using the most reliable astronomical information
available, and derived another table of mistakes in the projected dates of
Easter from 1477 to 1532. This sort of thing must have raised appreciably
the general level of awareness of the calendar's faults. At the same time,
the calendar was occupying an increasingly important place in the consciousness of people other than the clergy. The introduction of printing shifted
the centre of gravity of learning, especially through the agency of vernacular

texts, of which an excellent example is the *Compost et Calendrier des Bergiers*, the "shepherd's calendar" so often printed, from 1493 onwards, not only in French but in English, German, and Dutch.

It was no doubt Regiomontanus' ephemeris that led Pope Paul Sixtus IV to invite him to Rome with a view to recommendations for calendar reform; but the death of the astronomer, in his prime, occurred on 6 July 1476, shortly after his arrival.[45] Although reform tracts continued to appear sporadically, no serious papal initiatives were taken for more than thirty years — and then at last with the Fifth Lateran Council (1512) it seemed that reform was at hand. The Council had its origins in strife reminiscent of that at Basel, and earlier at Constance. Louis XII of France organized a schismatical "General Council" at Pisa in 1511, as a stick with which to chastise Pope Julius II, who responded with his own Council. This was continued after his death, by Leo X his successor, until 1517, the year of Luther's revolt. Politically successful — witness the concordat with Francis I of France — and also successful from the point of view of the papal party — since it established papal ascendency over General Councils — the Fifth Lateran Council did little to silence demands for the suppression of abuses within the Church, but in regard to the calendar it produced a polemicist in the tradition of Roger Bacon, Pierre d'Ailly and Nicholas of Cusa, namely Paul of Middelburg.

Paul of Middelburg was an old hand at astronomy, at least that of an astrological cast. Coming from Middelburg on Walcheren in Zeeland, and educated at Louvain, he moved nearer the orbit of the papal curia in 1479, when he went to teach astronomy at Padua. In 1480 he became physician to the duke of Urbino, and from 1494 until his death in 1533 (when called to Rome by Paul III with a view to making him a cardinal) he was bishop of Fossombrone. Most of his fourteen printed works, all published after his move to Italy, are in some way concerned with prognostication, but it was he who became bishop, and Savonarola — as Thorndike pointed out — who wrote against astrology and who yet went to the stake.[46] In 1491 he wrote a tract "mirum tibi fortasse in debitum...",[47] in which he exorted Pope Innocent VIII to reform the calendar, and his interest in the subject seems to have been of long standing, for Petrus de Rivo (1442-99), an old adversary at Louvain, had in 1488 written on this topic a work directed against him: *De anno, die et feria dominicae passionis et resurrectionis.*[48] In 1513 Paul of Middelburg published his most ambitious work, *Paulina de recta Paschae celebratione* [etc.], printed at Fossombrone. It opened with letters to Leo X, Maximilian, the College of Cardinals, and the

Lateran Council. On 16 February 1514 Leo invited him to the Council. He went to Rome, and was duly given presidency over the commission set up to produce a new calendar.

In his *Paulina*, and especially in his letter to the Council (printed, for instance in the monumental collections of Council proceedings by J.D. Mansi),[49] Paul of Middelburg offers a rather aggressive and often unjustified critique of his predecessors, but he does put his *laissez-faire* arguments reasonably well. In brief, he was in favour of leaving the solar calendar as it stood, of simply recognising the fact that the equinox had moved to March 10 (as he thought — in fact wrongly for two out of four years of the leap-year cycle in his time), and of allowing it to drift through the calendar by a day (as he thought) every 134 years (i.e. the Alfonsine figure). He was afraid that dropping a number of days from the calendar would upset too many people, and he criticised Marcus Vigerius (cardinal from Sinegaglia) in particular for wishing to drop no fewer than 14 days, thus bringing the calendar in line with Pliny's. (He had also implied $J = 100$.) Paul criticised his old adversary Petrus de Rivo for an "Egyptian calendar according to Bede"; and Pierre d'Ailly for flirting with the purely lunar Muslim calendar, and for certain opinions as to Easter limits. As for Paul's second constructive proposal, for the lunar calendar, and for the "Golden Number which had turned into lead", he would effectively leave the cycle as it was, except that the month containing the *saltus lunae* should be taken with one of the embolistic months as a pair, one being always full and one hollow. This would reduce the number of embolismus from seven to five. Finally, he wanted the lunar calendar to be set back a day every 304 years (cf. Cusanus).[50] He wanted the lunar months to be named after the Egyptian months, perhaps thereby hoping to remove traces of Muslim astronomy from Christian writings, and surely not consciously flattering his now deceased opponent Petrus de Rivo by imitation. This would do for Rome: he could not worry about the Greeks, who were already deep in error. Things had changed since the time of Cusanus and Eugenius IV, when hopes of reconciliation ran high.

The proposals were to be considered in the tenth session of the Council, planned for 1 December, 1514. (If successful, the cycles would have been declared to have started on 1 January of jubilee year 1500, when — a sign from God — the mean conjunction of the Sun and Moon at the Rome meridian occurred, at mid-day.) Letters were sent from the papal curia to all important Christian monarchs asking them to get the opinions of their astronomers. Thus Maximilian in turn made enquiries of those at Vienna, Tübingen and Louvain — and since the letter to Vienna was dated 4 October, 1514, less than two months ahead of the decisive meeting, the time margin was clearly being cut very fine.[51] Too fine for the English, I regret to say. In the *Letters and Papers, Foreign and Domestic, of the Reign of Henry VIII* there are four letters from Leo X, and no evidence of any reply from Henry.[52] On 21 July, 1514 the original request was sent, couched in conventional terms (errors in Easter giving cause of ridicule to Jews and heretics), and asking that Henry send either his best theologian and astronomer or the opinions of the same. In a letter of 1 June of the following year Leo laments that the greater number of requests, like that to Henry, have remained unanswered, and hopes that answers will come in time for a discussion at the eleventh session. On 10 July, 1516, and again on 8 December, 1516, the request was repeated in the same terms. One can hardly accuse the curia of not trying, or the English of enthusiasm for a subject they had once made very much their own. Nearly seventy years later, writing for Henry's daughter Elisabeth, in a highly nationalistic vein, John Dee noted that "none had done more skilfully than another subject of the British sceptre Royal, David Dee of Radik", by which he meant Roger Bacon, whose letter to Clement he transcribed. Paul of Middelburg, Dee added, followed the precepts of Bacon but was loth to acknowledge his debt, even reprehending his teacher over the date of Christ's passion. It was as a result of Bacon's letter to Clement V, said Dee, that Paul set to work and *called on Copernicus to advise*.[53]

9. *Copernicus and His Supposed Influence*

Here is one of the thornier problems of the history of calendar reform — or at least it has been treated as such. Here I can only discuss it in a peremptory way, taking as my starting point the fact that among those listed by Paul of Middelburg as having actually replied to the papal requests is "Nicolaus Copernicus Warmiensis".[54] The two could well have met in Italy, for Bernardino Baldi says they were friends.[55] The story that Copernicus was asked to *attend* the Council might stem from these small beginnings, but the earliest statement to that effect occurs in two letters written by Galileo to defend his Copernican opinions.[56] (Perhaps his misunderstanding, if such it was, arose from a misreading of the dedicatory epistle, to Paul III, of *De revolutionibus*.) A third issue, on which opinion has been divided in recent years, and one not without relevance here, is whether Copernicus had anything substantial to offer by way of new data, and in particular, whether the Gregorian reformers used his data for the year and month. That they did so is a conclusion drawn from so authoritative a writer as Christopher Clavius. Broadly speaking, the counter-argument is based on the fact that the Gregorian reform rested on proposals by Aloisio Giglio, who espoused Alfonsine parameters.[57] The last question is not at all as simple as it first appears to be. Clavius, whatever his views about Copernican cosmology — and they were extremely sceptical — did acknowledge with reservations that Copernicus was an authority on astronomical measurements such as are needed for the calendar, and it is not surprising that he and the commission should have wished to heed the Prutenic tables. For the Moon, Giglio designed his system of epacts around the assumption that $M = 312.5$, and the Gregorian followed suit. (Eight days are dropped in 2500 years.)[58] This is the Prutenic (Copernican) figure sure enough, corresponding to a lunation of 29.530592361 days. The Alfonsine figure of 29.530590860 yields $M = 310.7$. We recall the fearful monotony with which M was taken as 304; and indeed J.J. Scaliger once explained how epact systems could be made to fit that old favourite figure, as though it wall still relevant (in 1583). Thus far it seems that our modern "Copernicans" have the right of the argument; but when it comes to the solar year, Giglio's and Gregorian sources are harder to settle in this way. As already pointed out, J is for them simply 400/3, or 133.3. An approximation to the Alfonsine 134.2? In fact if the Giglio biographers tell us that he favoured that figure, took it from the Alfonsine tables, and was instrumental in its being accepted by the Gregorian reformers, then there is, from this point of view, an end of the matter. But Clavius wanted to cover other possibilities. He quoted the maximum ($365^d\ 5^h\ 55^m\ 37.7^s$) and minimum ($365^d\ 5^h\ 42^m\ 55.1^s$) lengths of the tropical year according to Copernicus, and took the mean value. If we calculate the value of J corresponding to the mean, we find 134.24. Perhaps this "Alfonsine" figure tells us something about Copernicus. It certainly tells us that calculation alone is not a sure guide to historical influence. It also tells us that great reputations can be a delusion and a snare. As far as my own theme is concerned, I can leave Copernicus' influence to my colleagues with the observation that if indeed he affected attitudes on the character of the needed reform, this does not show clearly until long after his death, and then in two respects: in the matter of the lunar calendar, and in his having shaken confidence (by discussing variable procession) in the *constancy* of the tropical year over long periods of time. Another interesting hypothesis regarding the origin of the Gregorian year is founded on a comparison of the lengths of the tropical year as given or inherent in the Alfonsine tables, the *De revolutionibus*, and the Prutenic tables respectively. If these values of the year are expressed in days and sexagesimal fractions of a day they all begin with 365; 14, 33, differing only from the third sexagesimal onwards. Consequently, all three sources agree if the length of the year is rounded off to 365; 14, 33 days — a value which gives *exactly* the Gregorian intercalation.[59]

10. *The Outcome of the Fifth Lateran Council*

At the beginning of June 1516 the Pope summoned the Commission and certain cardinals to discuss the whole question, and Paul of Middelburg prepared a report, his *Secundum Compendium correctionis Calendarii*, impressum... Romae pridie nonas Junii 1516. (It includes the list of some, who had responded to the request for criticism, one such name being that of Copernicus, as we have seen.) With an 8 July letter to heads of state Leo sent a modified *Compendium*, since printed by Mansi, and also Marzi.[60] Another appeal followed on 10 July. Some opinions were obtained, as already explained, but Paul's and Leo's efforts ended in failure. The last session of the Council (16 March, 1517) was preceded by a congregation of 13 March. As the pope was informed by Cardinal Giulio de' Medici, the Bishop of Fossombrone wanted the question of the calendar to be settled, and tried hard to persuade each of his colleagues, but in the end was denied his wish.[61] The topic was not introduced to the final session, and I suppose that in the end one may possibly ascribe the failure to a strengthening of conservatism, but certainly to a measure of scepticism in the ranks of the non-experts, confronted by experts who could not agree among themselves.

Their reasons for disagreement were various, and not very startling. Georg Tanstetter and Andreas Stiborius, for instance, answering from Vienna, liked a *fixed* date for the vernal equinox, and wanted an adjustment of a day every 134 years.[62] They wanted a printed table for 1500 years, with equinoxes and spring new moons: it was, they said, foolish to get so attached to the Golden Numbers that they were preferred to the truth. They noted the geographical problem: should Lisbon and Canton [Catigara] celebrate Easter according to local phenomena, even if — as might happen, in consequence — this were to mean doing so eight days apart? Should the Rome meridian be accepted as standard? Johannes Stöffler took a similarly severe astronomical approach: we should stick to astronomical tables, for truth is better than fiction. He tabulated Easters between 1518 and 1585 according to the usage of the Church, of the Fathers, and his local "truth" (the meridian of Tübingen and his Alfonsine tables serving to establish this). Like Paul of Middelburg, he claimed rationality for his principles, but of a different sort, for he would have had Easter of 1571 celebrated on 18 March, while Paul would have had it on 15 April.[63] As for the meridian, why Rome? (Do we detect the stirrings of German independence in all this?) Why not the Western Isles (the Fortunate Isles) of Ptolemy's geography, or even the islands newly discovered by the kings of Castile and Portugal? So much for the claims of Jerusalem. From Paris a certain (unidentified) doctor sent strong objections, known only from Paul's answer.[64] Broadly speaking it seems that even the expert astronomers were divided, not about the need for accuracy, but as to whether or not it was achievable by the use of the old techniques using simple cyclical relationships and procedures.

One of the more important consequences of the initiatives taken at this time is that they stimulated interest in the astronomical problems of the calendar, especially in Italy. The letter to Florence from the curia was for some reason more detailed than at least some of the others sent out on 10 July, 1516, and the Republic responded in a very laudable way, posting up a printed document (September 1516) on all principal churches and other buildings in its territory, inviting competent persons to occupy themselves with the problems of calendar reform.[65] I mentioned earlier a work by a Florentine monk Jo. Lucidus (1525, printed 1546), which was a symptom of Florentine enthusiasm, a phenomenon well charted by Marzi. It is as well to remember that at the Council of Trent, not far in the future, and at later Councils, questions were not decided by nations as in the past, but by a majority vote, and that the Italians would always have a numerical preponderance. In a strong sense, Italian interest in this matter was crucial.

11. *The Council of Trent and a "Forgotten Reform"*

Pressure to hold another General Council came from many quarters, but most importantly from the Emperor Charles V, confronted with the task not only of defending the Faith, but of keeping the peace in his own territories. A Council was planned for 1537, in Mantua, but the Protestants invited would not take part in an assembly in Italy, added to which there were political complications and delays, so that the Council summoned by Paul III to meet at Trent on 1 November, 1542 did not begin there until 13 December, 1545. (Trent, though south of the Alps, was on Imperial territory.) Even then there were further delays. A plague led to a temporary removal to Bologna, although the German bishops generally refused to move; proceedings were suspended by Paul III (1549) and recommenced by his successor Julius III (1551-2), to be ended when the troops of the Elector of Saxony seemed to threaten them. At last Pius IV summoned the bishops to Trent for the third time in 1560, the work of the Council beginning in January 1562. The 25th and last session was held on 4 December 1563. This most important of Councils — whose decrees and canons are so well known — was thus a quarter of a century in the convening.

At the last session of the Council the Pope gave orders that the Breviary and Missal should be reformed — which meant that attention be paid to the calendars that were an essential part of them. The new Breviary was ready by 1568, with one small change in the calendar: the Golden Numbers were moved up by four places (so that, for example, the first occurrence of I is no longer 23 January but 19 January). In a short explanatory passage the reader was told that it would in future be necessary to move the Golden Numbers up a place every 300 years beginning in 1800.[66] (This, in short, is to accept $M = 300$.) As van Wijk comments, this forgotten reform of the calendar was scarcely ever mentioned in later professional discussions. It was emphatically not the reform for which astronomers had been clamouring.

The Council of Trent led to the printing of a number of works on the calendar, some of them prepared at an earlier date. One such example that is notable for its contents was by Petrus Pitatus (Verona, 1564, written 1539).[67] He discussed the rule of Easter — he favoured continuation of the existing rule, on moral grounds — and drew up a lunar calendar for 1539-1805. He argued for the same sort of lunar calendar adjustment as in the Breviary (now with $M = 304$), but he also wanted to have no fewer than 14 days dropped from the solar calendar, to bring it back (as he thought) to that of its founder, Julius Caesar. Most significantly for later history, accepting a value of 134 for J, he compromised and pleaded for the rule whereby three out of four centennial years be ordinary (non leap-years). This is, of course, the Gregorian rule. As when Columbus stood the egg on its end, he broke the rules, but did so very gently. More reminiscent of Alexander's handling of the Gordian knot than of Columbus' egg was Luther's comment on the Easter problem. He would have had it a fixed day of the month, and was very dismissive of the ancient rationale. Calendars, he thought, have nothing to do with faith; they are a question only of worldly authority. It is obviously a very fine line that separates reformers from revolutionaries.

12. *John Dee as a Historian*

By its very nature, calendar reform turned a large proportion of its practitioners into chronologists, if not historians. In an essentially scientific subject this produced — and we have seen many examples — an unfortunate brand of eclecticism in all but the best writers, and where there should have been a coherent system, sectarianism and petty nationalism often prevailed. As we know, by something of a miracle, and notwithstanding the conservatism of many of its members, the Gregorian commission somehow managed to agree upon, and obtain acceptance for, a simple and even elegant reform — not such that it would have astonished Conrad or Grosseteste, but an excellent solution to those who refused to accept Luther's axiom. We know how national and religious discord stood in the way of its universal and immediate acceptance. What I should like to append, as a tailpiece to my story, is an example of a way in which those same forces led to a mildly distorted history of the subject, if "history" is the right word. I have in mind John Dee, graduate of Cambridge, editor of Euclid, and familiar with entities as diverse as angels and navigational instruments. Dee was a frequent visitor to the continent, where he had many friends, and yet his advice to Queen Elizabeth in *A Playne Discourse* (1582) [68] was excessively insular. It is of interest for another reason, however, namely the way in which it reveals Dee's use of some of the many medieval manuscripts he owned. And it is worth remembering that it was revived as one of the documents thought relevant at the time Lord Chesterfield finally brought his country into line with Gregorian Europe, 170 years after 1582. (The British parliament is performing a rather similar historical exercise at this very moment, in connection with islands first sighted by one of Dee's acquaintances.)

The bull of Gregory XIII ordering the use of the reformed calendar (24 February, 1582) was followed very soon afterwards by a proclamation by the English Queen (28 April, 1582) "declaringe the causes of the reformation of the Calendar and accomptinge of the years, hereafter to be observed, to accord with other countryes next hereto adjoyninge beyond the seas". By 25 March, 1583 Dee had produced his report, and — as already intimated in connection with the Copernicus problem — the Lord Treasurer's advisers in turn reported (for the benefit of the Lords of the Council), in essence, that although they liked Dee's ideas they thought it more convenient to go along with the Gregorians. This did not happen. Liberal as had been the mathematicians, the clergy — Archbishop Grindal and Bishops Aylmer, Piers, and Young — were unanimous in recommending further discussion, amounting to a rejection of the scheme originating in the see of Rome.

Dee's *Playne Discourse* opens with an elementary introduction to the problem, after which it has a curious historical dial around which are the great names in the history of the calendar — ending prematurely with "Regina Elizabeth Reformatrix anni civilis juxta epocham Christi" (10-11 for the explanation).[69] Among the names is that of Simon Bredon, one of Dee's national heroes, and I am reasonably sure that Dee got all his information about Bredon from what is now MS Digby 178, at least half of which he owned.[70] Bredon was

> a naturall subiect to the Brytish scepter Royall, with other Mathematicall students and Skilfull masters in Astronomy, being in his company assistant, at Oxford, Anno Christi 1345... [and] made diverse excellent observations, which now, in due tyme, being publyshed, may be very profitable for the veritie Astronomical. (15)

Dee then set down in Latin the record as he found it "many years since". I will not reproduce it, but point out that it comes fairly certainly (with one small copying mistake) from the manuscript I mentioned and that it concerns a series of noon positions of the Sun around equinoxes and solstices, together with interpolations to derive the times of entry into the four signs: Aries, Cancer, Libra, and Capricorn. Dee gave no year (!) but this was 1341. He then went on to express the pious hope that "we had many such records of observations so diligently and precisely made", to lead us to a certain knowledge of heavenly motions But of course the positions he recorded were not observations at all, but positions calculated by William Reed on the basis of the Alfonsine tables.[71] Dee was on safer ground when he set down observations by Copernicus (17) and himself (for the years 1553-5), and indeed a considerable part of his tract is devoted to explaining along Copernican lines how the year from vernal equinox to vernal equinox is not equal to the year from autumnal equinox to autumnal equinox, "which point (perhaps) to some young Mathematiciens maie seeme halfe a Paradox" (30). He relied heavily on the Prutenic tables. (As with Clavius, this does not make him a Copernican from the point of view of his accepted cosmology, a question that has occupied the thoughts of students of Dee.) He had a touch of the Holbrook disease, converting true longitudes of events in the ecliptic as between different meridians, such as London "about $1^h 48^m$ west of Koningsberg", and expressing the results to *sexagesimal fifths*. (32 ff.)   His favoured meridians were London, Königsberg, Jerusalem, and Bethlehem, but not, of course, Rome.

I mentioned elsewhere Dee's transcripts of Bacon (whom he seems to have claimed as a kinsman) and Grosseteste.[72]   When he cited Ptolemy and Albategni, he was taking his information from his compatriots' works, and this goes for his values for the parameter I called J (he noted values of 300, 106, 128, 115, 125, 120).   (See pp. 31, 53, 54.)   What he was most anxious to do was establish the correct time and date of the vernal equinox, and for all his talk of observations, he rested his answers for the most part on the calculations of others — for instance of Michael Mästlin of Heidelberg (38).   Much of his text is occupied with showing that the "Romanists" were mistaken to dock 10 days, rather than 11, from the calendar.   But *how* to remove the eleven days?   He favoured dropping the last day of each month from January to September, and the two last from October (p. 48).   It is interesting to see that in his little calendar for 1583,[73] commented on in the Lord Treasurer's report, he had a scheme for dropping only *ten* days, and this in such a way as not to interfere with Trinity Term or any Feast Day!   It seems that he was prepared to compromise with Rome, at last.   In his *Playne Discourse* he had been more confident.   "He rejoiced greatly that although [Elisabeth] was not consulted by the Roman Bishop" he will be obliged to acknowledge her Majesty's act "to be the most bewtifull flower of opportunity to bowlt owt and ymbrace the veritie..." (48).   The bad verses later in the Ashmole codex do not match the prose, and I regret to say that the overall quality of the work, which left aside the more problematical lunar question, was a pale shadow of the Gregorian recommendations.   But to be thwarted by four English bishops!   His suspicion of church councils was justified.   What was so special about the Nicene Council, that the Romanists should have chosen it for their "principall marke"? (43)   He wished to work from the *Epocha Christi*, he said, for we should prefer the "Trinitie Cownsell", that is, Father, Son, and Holy Spirit.   A very refined form of nationalism, of course.

13. *Some Conclusions*

With the Gregorian reform of 1582, more than four centuries of disenchantment came more or less to a close. Of course there would be future discontent, for the calendar in one form or another is an institution that will last as long as humanity itself; but the motives would no longer be those of the late middle ages, or of that period of Church reform whose historical title — the Reformation — should remind mere historians of the calendar to be modest. The calendar was never more than a straw in an increasingly turbulent wind. It is perhaps not impossible to draw one or two interesting conclusions as to the component forces of the greater Reform movement from the calendar's history, but this might well be thought a rather perverse way of approaching the larger, perfectly accessible problem. It is of some interest, even so, to ask what sort of people were our calendar reformers. In doing so we have to bear in mind a not uncommon historical difficulty: of many reformers we know virtually nothing, and if we make generalizations from the cases of scholars known independently for their works or deeds, notable or not, we are likely to produce a distorted picture. There is no escaping the fact, however, that the calendar occupied the thoughts of a sizeable number of scholars who would be judged important to the scientific movement on quite independent grounds. Consider the very diverse roles of Grosseteste, Sacrobosco, Campanus, Bacon, John of Murs, Geoffrey of Meaux, Pierre d'Ailly, Nicholas of Cusa, John of Gmunden, Regiomontanus, Paul of Middelburg, Copernicus, Reinhold, John Dee, and Christoph Clavius, fifteen scholars taken from my account whose selection for an intellectual history of western Europe would certainly not rest primarily on their contributions to the calendar debate. It would be a mistake to try to chart the growth of a scientific consciousness in Europe from that particular debate; there are better and more direct ways of doing this. From a scientific point of view the calendar debate is disappointing. I have tried to emphasize the eclecticism of the scholars I have discussed, an uncritical eclecticism, in fact. Even after the introduction of printing, and the dating of colophons, writers were often extremely vague about the chronological ordering of the texts on which they relied. Astronomical writers all too often reported a mixture of tradition (and calculation based on it) and genuine observation — a cynic might say in the ratio of a hundred to one — which their readers were often incapable of distinguishing. I have already cited John Dee as an excellent example of a man who had no excuse for his uncritical confusion of the one with the other.

From an astronomical and arithmetical point of view, that is to say, granted all the ecclesiastical and theological presuppositions, the calendar was a simple enough exercise in the invention of rules, rules that would harmonize incommensurable movements as closely as possible. It was an exercise well within the capabilities of innumerable clerics, and the fact that a large social group existed with competence in both the religious and the scientific aspects of the problem at issue makes it all the more surprising that they took so long to achieve success. Why were they not successful sooner? One answer has been given as soon as we have listed the ways in which specific attempts at reform were thwarted, one by one, by external circumstance. It is often said that institutional conservatism was at the root of successive failures, but the evidence does not support this view. (The picture changes as regards Protestant countries after the Gregorian reform, but that is not my theme.) Of course there were conservatives within the Church, who no doubt rejoiced when the experts could not agree among themselves. On the other hand there was, as we have seen, a considerable degree of papal sympathy with, and interest in, reform. Whenever we do have clear evidence as to the reasons for the abandonment of one plan or another, they usually have to do with the need to solve more momentous, or simply more urgent, political and religious problems.

It cannot be said too often that the calendar problem in the period under discussion was a church matter. Secular interest in structuring the calendar (as opposed to enjoying its holidays) was minimal. In any case, in the milieu I have been considering there was hardly a man who would have put the common law above the law of the Church. This much having been said, there is no doubt that different reformers had different motives. There was the astronomer who, having detected error, could not forbear to point it out, whether to prove his cleverness, to save his church from derision (and Bacon was afraid that even the rustic might laugh at the lunar calendar's faults), or simply out of a love for the truth. There is no reason why the three sorts of motive should not have been combined in one man, although the proportions clearly differed appreciably from case to case. It would not be wrong, for instance, to describe d'Ailly, Cusanus, and Paul of Middelburg as "careerists", for all their undoubted virtues, while Regiomontanus, whose short-lived career had much in common with theirs, surely had a purer concern for truth, at least of the textual and astronomical variety.

This introduces a question I have not touched upon in my all too cursory survey, namely the biographical element. It seems to me that a closer study of those most concerned with calendar reform would reveal a common character trait, a certain critical unease in their outlook on the Church — a body of which they were all loyal members. It is tautologous to say that they were not wholly conservative, but there is no *a priori* reason why, in a wider ecclesiastical context, they should not have turned out to be something of the sort. My impression, at least, is that as far as the evidence goes this was never the case. It would be foolish to pretend that one man can stand proxy for all, but since Robert Grosseteste in a very real sense set the stage for our four centuries of calendar history, I will take him as an example of one who had a deep concern with the need for a wider, political and spiritual, reform of the Church. He used to be presented by English historians as a radical leader in the movement towards the liberty of the English church, a very misleading view. On the contrary, he held to a very clear view of the Church as a tightly bound organism with a strongly hierarchical structure. It was precisely this view that enhanced his awareness of the disorder and corruption which he saw threatening it. He was no anti-papalist. He could at one and the same time deplore the exactions of the pope and appeal to the pope for support in the struggle to prevent secular interference in the ecclesiastical system. He went to Rome in 1250, when he was at least eighty years old, to protest a number of iniquities. In Sir Maurice Powicke's words:

> Pope Innocent IV sat there with his cardinals and members of his household to hear the most thorough and vehement attack that any great pope can ever have had to hear at the height of his power.[74]

Grosseteste's vitality was a rare quality. (His fourth and last work on the calendar, the *Computus minor*, was written when he was in his seventies, administering England's largest diocese.) Those who followed in his footsteps in the cause of calendar reform could not hope to share it; but I suspect that many of them, perhaps most, shared his belief that the Church was at the mercy of the imperfect human beings holding office within it, and that they, together with the institutions by which they cared for the souls in their charge, should be subordinated to this greater responsibility. They seem, in short, to have been men of a rather independent frame of mind, reformers in a wider sense.

REFERENCES

1. And when it comes to writing on their history we are most of us plagiarists, for here is a theme on which some of the Church's most erudite scholars have written. A comprehensive bibliography would occupy some hundreds of pages. I will single out one or two titles that cover my own particular topic in a useful way. The vocabulary of calendar and computus is generally unfamiliar, and can be found very clearly explained (although in connection with an early work) in W.E. van Wijk, *Le Nombre d'Or: Étude de chronologie technique, suivie du texte de la Massa Compoti d'Alexandre de Villedieu* (The Hague, 1936). An equally lucid but short monograph on the Gregorian reform by van Wijk is *De Gregoriaansche Kalender* (Maastricht, 1932). Two works on reform proposals in the late middle ages (1876) and "Gregorian" period (1881), both by F. Kaltenbrunner, are likely to remain standard for many years to come: "Die Vorge-schichte der gregorianischen Kalenderreform", *Sitzungsberichte der phil.-hist. Classe der kaiserlichen Akad. der Wissenschaft, Wien*, lxxxii (1876), 289-414; and "Beiträge zur Geschichte der Kalenderreform", *ibid.*, xcvii (1881), 7-54. I shall constantly refer to these as "Kaltenbrunner (1876)", and "Kaltenbrunner (1881)". Kaltenbrunner's ordering of his subjects is sometimes very misleading, especially in the early centuries, and he is occasionally over-enthusiastic (for example, about Master Conrad and Nicholas of Cusa) and repetitious, but he had a better command of his material than most of his contemporaries. His work is well supplemented on the sixteenth century by D. Marzi, "La questione della riforma del calendario nel 5° Concilio Lateranense", *Pubbl. del R. istituto di studi superiori pratici e di perfezionamento di Firenze, Sezione di filos. e filol.*, ii (no. 27) (1896), 1-263; and Marzi surveys then recent literature in his "Nuovi studi e ricerche ... secoli XV e XVI", *Atti del congresso internaz. di scienze storiche, Roma, 1903*, iii (1906), 637-50.

   My dogmatic statements about currently accepted astronomical parameters are based on the *Explanatory supplement to the astronomical ephemeris ... etc.* (London, 1961) or derived from data given there, without further reference.

   For a good short bibliography, including such classical authorities as Clavius, Peta-vius, and Scaliger, see van Wijk (1936), 135-41. Note that Marzi's bibliographical references are frequently unreliable.

2. Several dates for the equinoxes were in use in late antiquity and the early middle ages. The matter is somewhat complicated by the fact that the Babylonians left behind them a tradition of reckoning the "zodiacal signs" from points such that the vernal point came out at Aries 10 (system A) or Aries 8 (system B). Pliny (*Nat. hist.*, lib. xviii) places the Sun at the 8th degree of the appropriate signs at the beginnings of the seasons. This gave rise to confusion among medieval computists, although probably never serious; and has muddied certain historical discussions of the dates assigned to the vernal equinox. The literature is vast, and need not detain us here.

3. L. Baur did not include the *Compotus* in his edition of the philosophical writings of Grosseteste, but he commented extensively on it in *Die Philosophie des Robert Grosse-teste* (Beiträge zur Gesch. der Phil. des Mittelalters, Bd. xviii, Heft 4-6), Münster i.W., 1917. See esp. pp. 46-33. There seem to be at least four works by Grosseteste relating to the calendar: the first to have been printed (Venice, 1518) was his *Compotus cor-rectorius* (1215-19?). A *kalendarium* with associated canons was edited by A. Lindhagen (1916). For further details, see S. Harrison Thomson, *The Writings of Robert Grosse-teste*, Cambridge, 1940.

4. I say "day divisions" rather than "midnight" since on some astronomical conventions the day is taken to begin at mid-day.

5. Kaltenbrunner (1876), 402.

6. "*Computus est scientia distinguendi tempus certa ratione...*" Bruges MS. 528, ff. 1r-6v.

7. For Grosseteste, see n. 3.

8. Sacrobosco's *Computus* was one of at least a dozen medieval treatises (cf. n. 6) beginning "*Computus est scientia...*" His was: "*Computus est scientia considerans tempora ex solis et lune motibus*" (see Thorndike and Kibre, col. 243). Many commentaries were written on it, and it was frequently printed in the sixteenth century: *De anni ratione, seu De computo ecclesiastico* (that issued with his *De sphera mundi*, Wittenberg, 1540 might be the *princeps*, although a Paris edition without date has been claimed for 1538). Even this is twenty years, however, after the Campanus *princeps* (see n. 9).

9. "Rogavit me unus ex hiis quibus contradicere nequeo..." (Thorndike and Kibre, col. 1365; printed with Sacrobosco's *De sphera*, Venice, 1518).

10 There are numerous editions of the *Speculum maius*, beginning with the seven volume edition of Strasburg, 1473-6; see that of Douai, 4 vols, 1624 (repr. Graz, 1964), vol. 1. lib. 15, *de formatione coelestium luminarium.* This whole enormous encyclopedia was written between c. 1244 and some time in the 1250s.

11. F.S. Benjamin, Jr. and G.J. Toomer, *Campanus of Novara and medieval planetary theory* (Madison, 1971), 7.

12. For comment on the reading of the text of the mandate (without the calendar interpretation) see L. Thorndike, *A history of magic and experimental science*, vol. ii (New York, 1923), 625. He settles on a reading which makes the pope ask for those "remedies you think should be applied in those matters which you recently intimated were of so great importance", this to be complied with "without delay as secretly as you can".

13. Ed. Brewer, 271-92.

14. Ed. Bridges, i. 281.

15. L. Baur, "Der Einfluss des Robert Grosseteste auf die wissenschaftliche Richtung des Roger Bacon", in A.G. Little, *Roger Bacon: Essays* (Oxford, 1914), 45.

16. Bacon admittedly groups Thābit with al-Battānī and others, who argued for $J = 106$, giving 131 to Asophus ('Abd al-Rahmān ibn 'Umar al-Sūfī). See Steele, ed., fasc. 6, 12-18.

17. For more information see my "The Alfonsine tables in England", in *PRISMATA - Festschrift für Willy Hartner* (Wiesbaden, 1977); *Stars, Minds and Fate*, ch. 21, pp. 327ff.

18. "Cum sit intentio ostendere falsitatem kalendarii nostri..." (Thorndike and Kibre, col. 342); expl.: "Explicit compotus novus phylozophycus compositus per fratrem Iohannem de S.". Probably South German. His radix for tables of the Moon: 21 March 1273 (the calendar being supposed corrected), fer. II, $6^h 25^m$, golden no. I. His equinox: 14 March 1273, 20^h (but correctly: 13 March, $2^h 43^m$ a.m.). His value for J: 120. (Compare Vincent of Beauvais!) See Kaltenbrunner (1876), 307. Whether rightly or wrongly I do not know, but Kaltenbrunner implies that 14 March meant 14 days *complete*, and that we should write 15 March, 8 p.m.

19. Peter of Dacia (Peter Nightingale) calculated his calendar for a 76 year period beginning 1292. It included the same features as that of William of Saint Cloud, such as the altitude of the sun at noon, and the time of the lunations to a quarter of an hour. It is known in more than fifty MSS and was printed (from one of them) in the *Bibliotheca Casinensis* iv (1880), 232-47. A critical edition by F. Saaby Pedersen is in preparation for the *Corpus Philosophorum Danicorum Medii Aevi*. Cf. O. Pedersen: *Petrus Philomena de Dacia*, Copenhague, 1976 (= *Cahiers de l'Institut du Moyen-Âge Grec et Latin*, No 19).

20. The *almanach* begins: "Cum intentio mea sit componere almanach..." (Thorndike and Kibre, col. 310) and the calendar "Testante Vegetio in libro suo de re militari antiquis temporibus..." (ibid., col., 1568).

21. For quotations from John of Murs and discussion of authorship, see Lynn Thorndike, *op. cit.*, iii (1934), 297-8. On the general problem, and further examples of conservatism, see the article referred to in n. 17.

22. For the text of the letters see E. Deprez, "Une tentative de réforme du calendrier sous Clement VI: Jean de Murs et la chronique de Jean de Venette", *École française de Rome, Mélanges d'archéologie et d'histoire*, xix (1889), 131-43. There is an excellent summary of John of Murs' work generally in the *Dictionary of scientific biography* (by Emmanuel Poulle), vii (1973), 128-33.

23. For MSS, and discussion of a longstanding confusion over the authorship of the document — was it John of Murs or John of Lignères? — see Thorndike, *ibid.*, iii (1934), 268-9.

24. The value for J comes explicitly in tract I, and for M in tract II, but some have mistakenly accepted 210^y 260^d as their figure for M, drawn from the Alfonsine tables, rather than the correct 310^y 260^d (310.71 years).

25. (Holcot) "Utrum stelle sint create..." (Thorndike and Kibre, col. 1674); (John de Thermis) "Ad honorem domini nostri..." and "Itaque me videar aliquid ex me..." (ibid., cols. 44 and 797). Kaltenbrunner (1876), 322, says that John was writing for Clement VI; but he was dead in 1354, and the MSS are explicit.

26. The tract by Nicephoros Gregoras (1295-1359) was included in his *History*, of which there are several editions in Greek and Latin. A partial translation into French by Louis Cousin was published in Paris in 1685.

27. For more details of the tables by Levi (1288-1344) see B.R. Goldstein, *The astronomical tables of Levi ben Gerson* (Hamden, Conn., 1974), *passim*, and for an expansion of these brief comments, *ibid.*, 27.

28. See my "Astrology and the fortunes of churches", *Centaurus*, xxiv (1980), 181-211, esp. 200-01. See, North, *Stars, Minds and Fate*, ch. 8, esp. pp. 78-9.

29. To give only one of several possible examples, in his *Tractatus contra astronomos* he took long sections unacknowledged from Oresme's *De commensurabilitate*. See E. Grant, *Nicole Oresme and the kinematics of circular motion* (Madison, etc., 1971), 130-1, with references also to G.W. Coopland, who first realised the borrowings from Oresme both there and in two other tracts.

30. Three works by Pierre d'Ailly, with their references in Thorndike and Kibre, are: "Astronomice veritatis viam sequentes quam sapientes...", namely the *Concordantia* (col. 158); "Sanctissimo domino pape Johanni vicesimo tertio..." (col. 1372); and "Non modica diligentie cura...." (col. 921). Cf. C. v. d. Hardt, *Magnum Concilium Constantiense* (Frankfurt & Leipzig, 1679), ii, 72, and further references in Kaltenbrunner (1876), 328, n. 2. For general material on the cardinal, see *Lexikon für Theologie und Kirche* (Freiburg, 1963), viii, 330.

31. Kaltenbrunner (1876), 330, 334.

32. *Ibid.*, 333.

33. *Loc. cit.* (following v. d. Hardt).

34. His proof of the spurious nature of the Donation of Constantine and his arguments for its having been an eighth century forgery are justly famous.

35. *Historia gestorum generalis synodi Basiliensis* ed. Birk in *Monumenta conciliorum generalium*, ii (Vienna, 1873); see cap. xix, lib. viii, pp. 708 ff. Paraphrased by Kaltenbrunner (1876), 336-7.

36. Perhaps Hermann Zoestius, author of a tract on the calendar in MS. Melk K. 24, where he says he had written earlier, in 1432, on the subject. The advantage of omitting the seven days (which was not enough to put the equinox right) was that the dominical letters remained the same — a trifling advantage. Hermann noted four ways of dropping the days: by reducing all 31-day months to 30-day months, for one year; by dropping a day out of each of seven consecutive years; by dropping seven days from May 1437.

37. See his *Opera* (1565), 1155-67 and Thorndike and Kibre, cols. 53, 1536.

38. For John of Segovia's comments see Kaltenbrunner, 338, n. 1.

39. See the volume for 1435-41, pp. 347, 349-50.

40. Actually the entries are vacated, since they are repeated below, with the stipulation that the award be made before Whitsunday next. I hope they could agree on when Whitsunday was to be celebrated.

41. This occurs on the last of eight fly leaves (8r) prefacing what is a fine early (12-13 c.) astrological and astronomical collection. The statement is followed by what looks to me standard John Holbrook material (Cambridge based data; length of the year to sexagesimal fourteenths, etc.) very probably deriving from Richard Monk's tables in some measure, as when he gives the entry of the Sun into Aries at the Cambridge meridian for 1434 (March $11^d 5^h 20^m$; 14, 28, 19, 25!). See also the note at end of References.

42. See my "Chronology and the age of the world" in *History, cosmology, theology*, ed. W. Yourgrau (New York & London, 1978), 307-33. The tradition had of course much earlier roots. See North, *Stars, Minds and Fate*, chapter 9, pp. 91-117.

43. It is hardly necessary to give full references to the tables, which begin on f. 14v and end on 21v. The Laud catalogue distinguishes three items, but rather arbitrarily, and I shall assume that these pages were meant to comprise a coherent whole, inconsistencies notwithstanding.

44. *Dictionary of scientific biography*, vii, 324b (art. by A.P. Youschkevitch and B.A. Rosenfeld).

45. See further the article by K. Ferrari d'Occhieppo in *Regiomontanus Studien* (ed. G. Hamann), *Oesterreichische Akad. der Wiss., phil.-hist. Klasse, Sitzungsberichte*, ccclxiv (1980).

46. *A history of magic and experimental science*, iv (New York & London, 1934), 561.

47. Bibl. Apostolica Vaticana, MS 3684, ff. 2r-8v.

48. H. De Jongh, *L'Ancienne faculté de théologie de Louvain au premier siècle de son existence* (1432-1540) (Louvain, 1911), 83-7.

49. *Sacrorum conciliorum nova et amplissima collectio* [etc.] (Florence, Venice, Paris, etc., 1759 onwards; with reprints etc.), suppl. vi (Lucca, 1852), 462.

50. For more details see Kaltenbrunner (1873), 375-86, who notes that Paul did not see all the consequences of his plans.

51. On Vienna especially see Kaltenbrunner (1873), 386-98.

52. Vol. i, pt. 2 (for 1513-14; enlarged ed. 1920), 1320 [21.07.1514]; vol. ii, pt. 1 (for 1515-16; 1864), 151 [1.06.1515], 647 [10.07.1516], 829 [8.12.1516].

53. Unless I remark to the contrary, I shall quote John Dee's views from his *A Playne Discourse and Humble Advise, for a Gratious Queen Elizabeth* [etc.], Bodleian Library, MS Ashmole 1789, pp. 3 ff. There is much ancillary material, which I shall not discuss in detail here; but note the petition to the Queen at p. 61 (she is to press for dropping eleven days among Christian kings, hoping that they will see sense — this of course was

after the Gregorian reform), and note the draft calendar for May 1583-December 1583, and letters to Dee, later in the MS. For a transcript without the tables filled in, and ending before the petition, see Ashmole 179, item vii. There is a fine transcript in Oxford, Corpus Christi College MC (C) 254, ff. 147v-161v, followed by several valuable documents, beginning with f. 161v: "To the Lordes of the Councell. The Lord Threasurers report of the consultation had and examination of the plaine and brief discourse made by John Dee for the Queenes Majestye, etc. 25° Martii 1582". (See also n. 67 below). We are told that Digges, Savile, and Chambers agreed in principle with Dee, but thought it too late, and that Britain should now go along with adjacent countries in dropping only *ten* days. They refer to a scheme drawn up by Dee for doing just this — which is his calendar of Ashmole 1789 (see above), of which there are two copies here (163v-176v). There follow excerpts from Grosseteste's *Computus* (176v-177r) and actual letters about the calendar to Dee from Walsingham (1582), as well as other transcripts. Much of this MS is in Dee's hand.

54. D. Marzi (1896 - see my n. 1), 174.

55. *Ibid.*, pp. 233-50 contains an edition by Marzi of a MS life of Paul of Middelburg written by Bernardino Baldi (1553-1617).

56. For this whole episode see D.J.K. O'Connell, "Copernicus and Calendar Reform", *Studia Copernicana* (Warsaw, 1975), xiii, 189-202, esp. pp. 196-7.

57. Thus E. Rosen, in the discussion appended to O'Connell's paper (see n. 56), says that Giglio accepted the length of the *year* given by the Alfonsine tables. See the following note.

58. The system of epacts devised by Giglio (Lilius, in its Latin form) is explained in van Wijk (1932), 25-9. He accepts Giglio's use of the Copernican figure for the length of the month, but I note that H.J. Felber, in a paper preceding O'Connell's, p. 187, says that the Lilian new moons were computed by the use of the old Alfonsine dates. I leave the question to my colleagues.

59. This hypothesis was proposed by Professor Noel Swerdlow: "The origin of the Gregorian civil calendar", *Journal for the history of Astronomy*, v (1974), 48-9 (as pointed out to me by Professor Bernard Goldstein). This very acute observation is lacking in historical force, at least as far as settling the question of who used which authorities. It suggests, first of all, that those concerned with the reform were not making a conscious compromise. (We know several examples of would-be reformers who advocated approximative rules in the interests of simplicity). The Gregorian rule could easily have been seen as a reasonable approximation to any or all of the "best" authorities, unrounded. But as Swerdlow shows, it could just as well have been seen as an exact rule, accepting any of the "best" authorities, suitably rounded. He opens our eyes to a historical possibility, without really venturing into historical territory. From a theoretical standpoint, indeed, he opens our eyes to more possibilities than he mentions. It may be very easily verified that (with our modern way of rounding, at least) *any* author who accepted a value for J lying between 130.99 and 135.76, when working to the limits of accuracy usually claimed (say two fifths of a second or better), would have rounded sexagesimally so as to give the Gregorian rule. This would have included, for instance, anyone working from the Toledan tables of solar motion (sidereal coordinates, thus needing correction for precession) with a vernal point moving temporarily (I assume he would have accepted trepidation, but he need not have used it to get this figure) at the rate of a degree in 72 years. But this is just an example of how to miscalculate history. Swerdlow had in the background to his argument one important historical fact, namely that Pitatus (1560) remarked that the intercalation of 97 days in 400 years was "consistent with the tropical year of the *Alphonsine tables* and with the mean tropical year of Copernicus and Reinhold". And the *need* for such information brings me back to the point I was making in the text.

60. Mansi, *op. cit.* (see my n. 49), 702-6; and Marzi (1896), 191-5.

61. *Ibid.*, 211.

62. Kaltenbrunner (1876), 386 ff.

63. *Ibid.*, 392.

64. Marzi (1896), 206-8.

65. *Ibid.*, where the document is printed as a frontispiece.

66. W.E. van Wijk (1932), 22, correcting a misunderstanding by Kaltenbrunner. Note that the President of the Gregorian Commission (Cardinal Sirleto) had been involved in the preparation of the Breviary.

67. See n. 53 above. There is some doubt about the date, which might be in error for 1583 (the New Year for purposes of civil dating began on 25 March).

68. *Calendar of State Papers, Domestic... Elizabeth, 1581-1590*, ed. R. Lemon (London, 1865), 107.

69. The numbers in parentheses are page references to MS Ashmole 1789. See n. 53 above.

70. For an illustration from this MS, and some comments on it, see R.T. Gunther, *Early science in Oxford*, vol. ii (Oxford, 1923), 52, pl. See, however, the following footnote.

71. The vital parts of MS Digby 178 seem to have been copied by some mid-fourteenth century Mertonian, possibly Bredon. There is what has been taken as his autograph (on the grounds, in particular, that it uses the first person singular) on f. 13r. Between ff. 11r and 13r are tables for the Sun between 1341 and 1344 (a full leap year cycle), as well as a table of lunar latitude. There is another copy of these pages, with the "ego Bredon" (!), in MS Digby 176, ff. 71-2, and there it is stated that the tables were *calculated and written by William Reed.*

72. See n. 53.

73. See n. 53.

74. *Robert Grosseteste, Scholar and Bishop*, ed. D.A. Callus (Oxford, 1955), xxxiii.

Addenda

My reference to a Wittenberg, 1540 edition of Sacrobosco's *De anni ratione* is probably a mistake, as Mr Richard Priest points out to me. (Here I believe I followed the bibliography of M. B. Stillwell.) Mr Priest draws my attention to a British Library listing of a Wittenberg (I. Clug), 1538 edition, with a Melanchthon preface.

The value of 120 for the parameter J (at the end of p. 45) occurs in Grosseteste's first work. (I follow Sir Richard Southern's *Robert Grosseteste*, Oxford, 1986, pp. 128-9; and a reference to his pp. 127-31 generally could be added to my n. 3. Southern lists three Grosseteste works on computus, and puts them in a new order.) In his later work, Grosseteste accepted a figure of 100, seemingly following Battani and, as he has it, 'per experimentum nostri temporis'. Cf. n. 18. Miss J. Moreton notes that the 120 parameter in MS Trinity Coll. 441 is not Sacrobosco, but is in a passage substituted for one where the writer had erred in his arithmetic. She holds that the Trinity treatise is wrongly ascribed to Grosseteste.

Mr K. Snedegar (in an unpublished Oxford D.Phil thesis) notes that according to John Ashenden, *Exafrenon* drew on Grosseteste's *De impressionibus aeris*. This would provide yet another link with the value of 120 for J.

For an exhaustive new study of the problem adverted to in n. 59, see N. Swerdlow, 'The length of the year in the original proposal for the Gregorian calendar'. *Journal for the History of Astronomy*, xvii (1986), 109-18.

[*Note added in proof*: Essentially the same material as that referring to "R. M." in MS Ashmode 369 (referred to on page 55) is to be found in MS Gloucester Cathedral 21, f. 9r. This was pointed out to me by Mrs Hilary Carey. Sure enough, the manuscript contains Holbrook material. It belonged to John Argentine, chaplain to Henry V and Henry VI. See N.R. Ker, *Medieval manuscripts in British libraries* (Oxford, 1977), 952-5.]

5

THOMAS HARRIOT'S PAPERS ON THE CALENDAR*

Among the Petworth House manuscripts of Thomas Harriot, formerly with MS 241 but now separately bound, are twenty pages on calendrical and chronological matters. Although they are by no means the most original of his works, they do help to fill out the picture we have of a man who occupied a key position in early seventeenth-century science. An edition of them is appended to the present chapter. The papers are disordered: some of them refer back to the turn of the century, but most seem to have been put together around 1615 or 1616. The first dozen pages form a coherent little tract on the determination of Easter, seemingly intended for a popular audience, and beginning with rules as clearly set out as any you will find in comparable works aimed at those who are able to compute without recourse to counting on the joints of the hand. A thirteenth page is a fair copy of earlier material, and I do not think that it is of Harriot's writing. The last pages show Harriot's concern with the date and hour of the creation of the world, and with the real date of Christ's birth – a perennial problem for the Christian chronologist who was conscious of the errors in the Dionysian reckoning in common use. Harriot records inconsistencies as between the Roman and Alexandrian reckoning of Easter in the early centuries of Christian reckoning. Inevitably he introduces astronomical material concerning solar and lunar motions, but here it is always of an elementary character, and the later pages are more interesting for the gleams of light they throw on his historical sources.

The context of these papers has at least a twofold character. On the one hand, there was a concern with past chronology that was occupying so much of the scholarly thought of Harriot's contemporaries. Complementing this there was a long-standing preoccupation with the Easter computus. In those countries where the Gregorian reform of the calendar had been accepted (whether in 1582 or later), the computus was fast becoming something that could be left with equanimity to others – to a handful of clerics, that is, who could tell the rest of the Church when to celebrate Easter. For four centuries, say from the end of the twelfth to the end of the sixteenth, the standard method of calculation

* I should here thank Lord Egremont for kind permission to reproduce the contents of the Petworth manuscripts.

had been a matter for intense dissatisfaction on the part of those trained in astronomy, who had seen the shift in the equinoxes and the lunar phases in relation to the old calendars making a nonsense of the decrees of the Council of Nicæa. [1] Thomas Harriot was writing thirty years and more after the Gregorian reform of 1582, but in a country whose bishops had thwarted the efforts of its astronomers, even though they had the support of Queen Elizabeth herself in some measure, to bring the English calendar more or less in line with that of Rome. [2] Harriot was no longer living in an age when an Englishman could write polemics in the style of Robert Grosseteste, Roger Bacon, or even John Dee, for to demand reform was to support those steps now actually taken by the Church of Rome. He thus opens with a few critical observations expressed in a most matter-of-fact tone. "The Effect of the Decree of the Councell of Neece for the Observation of Easter day" is his title, and the first of his three rules is simple enough: "The next Sunday after the 14th day of the first moone is to be observed for Easter day". But what is the "first moone"? It is that (rule 2) whose 14th day falls on or next after 21 March, which "was supposed to be the day of the aequinoctium at that time, but is now about the 11th of March". As for discovering when it occurs (rule 3), we are to use the Golden Number appropriate to the year, for this marks the positions of the first days of lunations in the standard ecclesiastical calendars, that is, days of mean conjunction of Sun and Moon, "within a day more or lesse". Alas, "the media coniunctio or change day is now aboute 5 dayes before the place of the Golden Number". And Harriot adds merely that "The Church of England & some other Churches not subiect to the Church of Rome follow this decree still uncorrected". [3] In the last of four notes to the opening decrees he writes that "our Common Easter day is sometime the same day as the true Easter day intended, and if it differ, it wilbe later than the true by just 1, 4, or 5 weekes for these 700 yeares to come & upward; and after it wilbe more sometimes & more &c.". There is surely a touch of resignation in the calculation of future divergence between our Easters based on the old calendar and the "true Easter day intended". These last words are just about the strongest protest Harriot essays, although he brings home his point as to divergence when he tabulates Easters on both systems between 1614 and 1620, and shows that five out of the seven years involve disagreement.

On the technical side, Harriot's opening tract is reasonably straightforward. The main table he presents with the basic rules mentioned already is one he

---

1. See further my "The Western Calendar — 'Intolerabilis, horribilis, et derisibilis'. Four Centuries of Discontent", in G.V. Coyne, *et al.* (ed.), *Gregorian Reform of the Calendar: Proceedings of the Vatican Conference to Commemorate its 400th Anniversary, 1582–1982*, Vatican City, 1983, pp. 75–113; see above, chapter 4, pp. 39-78.

2. *Ibid.*, p. 66-8.

3. All quotations here are from p. 1 of the manuscript.

could have found in any medieval ecclesiastical calendar, listing days of the month (for March and April only, since Easter always falls within them), weekday letters, and Golden Numbers — these being placed opposite the days on which the change of the moon was supposed to occur for each of the nineteen years of the lunar-solar cycle.[4] The system is based on the assumption that 19 Julian years (of 365^d6^h) are precisely equal to 235 lunations, and that after 19 years new moons will occur on the same days of the month in the calendar as before. Harriot performs a short calculation of the error involved, and finds that 235 months fall short of 19 Julian years by 1^h 27^m 32^s 42^t, "which being so little the aunccient fathers semed to neclecte; or did thinke the astronomicall observations not so exact, or if that they were, the error that should rise upon that account might be mended when in time it should be found notable".[5] He notes that the amount builds up to a day in 312½ years — his value for a parameter that was much quoted (with such values as 300, 304, and 310) in the late middle ages. Harriot's figure is not peculiar to him, though, for it is at the heart of the Gregorian reform, which was based on Antonio Giglio's system of epacts designed around the figure 312.5. (Following the Gregorian rules, eight days are dropped in 2500 years.)[6] Not that Harriot breathes a word of the association.

For determining Easter, Harriot favours the system based on Paschal Terms ("termini Paschales", "fourteenth days of the first moon", etc.). Suppose we are in a year with Golden Number 17. This is found against 27 March in the standard calendar and the table already mentioned. The fourteenth day of the moon (counting inclusively) will then be 9 April, so in a table of Paschal Terms (that is, another sort of table, often just labelled "luna xiv") there will be a column of Golden Numbers ("Aureus Numerus") with 17 placed against 9 April. Suppose the Dominical Letter for the year is c, that is, all Sundays in the year (or at least in the months after February, this proviso being necessary since in Leap Years the Dominical Letter as judged by the standard calendar shifts by one after 29 February) have the day-letter c. By the rules for Easter already expressed, we look down the calendar for the next c after 9 April. In fact it is at 11 April, which will therefore be Easter day.

It was Harriot's ambition to reduce this standard procedure to a straightforward set of mathematical rules. In this aim he was only partly successful — and not until the mathematician C.F. Gauss (1777–1855) did anyone set down a complete arithmetical algorithm. The problem is that we are involved in arith-

---

4. The Golden Numbers go from 1 to 19, the cycle of years after which the lunations (numbering 235) repeat supposedly precisely. How the Golden Numbers are placed can be found from standard works on the calendar, or from the explanatory supplements to the *Astronomical Ephemeris*.

5. Petworth calendar manuscript, p. 3.

6. *Op. cit.*, note 1 above.

metic of congruences, of cycles of numbers. There are the cycle of lunations, the cycle of days, the leap year cycle, and even the cycles for calendar correction, for example, and they do not even share a common starting point. Such cycles had been incorporated in a number of different tables, devised, as Harriot says, "for the more simpler clarkes, that they might not be mistaken".[7] (He mentions the English Bible and larger Books of Common Prayer, the works of the computists, and the Roman missals and breviaries.) To manage without them, Harriot says, he has set down a memorable rule of his own. The rule is very simple: *the Paschal Term is found by subtracting the epact from 47, this giving the date in March*. This rule takes some things for granted, as I will explain, and it only takes us as far as the Paschal Term, leaving another step to come before Easter is found. It takes for granted a particular system of epacts, and some calculations on an isolated sheet[8] show that Harriot pondered the merits of this carefully.

Allocating the Golden Numbers is, as I hope is now clear, equivalent to making statements about the age of the moon. Golden Number 5 against 29 December, for instance, means that on the following January 1, that is, in the year after that with Golden Number 5, the moon will be in day 4 — "four days old", in traditional parlance. The epact is sometimes defined as the age of the moon on 1 January, but this is a dangerous definition, owing to the slip of the lunar calendar. (For Harriot's time we could take such a definition with 3 January, or 3 March. Note that the definition "the age of the moon on 1 March" is more or less equivalent to the definition "age of the moon on 1 January", since two lunations more or less equal the interval of 59 days.) Ordinary calendar techniques presupposed that at the end of every Julian year the moon had gone eleven days beyond 12 lunations. We find Harriot calculating the actual slip on the basis of an annual difference of $10^d 21^h 11^m 21^s 52^t 24^f$, for a complete 19-year cycle, and needless to say, the accumulated error is the $1^h 27^m 32^s 19^t 40^f$ already mentioned. Using the approximate rule, nevertheless, he is able to quote two basic arithmetical formulae for Golden Number (G) and epact (E) corresponding to the year (Y). (I denote the remainder when a is divided by b by 'rem[a/b]', which makes the following formulae look more like explicit definitions than would the mathematicians' notation 'a = r (mod b)'.)

$$G = rem[(Y + 1)/19],$$
and $E = rem[11G/30]$.

(The second follows because every new year of the G-cycle the age of the moon on any particular date chosen as norm advances by 11 days. We reject multiples

7. MS, p. 6.
8. *Ibid.*, p. 5.

of 30, the approximate length of a lunar month, and get rid of the constant by the suitable choice of a base date, bearing in mind the meaning of E, "age of the moon on such and such a date".) In short, by sacrificing his astronomical precision he can keep the old rule. We can now see the source of his rule of "$47 - E$". We are only interested in cases where day 14 of the moon comes after 21 March. Let us suppose that the moon is of age E on 3 March, then it will be of age 14 days on $(3 + 14 = E)$ March, and again 30 days later, that is, on $(47 - E)$ March. The first will never occur in the Easter calculation, for there we are only interested in Paschal Terms after 21 March, which with the first date would imply negative E. The remaining alternative is Harriot's rule. We may check it against his own example in these papers, for the year 1616: here $G = 2$, which comes against 12 March in any calendar. By the division rule, $E = 22$. Day 1 being 12 March, day 22 will be 2 April. With the help of the Golden Number against this date it is easy to check that the moon would have been judged to have the same age on 3 March, 2 February, and 3 January, for example. (The intervals are not all 30 days, of course, for the patterning of the calendar, with its 29- and 30-day months, must achieve an average of about $29\frac{1}{2}$ days to the month.)

To our two arithmetical rules of the last paragraph, Harriot has thus added a third, namely

$T = 47 - E$ if this is greater than or equal to 21,
and $T = 47 - E + 30$ otherwise,

where T is the March date (with an obvious rule if it takes us over 31 and into April) of the Paschal Term, luna xiv. The alternatives are simply to ensure that we do not look for an Easter before the conventional equinox. The addition of 30 takes us on a lunation. But what of the last step, to the Easter date itself? This is a step Harriot cannot take, without the help of a table; a simple enough table, admittedly:

1	2	3	4	5	6	7	8	9	10	11	12	13	14
g				b				d				f	
f	e	d	c	a	g	f	e	c	b	a	g	e	d
15	16	17	18	19	20	21	22	23	24	25	26	27	28
		a				c				e			
c	b	g	f	e	d	b	a	g	f	d	c	b	a

He indicates some years on this, but they are not of fundamental importance: for the record, 1600 and 1628 and 1656 come opposite 13 and ef. The table correlates the Dominical Letters a to g with what Harriot calls the "numbers . . . of the auncient Dionysian cycle and now yet commonly used". Denoting them by N, we can express his rule for N as follows:

$$N = \text{rem}[(Y + 9)/28].$$

As we all know, the year slips a weekday each year, and an extra weekday in leap year, running through a complete cycle in $7 \times 4 = 28$ years. The Dominical Letter, the weekday letter that in a particular year is to count as Sunday, shifts in the manner shown in the table, and it is the "doubling up" of the letters every fourth year that makes the translation of the table into arithmetical terms so awkward. (Note that the lower letter, in both senses of the word, serves for the latter part of the year, and thus for the Easter calculation. Note also that the interpolated day in a leap year was not always where it is now, at the end of February, and that Bede took it after 24 March, while Harriot took it after 24 February, St Matthias' day.) It is not difficult to turn the table into a rule, perhaps, making a fifth arithmetical rule, to which can be conjoined a sixth, along the lines "Search for the Dominical Letter (or a number you might use to replace the letter) in the calendar sequence next following T, remembering that 22 March has day-letter d, 23 March e, etc.". Not difficult, but Harriot did not do it, and I shall leave matters there, with the remark that the little table above is almost as easy to write down as Gauss's formulae, which themselves contain a number of awkward constants. But of course we must give Gauss the palm.

Harriot was not writing for a sophisticated audience, however. (For whom, indeed, was he writing? He could well have turned this document over to a printer without any difficulty, I would judge, but presumably had patron or friends in mind. Or was it merely lethargy that stood in the way?) He therefore added rules for calculation through counting on the parts of the hand. He gave a rule for the epacts by counting on the joints of a single finger;[9] another using four fingers and a verse he says he composed "many yeares past" for the Dominical Letter beginning 1600. "Fox eares dogs clawes brave a goose" might be thought harder to memorize than the alphabet that its initial letters embody, except that it runs backwards and does not terminate in a. There is included in these papers a sheet of jottings in the style of those other papers of his where he permutes words for secret or mnemonic purposes: thus "doc bis ac glis fiet" is meant to remind one of "dc b a g fe", a string of Dominical Letters, but for what years Harriot, in these jottings, does not say. (Note how he put the double letters into a single word.) Perhaps they come from the period of his life 1596–1600; they are not likely to be for 1624–1628, which they also cover. (He died in 1621.) Other rules systematically written up in his tract (as these jottings are not) are:

– To find what letter by the hand standeth agaynst any day of any month in the yeare as it is in the calendar.
– To know what day of the week any letter signifyeth.

9. *Ibid.*, p. 7.

– By knowing the day of the week, to know what day of the month it is, supposing I know allready within a week or thereabouts; otherwise we cannot do it well though we have a Kallender.

– To know how many dayes every month hath.[10]

All are in the tradition of rules taught (of course through Latin) to the medieval clerk. They are hardly worth expounding. The first requires a great deal of information to be carried on both the inside and outside of the hand, and uses this mnemonic:

Age drawes death, gold blindes eyes,
Goods choose frendes, ale drownes flyes.

Harriot tells us again that "some years before" he had written for the same purpose:

Ale dregs, dried guts, bulls eares,
Good cheare for a dog's feast.

The verses are essentially for the day-letters for the first of each month; the fingers are used to carry on counting in sevens. The other rules are rather more trivial, but it is noteworthy that the last is a very awkward finger rule for our well-known "Thirty days hath September, &c.". How old our own rule is I cannot say, but it is certainly older than Harriot.[11]

The last seven pages of the group concern, as I have said, ancient chronology. As in other parts of Harriot's literary remains, it is hard to say what he did and did not believe, for so much is taken from other writers, to be deposited here only as notes. Thus central to a group of calculations relating to the epoch of the creation of the world there is the hypothesis that this took place at 9 p.m. on September 24 (the autumnal equinox, supposedly) in a "year 4104", presumably before the birth of Christ (conventional rather than "correct"). The sun had then set. In working out the interval to the next equinox he made a couple of mistakes, one of which he corrected, so we can assume that he was not copying entirely from Jacob Christmann – although certainly he took the epoch from that author. Christmann's widely read *Epistola chronologica ad Justum Lipsium* [etc.] of 1591 (Heidelberg) and later, which appeared as *Disputatio de anno* [etc.] in expanded form in 1593 (Frankfurt), does not tally with Harriot's page references, and it turns out to be the ostensibly less probable

---

10. *Ibid.*, pp. 10–12. Note that the pages are out of sequence, and that 10 should follow 11, as at present.

11. *The Oxford Dictionary of Quotations*, 3rd ed., 1979, p. 8, suggests c. 1555.

Muhamedis Alfragani Arabis chronologica et astronomica elementa of 1590
(Frankfurt) on which Harriot was relying so much. He used Christmann again
in papers now in the British Library,[12] together with other authors I shall mention shortly in connection with the calendar papers. It is important to recognize
that many scores of dates were canvassed for the creation of the world, and that
there was considerable activity concerning biblical chronology generally from
the middle ages to modern times.[13] Archbishop Ussher's writings, although well
known, and indeed in their way remarkable for balanced scholarship, were by
no means unusual. Thomas Lydiat was another English contemporary of Harriot working on more or less the same assumptions as he (as regards creation)
in his *Defensio tractatus de variis annorum formis contra J. Scaligeri obtrecta-
tiones*, published with *Examen canonum chronologiae isagocicorum* in 1607
(London), and in *Emendatio temporum ab initio mundi . . . contra Scaligerum
et alios* in 1609 (London). The first work was dedicated to Sir Anthony Cope
and the last to Henry, prince of Wales, who gave Lydiat a position as chronog-
rapher and cosmographer for a time.[14] There is a good chance that Harriot and
Lydiat were acquainted, although I have been able to find no evidence of this.

There is a firmer connection with chronological affairs, of course, through
Harriot's old patron Walter Raleigh. There are few clear signs that Harriot
helped with *The History of the World*, although he is named as one of those
who occasionally did so.[15] Raleigh's work, composed in the Tower between
1607 and 1614, and published in that year, ends with a chronology. Even here,
though, Harriot's hand is hard to detect. The chronology runs from year 1 of
Adam and the world (said to be year 683 of the Julian period) to year 3867 of
the world, when Torquatus and Octavius were Roman consuls, that is, the year
165 B.C. The history itself ends at around 130 B.C. Raleigh, surprisingly, gave
no prominence to the date of the world's creation he was using, but this must
have been 4031 B.C., different from Harriot's (4104 B.C.), and indeed different from the many others of which I am aware. Simply because he had no opportunity, or spirit, to complete his *History*, Raleigh had no opportunity to include material assembled by Harriot in the papers I am here discussing, but
their apparent date (1614, 1615, or 1616) is consistent with the hypothesis that
they were meant for Raleigh's use. As an example of a passage that might well

12. See Add. MS 6788, ff. 497r and 499r, for example.

13. See my "Chronology and the Age of the World", in *History, Cosmology, Theology*, ed. W.
Yourgrau, New York, Plenum, 1978, pp. 307–33; North, *Stars, Minds and Fate*, ch. 9.

14. He went to Ireland at Ussher's invitation, but later returned to London. His hopes of prefer-
ment were dashed by Henry's death in 1612. There is a charming comment in the *Dictionary of National Biography* article (by A.F. Pollard) to the effect that his defective memory and utterance led
him to relinquish both the study of divinity and his fellowship (at New College, Oxford) in 1603,
"in order to devote himself to mathematics and chronology".

15. See the *D.N.B.* article by John Knox Laughton and Sidney Lee, at p. 645.

be thought to bear Harriot's mark, there is a reference to the "authoritie of that great Astrologer *Ptolomie*, from which, there is no appeale", this being here authority for identifying year 519 of Nabonassar with year 82 of "The [Seleucid] Kingdome of the Greekes".[16] A point worth making is that Raleigh, whose library was obviously large, though its size has often been exaggerated, was here quoting indirectly, from Gauricus. When he added a reference to Bunting, who had calculated Saturn's position in agreement with a position given by Ptolemy, we may imagine that Harriot had guided him to what − as I shall show in a moment − was a favourite source of reference. Whether or not this was so, it is the easier to believe when we read further that

> These observations of the Celestiall bodies, are the surest markes of time: from which he that wilfully varies, is inexcusable.

To return to the calendar papers: on another sheet we find a "Collatio annorum à Christo secundum Baronium . . .", best illustrated by a specimen entry, showing that Julian year 46 has Golden Number 2, Dominical letter b, was year 1 in the Dionysian cycle we use for reckoning time (i.e. was A.D. 1), but was year 3 of Christ according to Baronius. Cesare Baronius (1538−1607) was a widely read Catholic writer, the compiler of the quasi-official *Martyrologium Romanum*, done at the instance of Pope Gregory XIII (who instituted calendar reform), and was a cardinal. Lydiat had done the safe English thing and attacked Baronius vigorously. Harriot seems to be citing the first two of the twelve volumes of the Cardinal's *Annales Ecclesiae* (1588−1607), which rest for relevant early chronology on Irenaeus, Tertullian, Clement of Alexandria, Eusebius, and others. In other manuscript papers (in British Library MS Add. 6789) Harriot again shows that he leant heavily on the same writer, taking from him lists of consuls, errors in feasts as calculated by other writers, and other chronological matter.[17]

Again, in both sets of papers we find numerous references to the would-be calendar reformer Paul of Middelburg, who had exhorted first Innocent VIII (in 1491) and later, in an ambitious printed work, Leo X, Maximilian, and the College of Cardinals, not to mention the Lateran Council, to reform the ailing calendar. It is this last printed work, *Paulina de recta Paschae celebratione* [etc.], that Harriot was consulting.[18] It is interesting to see that some Paschal

16. Book 4 "of the first [and only] part", ch. 5, sect. 7, at p. 262 in the second pagination of the first edition of 1614.

17. See ff. 476r−486r, *passim*.

18. 1513 and 1516 (Rome), with a slight change in the later title (*Compendium correctionis calendarii pro recta* [etc.]). Cf. British Library MS Add. 6788, f. 494v. See further my article referred to in note 1 above. The Petworth ref. is p. 20, which I should in any future rebinding place after p. 17 and before 15, 16, 14, 18 and 19, in that order. For further references to Paul see Add. MS 6788, ff. 494v and 556v.

Terms calculated by Paul and reported by Harriot (AD 17 and 25) do not agree with Harriot's tables, but the whole subject is hard to reduce to consistency, with its frightening network of inter-related presupposition. There is a couple of eclipse records (AD 14 and 17). For the rest, as already mentioned, there are lists of years taking us up to the year 470, and for the most part concerned with conflicting calculations.

Apart from Christmann and Paul of Middelburg there are several other authorities named, but only two of these does Harriot know from their own works, namely "Bunting" and "Pascasinus in Leone". Second-hand are his references to Ambrosius, Valentinianus, Marcellus (all from Christmann, pp. 435, 445–8, 452), and to the "first, second and third controversies" about Easter (for the years 330, 341, 349, in which he follows Paul of Middelburg, p. 56). Broadly speaking, the Roman Church followed the Alexandrian except when Easter was fixed in the period 22–25 April. [19] By the time of Ambrose, Bishop of Milan, at the end of the fourth century, there was a Milanese rite developing, differing from the Roman. Marcellus was a 4th-century Bordeaux physician, [20] and Valentinian II was simply Ambrose's emperor, whose death by strangulation is a rare historical date among the many Easter dates listed by Harriot.

The published work Harriot had to hand by the pastor of Grunow, Henricus Bunting, seems to have been his *Chronologia, hoc est omnium temporum et annorum series, ex sacris bibliis* [etc.] (1590, etc.). Bunting is cited in the British Library papers; there he is used for the date of the death of Constantine. [21] As for "Pascasinus", qualified as "in Leone" this is a reference to Bishop Paschasinus of Lilybaeum, who was consulted by Pope Leo over the disagreement between Roman and Alexandrian reckoning for the year 444. The bishop replied. In 451 Leo wrote again to him, to consult about certain heresies, and so as not to be caught out a second time raised the matter of a potential Easter conflict to come in 455. No reply is known. [22] "Paschasinus in Leone" was used by Harriot, finally, to give the consulates covered by the table of Easters for a hundred

19. Here I follow C.W. Jones, *Bedae Opera de Temporibus*, Cambridge, Mass., 1943, p. 55. According to Paul, however, in 330 and 341 the Romans took 22 March while the Alexandrians took 19 April.

20. *Ibid.*, pp. 334, 365–6.

21. Bunting's best-known work was his *Itinerarium* [etc.], first published in Magdeburg in 1597, with woodcut maps, and enormously popular thereafter, judging by editions and translations. Harriot was certainly using the less well-known *Chronologia, hoc est omnium temporum et annorum series, ex sacris bibliis* of 1590 (Servesta, from the press of Bonaventura Faber), a solid work with much astronomy that would have been to Harriot's taste, and interesting for its contemporaneous astronomical reference. Another edition, beautifully printed by Ambrosius Kirchner in Magdeburg, has different pagination from that quoted by Harriot. Note that Bunting puts Creation at c. 3968 B.C.

22. Jones, *op. cit.*, pp. 55–6, for references to MSS and printed editions, updating a study by Krusch.

years as compiled by Theophilus, bishop of Alexandria. (Note here the mention of Pope Zozimus, also referred to by Bede.) Just four years were abstracted from Theophilus' century of years, two of them (444 and 455) being the ones that concerned Pope Leo.

It is, after all, an accident of history that the Petworth papers are separated from the collections in the British Library, and a brief survey of similar chronological materials in the latter is in order. Harriot's concern with biblical history begins logically with the opening of the book of *Genesis*, "in the beginning God made heaven and earth", and his having cited this has attracted some attention for the light it throws on his supposed atheism. Consider MS Add. 6785, for example, at f. 30, where it is followed by the Hebrew equivalent of the English, and this in more senses than one: for the Hebrew is unpointed, and the English is written "n th bgynng gd md hvn nd rth"! This *jeu d'esprit* can hardly be treated as a devout declaration of belief; but I am sure the evidence for his belief is explicit enough, elsewhere.

In MS Add. 6789, ff. 468–96, we find copious reference to Josephus, with numbers that might easily be taken as years, but that are in fact page numbers to the edition of *De bello Judaico* Harriot used. At one point Harriot in the tradition of J.J. Scaliger explores the behaviour of three calendar cycles, namely the cycles of 19 years (the Metonic), of 28 years (the solar or weekday cycle), and of 15 years (the old Roman Indiction cycle). He deduces a "Golden Year", a year which, when divided by 19, 28 and 15, yields the same remainders (1, 1, and 3) as the year of Christ's incarnation. The Golden Year is 5853, not exactly in the immediate future.[23] It was Josephus who, as evidenced by another British Library MS (Add. 6788), set Harriot off calculating not only the years spanned by biblical generations, but *populations* resulting by geometric progression.[24]

Josephus, Bede, Paul of Middelburg, Clavius, Bunting, Christmann, Baronius, Scaliger, and the relevant astronomers for their solar and lunar motions, namely Ptolemy, Vieta, Copernicus, Tycho, and Kepler; in referring to so many authorities Harriot gave evidence of serious interest in the problems of chronology and the computus. This raises some important questions. Why on earth did Harriot take so many pains over this material? The explanation can-

23. See f. 473. The period for complete repetition of the three cycles is evaluated as 7980 years. This is the period that Joseph Justus Scaliger had introduced in 1583. His "Julian day" system of reckoning time has of course nothing to do with the date of Creation, but begins with 1 January 4713 B.C. (as day 0), since that year is the last in which the solar, Metonic, and Indiction cycles began together.

Harriot added more in the same vein, for instance dating events according to the three cycles. Note that at the end of the Petworth calendar papers he made use of the standard (19 × 23 =) 532 year cycle, that cycle after which Golden Numbers and Dominical letters recur jointly.

24. Beginning f. 507r and ending (with interpolated material from Viète, Copernicus, and others) at f. 549v. Note Harriot's emendations to Josephus at f. 496r.

not simply be put down to his love of calculation, for by any standards the calculations involved were very tedious stuff. He took the task seriously. One ought not to disagree lightly with Scaliger — although there were plenty of precedents for doing so — but Harriot's calculations sometimes failed to confirm the great scholar's findings.[25] Why did he bother? Surely the answer is that Harriot took his faith seriously. Easter is for a Christian the celebration of the most important event in history, and it took a no-nonsense man like Luther to dispense with what were roughly the Jewish principles for establishing its date. Confronted though he was by the knots into which Christian computists had tied themselves in trying to apply their rules consistently, Harriot gives not the slightest hint of impatience with those rules. In matters of biblical history he seems to have been no less of a conformist. In its early stages, at least, Christian history connects with Old Testament chronology, and hence with creation — and how pleasant it was to feel that one could put a definite date to that. There are no signs, in the papers I have been discussing, of millenarianism of the sort that often turns up in biblical chronology. Harriot was simply caught up in scholarly discussions of a more genuinely historiographical kind. Luther's challenge to the papacy had, among so many other things, turned scholars' thoughts towards early scriptural history and patristic tradition, with documentary evidence replacing Church authority as an object of trust. Had not Raleigh, Harriot's former patron, written a history of the world in the same critical tradition? It is hardly surprising that Harriot gave the subject serious attention. What is worth mentioning in Harriot's case is his use of Protestant and Catholic authorities indifferently. He was by nature a computist, not a polemicist.

25. Add. MS 6788, ff. 499r–506v, and especially 556v–7v.

TEXT

(The bold numbers represent the pages of the manuscript.
All diagrams are redrawn)

The Effect of the Decree of the Councell
of Neece for the Observation of Easter day

1) The next Sunday after the 14th day of the first moone is to be observed for Easter day.

2) The 14th day of the first moone is the fouretenth day of that moone that falleth upon, or next after the 21th [sic] day of March. (Which was supposed to be the day of the aequinoctium at that time, but is now about the 11th of March.)

3) The first day of the first moone is to be reckoned that day of the month in March agaynst which the golden number standeth at that yeare [then inserted here:] if the 14th reaches unto or beyond the 21 day. Otherwise in Aprill seek it, as when the golden number is, 11, 19, or 8. [Insertion ends.] (Which day was supposed at that time to be the day mediae coniunctionis Solis et Lunae or within a day more or lesse. But the media coniunctio or change day is now aboute 5 dayes before the place of the golden number.)

The Church of England & some other Churches not subiect to the Church of Rome follow this decree still uncorrected. As for example, 1616. The golden number is 2. And being leap yeare there are two Dominicall letters gf. But f the later is for March and the monthes following. Look the golden number 2 in March & it standeth agaynst the 12th day of the month. Then ad 13 & it maketh 25 or reckon that day for one and tell on 14 which will fall on the 25th of March (which is after the 21) agaynst which standeth g which is Munday. The next f following which is Sonday is the 31th [sic] of March, & therefore Easter day. And so in like manner universally for other yeares.

Aureus numerus	Martius		Aureus numerus	Aprilis	
3	1	d		1	g
	2	e	11	2	a
11	3	f		3	b
	4	g	19	4	c
19	5	a	8	5	d	+
8	6	b	16	6	e
	7	c	5	7	f
+ 16	8	d		8	g
5	9	e	13	9	a
	10	f	2	10	b
13	11	g		11	c
2	12	a	10	12	d
	13	b		13	e
10	14	c	18	14	f
	15	d	7	15	g
18	16	e		16	a
7	17	f	15	17	b
	18	g	4	18	c	+
15	19	a		19	d
4	20	b	12	20	e
+	21	c	1	21	f
12	22	d		22	g
1	23	e	9	23	a
	24	f		24	b
9	25	g	17	25	c
	26	a	6	26	d
17	27	b		27	e
6	28	c	14	28	f
	29	d	3	29	g
14	30	e		30	a
3	31	f			

There are other tables to find Easter day for Ever (knowing the golden number & dominicall letter which are easily had by rules common) yet this is the foundation of all; & playne inough of it self, if that be well marked which I have here set downe.

	Aur. num.	Lit. Dom.	Easter day	True Easter day
1614	19	b	Ap.24	A.20
1615	1	a	Ap.9	A.9
1616	2	gf	M.31	M.24
1617	3	e	A.20	M.16
1618	4	d	A.5	A.5
1619	5	c	M.28	M.21
1620	6	ba	A.16	A.9
&c.				

(2) ———

Note:
1) If the golden number be 16. The first day of the moone is the 8th of March & the 14th day of the moone wilbe the 21th of March. And then if c be dominicall letter, Easter day by the Canon wilbe the 28th of March. But if d be dominicall letter Easter day wilbe the 22th of March. So that the 8th day of March is the least number that may be the first day of the first moone. And the 22th day the least [for: last], that may be Easter day.
2) If the golden number be 8. The 14th day of the moone in March wilbe the 19th of March, which is short of 21. There 8 in Aprill sheweth the 5th day to be the first day of the first moone: & the 14th day of the moone wilbe the 18th of Aprill. And if c be Dominicall letter, Easter day wilbe the 25th of Aprill. But if any other letter be dominicall, Easter day wilbe short of the 25th day. So that the 5th day is the greatest number in April for the first day of the first moone, and the 25th the greatest, for Easter day.
3) The days therefore that may be Easter dayes, are all from the 22th of March to the 25th of Aprill (inclusive) & no other, the which I have marked all with prickes.
4) And this I thought good to note withall. That our Common Easter day is sometime the same day as the true Easter day intended. And if it differ, it wilbe later than the true by just 1, 4, or 5 weekes for these 700 yeares to come & upward; and after it wilbe more sometimes & more &c.

(3) ———

[The following statements are abbreviated with the help of present day notations. – J.D.N.]

In one Julian yeare, is 365^d 6^h

365^d × 19 = 6935^d = 19 common yeares.

6^h × 19 = 114^h = 4^{d}18^h.

Therefore 19 Julian yeares = 6939^{d}18^h.

———————

Astronomice computa:

One month secundum medium motum = 29^d 12^h; 44,03,10,48.

235 such astronomicall monthes = 6939^d 16^h; 32,27,18,00.

19 Julian yeares (as above) = 6939^d 17^h; 59,59,60.

Therefore 235 monthes is short of 19 Julian yeares only 1^h; 27,32,42.

Ergo. Therefore upon what day & houre of any monthe & yeare soever the moone doth chaunge: 19 yeares after it will chaunge the same day of the month, but before the former hower, 1^h; 27,32,42 only, which being so little the auncient fathers semed to neclecte; or did thinke the astronomicall observations not so exact, or if that they were the error that should rise upon the account might be mended when in time it should be found notable.

But suposing the astronomicall account just; that 1^h; 27,32,42 in 312½ years will mount unto 24^h or one day.

In 625 yeres, 2 dayes.

In 1250 yeres, 4 dayes, &c

It followeth also that if we have noted or marked in Julian kalendars in every month the chaunge day for 19 years together. As namely for the first yeare to set downe the figure of 1, agaynst that day of every month where the mone changeth. And the second yeare the figure of 2. And the third yeare 3. And so forth till all the nineteene yeres have received there severall markes by numbers in such manner as the other before. Then for the next 19 yeares following the dayes of chaunges are all knowne and had allready; as where the figure of 1 was written there are the chaunge dayes for the first yeare. And where the figure of 2 was written, there is the chaung day for the second yeare; and so forth for all the rest of the 19 yeres. And so likewise for other 19 teene of yeres, following one after an other; you have the chaunges of the mone according to such perenes [?] of truth as may be concluded by that which I before noted.

(4) ———

How to make the table called the table of Termini Paschales; or the foreteenthe dayes of the first moone: for all the yeares of the cycle of the mone, which is all the yeares of the golden number.

Seing that the 21th of March canbe the soonest day for the 14th day of the first moone and the 25th of Aprill the farthest day for Easter day (as by the decree of the Councell of Neece in the 1 paper, & the notes in the second paper); therefore first set downe all the dayes of those monthes from the 21th of March to the 25th of April as is here in this table with there hebdomadall or week letters agaynst them. Then for that yere whose golden number is one, looke the 14th day of the first mone in the table (according to the cautions of the decree) set down in the first papaer. There 1, standing agaynst the 23th of March which noteth that to be the first day of the moone, then the 14th wilbe the 5th of Aprill. Therefore agaynst the 5th of Aprill in this table is placed 1, shewing the Paschal Terme or the 14th day of the first moone when the golden number is 1. Likewise when the golden number is 2, looke in the table of the first paper & it standeth agaynst the 12th of March which noteth that day to be the first day of the first moone (agreing with the conditions required). Therefore the 14th day wilbe the 25th of March. And agaynst the 25th of March is placed in this table 2, signifying that when the golden number is 2 that day is the 14th day of the first month. And so in like manner of other golden numbers till you see how all are placed. One example more when the golden number is 8. That in the first table standeth agaynst the 6th of March, whose 14th day wilbe the 19th of March, but that 14th day cannot be of the first moone because it is short of the 21th day (agaynst the decree). Therefore look the next 8 following, which standeth agaynst the 5th of Aprill, which noteth that day to be the first day of the first mone. And the 14th day wilbe the 18th of Aprill. Therefore in this table is placed 8 agaynst the sayd day signifying the 14th day of the first moone when the golden number is 8. All the rest of the golden numbers so placed in this table shew the termini paschales, or 14th dayes of the first moone, which are no other dayes then agaynst which they are placed; & are in number only 19, but the Easter dayes may be any here written but the 21th of March, which in all are in number 35.

The use of this table is for the ready finding of Easter day. First for the yeare of our Lord larne the golden number. That number seeke in this table and the next Sunday after (known by the Dominicall letter) is Easter day. As this yere 1616, the golden number being 2 and the Dominicall letter f, sheweth Easter day to be the 31 of March.

Termini pasch. sive lunae pr. decimiquart.		
Mensis	Dies mensis / Lit. Dom.	Aureus numerus
Martius 21		16
22	d	5
23	e	
24	f	13
25	g	2
26	a	
27	b	10
28	c	
29	d	18
30	e	7
31	f	
Aprilis 1	g	15
2	a	4
3	b	
4	c	12
5	d	1
6	e	
7	f	9
8	g	
9	a	17
10	b	6
11	c	
12	d	14
13	e	3
14	f	
15	g	11
16	a	
17	b	19
18	c	8
19	d	
20	e	
21	f	
22	g	
23	a	
24	b	
25	c	

(5) ———

	d	h	'	"	'''	''''
	10	21	11	21	52	24
1	10	21	11	21	52	24
2	21	18	22	43	44	48
	32	15	34	05	37	12
	29	12	44	03	10	48
3	3	02	50	02	26	24
4	14	00	01	24	18	48
5	24	21	12	46	11	12
	35	18	24	08	03	36
	29	12	44	03	10	48
6	6	05	40	04	52	28
7	17	02	51	25	44	52
8	28	0	02	47	37	16
	38	21	14	09	29	40
	29	12	44	03	10	48
9	9	08	30	06	18	52
10	20	05	41	28	11	16
	31	02	52	50	03	40
	29	12	44	03	10	48
11	1	14	08	46	52	52
12	12	11	20	08	45	16
13	23	08	31	30	37	40
	34	05	42	52	30	04
	29	12	44	03	10	48
14	4	16	58	49	19	16
15	15	14	10	11	11	40
16	26	12	21	33	04	04
	37	08	32	54	56	28
	29	12	44	03	10	48
17	7	19	48	51	45	40
18	18	17	00	13	38	04
19	29	14	11	35	30	28
	29	12	44	03	10	48
	1	27	32	19	40	

(6) ———

How the Paschall tearmes, or the 14th dayes of the first moone, may be had without any table. And per consequence Easter day.

There be other tables for the finding of Easter day for ever, then I have set downe. But the   in the first paper is the ground of all, & easy enough if it be well marked. The other tables were devised for the more simpler clarkes, that they might not be mistaken. There is one speciall one in the Kallender of the great English bible & in the larger bookes of common prayer. The same is also in the Romane masse bookes & Breviaryes, and in many other Kalenders & in many of those authors which are called computists writing of this matter, where they may be seen by him that will. I for this time will have all other tables & set downe a memorable rule of myne owne, which will performe the thing desired without all tables or bookes, and exactly according to the decree of the Nicene Councell. But first suppose that for the yeare whose paschall terme or Easter day you seek, you have for that yeare his golden number and his epact, which in the papers following together with the dominicall letter are also taught by memorable rules. The epact therefore being supposed to be knowne, the rule is as followeth:

Subtract the epact allwayes out of 47. The remayne sheweth the 14th day of a moone from the beginning of March. And if it be more then 31 (the dayes of March) it is the 14th day of the first moone. Then out of the said remayne subtract 31, & this last remayne is the day of Aprill for the paschall terme.

But if the first remayne be lesse then 31, and be just 21, or greater, that number is the day of March for the paschall terme or the 14th day of the first moone.

But if the sayd first remayne be lesse then 21, it is agaynst the decree to make it the 14th day of the first moone; therefore adde unto it 30, and from the summe subtract 31. The remayne is the day of Aprill for the paschall terme. Or, without any addition of 30 subtract only 1, & the remayne is the same day of Aprill for the sayd paschall terme.

Golden number	Epact	All the paschall termes by rule, agreeing with the table before
1	11	$47 - 11 = 36$　$36 - 31 =$　5 April
2	22	$47 - 22 = 25$　March
3	3	$47 - 3 = 44$　$44 - 31 = 13$ A
4	14	$47 - 14 = 33$　$33 - 31 =$　2 A
5	25	$47 - 25 = 22$　M
6	6	$47 - 6 = 41$　$41 - 31 = 10$ A
7	17	$47 - 17 = 30$　M
8	28	$47 - 28 = 19$　$19 - 1 = 18$ A
9	9	$47 - 9 = 38$　$38 - 31 =$　7 A
10	20	$47 - 20 = 27$　M
11	1	$47 - 1 = 46$　$46 - 31 = 15$ A
12	12	$47 - 12 = 35$　$35 - 31 =$　4 A
13	23	$47 - 23 = 24$　M
14	4	$47 - 4 = 43$　$43 - 31 = 12$ A
15	15	$47 - 15 = 32$　$32 - 31 =$　1 A
16	26	$47 - 26 = 21$　M
17	7	$47 - 7 = 40$　$40 - 31 =$　9 A
18	18	$47 - 18 = 29$　M
19	29	$47 - 29 = 18$　$18 - 1 = 17$ A

All examples are conteyned in the table annexed, the which shewing all the paschall termes, & agreing with the table before, do verify the rule & justify it as perfect.

(7) ———

The rule for the golden number

Adde to the yeare of the Lord 1 & devide by 19. The remayne is the golden number.

For the yeare 1600 & after, this rule I make. The golden number was then 5, which may be remembered by the fitures of the yeare being 1 & 6, the difference being 5, the golden number. Then for any yeare after, ad 5 to the year above 1600. From the summe cast away the 19tenes of it be above 19. The remayne is the golden number. As if this yere, 1616. Take 16 & ad 5, that is 21. Take away 19. There remayneth 2, which is the golden number. It may be done also by the hand, but this rule is easy enough.

For the Epacts

Having the golden number, multiply it by a 11, and devide by 30. The remayne is the Epact. Or you may do it by your thumb in this manner, counting the first joynt 10, the second 20, the top 30 or 0. Then for any golden number beginning the reckoning at the first joynt, and telling on, as far as your golden number, then the number of the joynte with the golden number is the Epact if the summe be lesse then 30. If it be greater then take away 30 & the remayne is the Epact.

10	20	30
1	2	3
4	5	6
7	8	9
10	11	12
13	14	15
16	17	18
19		

As this yeare 1616, the golden number being 2 & the number of the joynt 20, the summe is 22 for the Epact.

A table of the cycle of the sonne, which is of 28 years for the dominicall letter. Beginning 1600 & ending 1627, it will begin also 1628, &c.

1656
1628
1600

f				a				c				e				1615
e	d	c	b	g	f	e	d	b	a	g	f	d	c	b	a	
13	14	15	16	17	18	19	20	21	22	23	24	25	26	27	28	
				g				b				d				
				f	e	d	c	a	g	f	e	c	b	a	g	
				1	2	3	4	5	6	7	8	9	10	11	12	
				1616											1627	
				1644											1655	

The numbers under written are of the auncient Dionysian cycle & now yet commonly used

The Generall rule for the dominicall letter

Ad to the yere of the Lord 9, and devide by 28. The remayne is the number of the cycle, which look in the table and it sheweth the dominicall letter, be it one or two. If two, the first serveth from the first of January to the 24th of February, being St Mathias day; the other all the yeare after. If there be but one dominicall letter it serveth from January all the yeare. Example for this yeare 1616. If it be devided having 9 added, by 28, that is 1625 by 28, the remayne wilbe 1, which looked in the table sheweth gf dominicall letters.

(8) ———

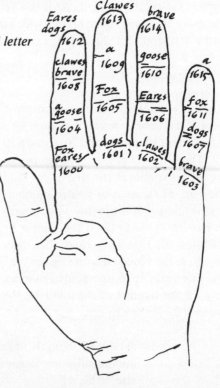

By the hand to find the dominicall letter beginning in the yeare 1600.

For this use I made many yeares past this verse following, which is memorable:

Fox eares dogs clawes brave a goose.

Then with your forefinger of your right hand beginne & reccon on the yeares of our Lord on the inside of the left hand in such order as is here pictured, beginning at the lowest joynt of the forefinger with the yere 1600, then on the next finger 1601, and so forth, till you come to the yeare of the Lord whose dominicall letter you seek; and mark upon what joynt or top of the finger it falleth on. As for example 1616, in this order it will fall on the back side of the forefinger upon the next joynt after the top. (And this recconing for other yeares is to continew to the last joynt of the little finger on the back side, which is the 28th place from the first of the forefinger on the inside. And then afterward you must begin with such a yeare as falleth on the first inside joynt on the inside of the forefinger for other 28 yeares following. And note this by the way, that all the yeares that fall on the forefinger are leape yeares, and therefore must have two dominicall letters.)

Then begin with your verse, and say two words on the forefinger, Fox Eares, & the next word on the next finger, & so onward as you se pictured (saying two words allwayes on the forefinger) till you come to the yeare you marked before, which was on the back side of the forefinger on the next joynt to the top. And on that place will fall these two wordes: Goose Fox. The first letters of the words, gf, are the dominicall letters for the sayd yeare 1616; and so for other yeares.

(9) –––

D   f
c   b   a   g   e

Doce bis ac glis fiet.
Dicat bis age fiet.
Docebit age fiet.

fiet docebit age.

glis fiet docebitur [erased]
grus   docebit age [erased]
grex [erased]
gres fiet dicat boa.

Adde de gleba et graecus fiat deflens. [erased]
Addo de gleba et graecus fiat deflens.
Adde   globo   fiam
                         fiat
                         fias
Adde de globo et graecus fias deflens.

(10) –––

I had made some years before these 2 following verses, or rather merry sentence for the same use, as the two others before serve for:

Ale dregs dried guts bulls eares
good cheare for a dogs feast

(11) ———

To find what letter by the hand standeth agaynst any day
of any month in the yeare as it is in the calendar

For this use I also made many yeares past two memorable verses consisting
of twelve monasillable wordes whose first letters are the letters that belong to
the first day of every month beginning at January. They are as followeth:

> Age drawes death, gold blindes eyes,
> Goods choose frendes, ale drownes flyes.

First I wold know what letter is for the first day of the month, as namely for
March. Say the months beginning with January in such order as you see pictured [2 figures of inside left hand, shown on opposite page] till you come to
March, and mark upon what joynt it falleth. Then say the wordes of the verse
till you come to that month of March and the word *death* will fall upon the
same, whose first letter d standeth agaynst the first day of the month in the Kalender. And so of any other month.

Then I would know what letter standeth agaynst the 25th day of March. For
this understand there is an other order of reckoning, which I call direct (the
other before being transverse). Understand that upon every finger, with inside
& outside together, are 7 places, which are fit for an intire week. I begin as I
did before, at the first joynt of the forefinger, & count that one; the next joynt
of the same finger 2; & so other joyntes & places of the same are 7. And on every
finger are 7, therefore 3 fingers make 21. Then the first joynt of the 4th finger
must be 22, & the top wilbe 25, which is the day of the month for which I seek
the letter. Then seing before is found that d is the first day, say forward e, f,
g, a, b, c, &c., untill you come where the 25th day was marked, and you shall
find that g wilbe the letter that falleth upon that day. And note that what letter
is on the first joynt of the forefinger, the same is on the first of all the rest, and
so all parallel joyntes have the same letters & are of the same day of the week.
And seing that before you found the 25th day to be on the top of the little finger,
in the beginning of the account, you needed not to reckon no farther by the letters then the top of the forefinger, & that being found to be g, it must follow
also that his parallel place the top of the little finger must also be b, the letter
that is sought for.

(12) ———

To know what day of the week any letter signifyeth

[See figure at bottom of opposite page] For this you need to use but the forefinger, placing the dominicall letter in the first place, as namely f for this yeare.

Inside of the left hand (1)

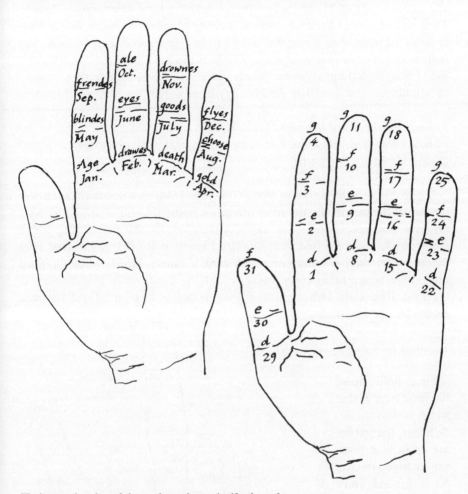

[To know what day of the week any letter signifyeth . . .]

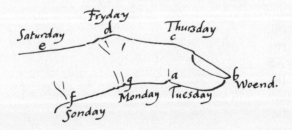

Then say on g, a, b, &c., till you come to the letter you seek, as namely d, which was the first day of March before found; and mark on what place it falleth. Then say, where you began, Sonday & forward Munday, Tuesday, &c., till you come to the place of d; & you shall find it to be Friday. And so by this you may understand that the first joyntes for March, in the 2 hand the next page before, were Fridayes, & the next rank of joyntes were Saturdayes. And so the tops were all Mundayes; and therefore the 25th day there sought for was on a Munday.

By knowing the day of the week, to know what day of the month it is, supposing I know allready within a week or thereabouts; otherwise we cannot do it well though we have a Kallender.

Suppose it were the month of March this yere, 1616, on a Sonday that I desire the day of the month. By the rules before we found that the first joyntes were Fridayes, the second rank Saturdayes, & therefore the third Sondayes, which methodically you must find by saying first Fryday on the first joynt, then after Saturday, Sunday, & marke on which rank it falleth, which wilbe the third rank. Then remembring every finger is 7, all the Sundayes of this month are: the third, 10th, 17th, 24th, or 31th; so that the day of the month I seek for must needes be one one these.

To know how many dayes every month hath.

First hold down the forefinger & that which is next the little finger, the rest being up in such manner as here you see. As if I wold know how many dayes October hath: begin with the tumb, & say on March, Aprill, May, &c., till you come to October, & mark on which finger it fallth, which wilbe the midle finger that standeth up; & therefore 31 dayes it hath, for all the

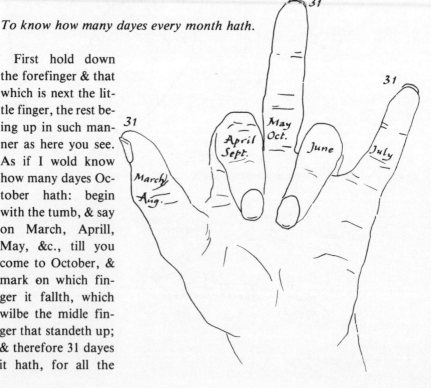

fingers that stand up signify 31 dayes, the rest, that be downe, 30, except for February, which hath 29 when it is leap yeare, but 28, other yeares.

(13) ———

Seeke for the goulden number of your yeare in the right syde of the table, and then looke amongst the Dominicall letters downeward, below the place where you finde the goulden number, for the Dominicall letter of your yeere, and right against it is that day of the moneth that Ester day shall fall on: reguard being had in the leap yeeres to take the latter letter.

[The table placed here is essentially that found on p. 4 above.]

As this yeere 1616 the goulden number is 2, and being leap yeare the latter Dominical letter is f: the goulden number is found against the 25 of March, and the next f downward is upon the 31 of March, which this yeere is Ester day.

[Note the many unusual spellings, not at all typical of Harriot. The words y^e, w^{ch}, y^t are expanded. These are not found in Harriot's work on the calendar. Finally, the script is not Harriot's usual script, but a careful fair book hand. – J.D.N.]

(14) ———

Alex. et Dyonis.	cyclus	Theophili	Christi		
	O	61	440	gf	
		62	441	e	
	6	63	442	d	
	7	64	443	c	
Pascasinus in –	5	8	65	444	ba	– pascha Lat. Martii 26 a
Leone 190	6	9	66	445	g	Alex. April 23 a
	7	10	67	446	f	
	8	11	68	447	e	
	9	12	69	448	dc	
	10	13	70	449	b	
	11	14	71	450	a	
	12	15	72	451	g	
	13	16	73	452	fe	
Leo –	14	17	74	453	d	– pascha April 12 d
Leo –	15	18	75	454	c	– pacha April 4 c
Leo –	16	19	76	455	b	* – pascha Lat. April 17 b
		1		456	ag	Alex. April 24 b
		2		457	f	———
		3		458	e	* hoc pascha confutat Christm-
		4		459	d	annum. Apud illum annus est
		5		460	cb	bisext: BA, quod fieri
		6		461	a	non potest.
		7		462	g	
Paul. Mid. 64 –		8		463	f	– pascha Lat. 24 Mart. f
et Christmann		9		464	ed	Alex. 21 April f Paul
prax. 453		10		465	c	– pascha Lat. 5 April c of Mid.
				466	b	Alex. 28 Mart. c 64
				467	a	
				468	gf	
				469		
				470		

[There is a computation of the Dom. Letter for 444 in the corner of the page. – J.D.N.]

(15, 16) ———

[These pages are in the style of p. 14, and take us up to AD 394. I combine two pages and abstract only the years with annotations. − J.D.N.]

Ambrosius. Christm. 447 P	− 13	373	f	− pascha 31 Martii f
Ambrosius. Christm. 448 P	− 17	377	a	− pascha April. 16 a
Valentinianus junior lagneo strangulatus				
in vigilia pentecost.				
Marcel Christm. 452 P	− 13	392	dc	− Pentecoste Mai 16 c
	26 3	325	c	− Concilium Niceum
	1 6	328	gf	− Concilium dimissum
Prima contraversia de paschate. Paul				
de Mid. 56	− 8	330	d	− pascha Lat. 22 Martii d Alex. 19 April. d
Christm. [5 characters] Bunting [2 characters]	− 15	337	b	− 12 cal. Jn. May 22 b pentecostes. Moritur Constantinus ergo pascha Ap. 3 b.
Paul. Mid. 56b 2ª contraversia	− 19	341	d	− pascha Lat. Mart. 22 d Alex. April. 19 d
3ª contraversia Paul. Mid. 56 b	− 8	349	a	− pascha Lat. 26 Mart. a Alex. 23 April. a
Ambros. Christ. 435 p 445 p	− 19	360	ba	− pascha April. 23 a

(17) ———

Initium anni Juliani

Jan: Novilunium   0^d 07;^{h} 06,35 P.M.
                 29   12;   44,03
                 ——————————
                 29   19;   50,38
                 28   12;   44,03
                 ——————————
                 58   32;   34,41
                 59   08;   34,41
                 31
                 ——
                 28   08;   34,41

Ergo:
Mart   08;   34,41 P.M.
————————————

13^d 01^h = 12^d 24^h 1080′   4104^a 216$^{cycl.}$
           12   20   204
           ————————
            0   4   876
            0   4   204
            ————————
            0   9   000   Tekupha

4104   6^f 23^h 0
        2   4   204   Tisri = 8^f 28^h 204′   Chri 344
        —————————————   Sept. 24
Differentia
        2   5   204   Tohu sive radix

Tekepha Tissri
Sept 24^d 9^h 0′   post occas. Chri.
   +   182   15
       ——————
       23   9
Sept.   7   15
N.   30
D.   31
J.   31
F.   28
    ——————
    127   15
O.   31

[table of planetary symbols, presumably for weekdays, which I give as initials. – J.D.N.]

    ——————
    158   .15   S   Th   Tu
     24   8   F   W
    ——————   M   S   Th
    182   15   Tu   8   F
                      W   M   S
                      Th   Tu   8
                      F   W   M
                      ——————
                      S   Th   Tu
                      8   F   W

(18) ———

Theophilus Episcopus Alexandriae scripsit canones de paschate pro 100 annis. Incipiunt in primo consulatu Theodosii. (Leo. 195.) Consulat Antonini et Siagrii paschasinus. In Leone 190

[☉ = Sunday]

443	Anno	$\frac{64}{100}$	7 cal. Ap. ☉ Luna 21ª occurrebat	(March 26 a.
444		$\frac{65}{100}$	9 cal. May ☉ Luna 19ª occur.	(April 23 a.

Tempore Zozimi Papri
    Honorio Augusto cons. 11°
           Constantio cons. 2°   8° cal. Ap. ☉

453	Anno	$\frac{74}{100}$	Opilione consule prid. Idus Apr. pascha ☉	(April 12 d.
454	Anno	$\frac{75}{100}$	prid. non. Apr. pascha ☉	(April 4 c.
455	Anno	$\frac{76}{100}$	8 cal. Mai. Alex. pascha ☉ 15 cal. Mai. Lat. pascha ☉	(April 24 b. (April 17 b.

Ætio et studio Co(anno post Opilionem): nulla dubitatio de pasc.

De dub. futuro   9° cal. Mai.   (April 23 a.

(19) ———

[Here there is some simple calculation; summarized, it amounts to:

      532 + 9 = 541
   and 541/28 = 19 rem. 9.

Harriot made two erroneous attempts at the second, before getting it right. There is also a scale:

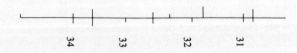

(20) ———

Collatio annorum à Christo secundum Baronium cum numeris Julianis annis Dionisianis cum cyclis, et incipiunt anni cal. Jan.

Paschata inventa per cyclum lunarem. Et consentiunt calculo Pa. de Midelburg.

	Juliani	Lun.	cycli Dyo-nis.	Anni Chr. Dion.	Anni Christi	Baronius
	44	19	e		1	Consules
	45	1	dc		2	Augustus XIII
	46	2	b	1	3	Plautius
	47	3	a	2	4	Siluens
	48	4	g	3	5	
	49	5	fe	4	6	
	50	6	d	5	7	
	51	7	c	6	8	
	52	8	b	7	9	
	53	9	ag	8	10	
	54	10	f	9	11	
	55	11	e	10	12	
	56	12	d	11	13	
Eclips. lun. totalis	57	13	cb	12	14	
Sept. 27, ho. 5 P.M.	58	14	a	13	15	
noct. B	−59	15	g	14	16	
	60	16	f	15	17	
	61	17	ed	16	18	
Eclipsis sol. nunq.	−62	18	c	17	19	
total. Feb. 15 hora	63	19	b	18	20	
meridiana fere	64	1	a	19	21	
Romae. Paul Mid. p303.	65	2	gf	20	22	
	66	3	e	21	23	
	67	4	d	22	24	
	68	5	c	23	25	
	69	6	ba	24	26	
	70	7	g	25	27	
	71	8	f	26	28	
	72	9	e	27	29	
Mar. 29 d [Monday]	73	10	dc	28	30	
Apr. 16 a [Sat.]	74	11	b	29	31	
Apr. 6 e [Thurs.]	75	12	a	30	32	
Mar. 26 a [Mon.]	76	13	g	31	33	
cal. 13 Apr. 14 f [Mon.]	77	14	fe	32	34 +	
Apr. 3 b [Fri.]	78	15	d +	33	35	
Ma. 23e[Tu] Apr. 21 f [Wed.]	79	16	c	34	36	
Ap. 11 c [Mon.]	80	17	b	35	37	
Mar. 31 f [Sat]	81	18	ag +	36	38	

6

THOMAS HARRIOT AND
THE FIRST TELESCOPIC OBSERVATIONS OF SUNSPOTS

One [mathematician] saith the Sun stands, another he moves, a third comes in, taking them all at rebound: and lest there should any Paradox be wanting, he findes certain spots or clouds in the Sun, by the help of glasses, by means of which the Sun must turne round upon his own center, or they about the Sun. *Fabricius* puts only three, and those in the Sun, Apelles 15. and those without the Sun, floating like the Cyanean Isles in the *Euxine* Sea. The *Hollander* in his *dissertatiuncula cum Apelle* censures all. . . .[1]

> Democritus Iunior [Robert Burton],
> *The Anatomy of Melancholy* . . . , Oxford,
> 1621, p. 329.

IT is a melancholy thought that Robert Burton, whose book was first published in the year of Harriot's death, should have known no better than to omit his compatriot's name from his list of paradoxers. Burton evidently knew of the controversy over priority in the discovery of sunspots, and he must have ignored the name of Galileo quite deliberately, for with Galileo's writings he was very familiar; but of Harriot's work he was in all probability totally unaware. What was true of Burton in 1621 was to remain true of many a later historian. Thus in a recent discussion of Galileo's observations of sunspots, running to more than three score pages, Bernard Dame was able to include Harriot's name only in a footnote, as one who 'a peut-être observé les taches avant Galilée'.[2] He observed that Harriot's observations were not known until 1784, and not published until 1833. In saying as much, although he made no mention of Arago, he reminds us of that passage where Arago, in a life of Galileo and in connexion with this very question of priority over the obser-

---

1. Notes to the text indicate that the discoverer of sunspots was 'Io. Fabricius de maculis in Sole. Witeb. 1611' (see below), and that the 'Hollander' published at 'Lugduni Bat. Ao 1612'. This is a reference to a rare and obscure tract by a writer calling himself simply 'Batavus', or 'Dutchman': *De maculis in sole animadversis, et, tanquam ab Apelle, in tabula spectandum in publica luce expositis, Batavi dissertatiuncula ad Amplissimum Nobilissimumque virum, Cornelium Vander-Milium Academiae Lugodinensis Curatorem vigilantissimum,* Ex Officina Plantiniana Raphelengii (Leyden), 1612, 19 pp, quarto. I have not seen this work referred to by any modern writer, but Burton's style of the previous sentence suggests that he had seen it. Raphelengius finished printing three years later with a book by the Italian astronomer Magini. 'Apelles' was part of a pseudonym of Scheiner, to whom Burton referred elsewhere, but apparently Burton was ignorant of the identity of the two.
2. 'Galilée et les taches solaires (1610–1613)', *Revue d'Histoire des Sciences,* xix (1966), pp. 306–70; reprinted in *Galilée: Aspects de sa vie et de son oeuvre,* Centre international de Synthèse, Paris, 1968, pp. 186–251. See p. 195, n. 3.

vation of sunspots—laid down his 'règle aux historiens des sciences'.[3]

Arago's guiding principle was one of public utility: 'Le public ne doit rien à qui ne lui a rendu aucun service', and 'en matière de découvertes, comme en tante autre chose, l'intérét public et l'intérêt privé bien entendu marchent toujours de compagnie'. Publication meant everything to Arago, and in reading him we are reminded of William Lower's plea to Harriot in a letter of 6 February 1610, where Lower has spoken of Harriot's discoveries passing to other men:

> Al these were your deues and manie others that I could mention; and yet to[o] great reservednesse hath robd you of those glories . . . Onlie let this remember you, that it is possible by to[o] much procrastination to be prevented in the honor of some of your rarest inventions and speculations. Let your countrie and frends injoye the comforts they would have in the true and greate honor you would purchase your selfe by publishing some of your choise workes.[4]

But Lower's lament was not quite the same as Arago's principle of the historical excommunication of writers without influence. Of course if a man were to have spent his life in a hermit's cell drafting political constitutions for his own amusement, his prospects of inclusion in a constitutional history would be quite properly slender. But even taking an extreme view of Harriot's isolationism, and supposing that he conceived of mathematics and astronomy as wholly personal intellectual enterprises, he was neither inconsistent nor foolish, unless it is foolish not to beg for admiration. Galileo wished to know, and also to be known. Harriot seems to have wished only to know. His surviving writings are far less extensive and discursive than Galileo's, and had they been otherwise they would not still be awaiting publication.

When we look to Harriot's solar observations, we find records made on two hundred or so days, with very few personal asides, a few rough calculations, and nothing more. Galileo, on the other hand, was by this time exposed to the public gaze. He gave glory to senates and to princes, in the expectation of rewards, and it was important that he not only reveal new truths, but also master the art of establishing priority in his discoveries. When in 1613 he published his 'history and demonstration relating to sunspots and their phenomena',[5] it was in the form of three letters to Mark Welser, a leading citizen of Augsburg, in which an attempt was made to refute a rival claim to the discovery. But the letters were much more besides: Welser, who had prompted them by his having published three letters on sunspots from Christoph Scheiner (under the long

3. *Oeuvres de François Arago: Notices Biographiques*, vol. iii, 1855, p. 271 sqq.
4. From the first part (now lost) of a letter of which the remainder is British Museum Add. MS 6789, ff. 427–8. See Rigaud (cited n. 11 below) at p. 43 where the missing part is printed from von Zach.
5. *Istoria e dimonstrazioni intorno alle macchie solari e loro accidenti*, Giacomo Mascardi, Rome, 1613. Translated in S. Drake, *Discoveries and Opinions of Galileo*, Doubleday, New York, 1957. Galileo's letters are dated 4 May, 1612, 14 August 1612, and 1 December 1612.

pseudonym 'Apelles latens post tabulam'),[6] had obliged Galileo to reply to a question about the nature of the spots. Scheiner was a Jesuit, and Welser a banker to the Jesuits. With great intuition and tact, Galileo in his replies helped to erode opposition to Copernicanism, and to the notion of an immaculate and lucid Sun, although it is doubtful whether Galileo's letters much affected Scheiner himself. In the course of the polemic, and with Welser's help, we may as it were interrogate Galileo closely. There is no such possibility with Harriot, for whom we must piece together what fragmentary evidence survives, and hope that something significant emerges.

THE DEBATE ON PRIORITY OF DISCOVERY
This is a tedious problem, but not entirely barren, historically speaking. It is now well known that spots were seen on the face of the Sun long before the advent of the telescope. Chinese observations go back at least to 28 BC,[7] and Humboldt has collected together a very useful list of references to sunspots in European and Islamic annals,[8] of which perhaps the most interesting are those in the *Annales Regi Francorum* (for 807 AD), and in Einhard's life of Charlemagne (before 814 AD).[9] The former was interpreted at the time as a transit of the planet Mercury, lasting for what we know to be an impossible eight days. In the same century, the philosopher al-Kindī explained away similar observations as a transit of Venus (lasting for a nonsensical 91 days), and before 1199 Ibn Rushd used the Mercury explanation once again. Even Kepler, who had observed a sunspot by projecting the image of the Sun through a small hole into a darkened chamber during what he believed would be a transit of Mercury on 18 May 1607, was totally but quite naturally misled by his expectations into thinking that he was indeed seeing Mercury.[10]

There are clearly two vastly different sorts of connotation for sunspot

6. *Tres epistolae de maculis solaribus* (published 5 January 1612) was followed in September by *De maculis solaribus et stellis circa Iovem errantibus accuratior disputatio*, both at Augsburg. The letters of Scheiner, Galileo, and of Welser to Galileo, are assembled in vol. V. of Favaro's edition of Galileo's *Opere*.
7. Two early papers are Alexander Hosie, 'Sunspots and Sun shadows observed in China BC 28—AD 1617', *Journal of the North China Branch, Royal Asiatic Society*, xii (1878), pp. 91–5; Joseph de Moidrey, 'Observations anciennes de taches solaires en Chine', *Bulletin Astronomique*, xxi (1904) 11 pp. The observations were often systematic, lasting for as much as ten consecutive days, and the spots were described in size and shape. Much further information and further references may be found in J. Needham and Wang Ling, *Science and Civilization in China*, vol. iii, Cambridge, 1959, pp. 434–6.
8. *Kosmos*, vol. iii, Stuttgart and Tubingen, 1850, pp. 412–6. See also the translation by E. C. Otté and B. H. Paul, *Cosmos*, vol. iv, Bohn's Scientific Library, 1852, pp. 379–86. This may be supplemented by G. Sarton, 'Early observations of the sunspots', *Isis*, vol. xxxvii (1947), pp. 69–71. See also B. R. Goldstein, 'Some medieval reports of Venus and Mercury transits', *Centaurus*, xiv (1969), pp. 49–59.
9. For Galileo's references to one of these early chronicles, see *Opere* (ed. Nazionale) vol. v, p. 138, and cf. his letters to Welser, op. cit., p. 117. Kepler had drawn attention to chronicles, and had actually 'restored' the text to its true form, in the light of his interpretation. See *Ad Vitellionem Paralipomena*, Frankfurt, 1604.
10. *Phaenomenon singulare Seu Mercurius in sole*, Leipzig, 1609.

records. Looking back with the advantage of our present knowledge that a sunspot is a relatively cool area forming a slight depression in the photo-sphere of the Sun, one might say that sunspots were not truly found until, at the very least, a proof was offered of their solar nature. Curiously enough, Stephen Rigaud took an almost equally severe historical position, and consequently refused to entertain the idea that with his earlier observation Harriot had discovered anything. He had not, said Rigaud, obtained 'more than a transient view of a phenomenon, which had been seen before his time, even with the naked eye', adding that it was 'a misapplication of terms to call such an observation a discovery'.[11] Speaking of Scheiner, Delambre said, not that he made no true discovery, but that if he had a telescope then it followed that he could see the spots: 'il n'y a pas grand mérite à cela'.[12] As Delambre pointed out, Scheiner was less happy in his conjectures than Galileo. He was an inferior geometer, and a less practised astronomer.

This was all part of a very proper reaction against the tactics used by the astronomers and their factions in the original dispute over priority, in which rhetoric had meant more than logic. The dispute had its repercussions on very many nineteenth century writers whom it would be pointless to list: it should be sufficient to say that in imposing some sort of criterion of historical validity which was not ostensibly designed to favour a chosen party, Delambre and Rigaud were untypical of their time. Nevertheless, with Rigaud's argument begins the slippery path to a very corrosive paradox of historiography. Why did he not go on to deny Galileo's discovery of sunspots, on the grounds that our knowledge of them suggests that they are different from the objects known to Galileo? If the historian defines his conceptual terms once and for all in order to escape this paradox, he runs the risk of being accused of not playing the priorities game impartially. Alternatively, should he emphasize that he writes of a conceptual world which is in a state of flux, he will be accused of writing with a grey historical uniformity. But a continuum is not necessarily homogeneous. Harriot left no indication of his attitudes to the physical nature of sunspots, let alone to their significance for theology and Aristotelianism, a matter which was uppermost in Galileo's thoughts. We have no explicit indication even of their supposed location. But slender as is the evidence, Harriot's observations show him to have been a remarkably good astronomer, the peer of Galileo in matters of observation, and superior in many small respects to Scheiner, Fabricius, and the rest.

Harriot recorded what we must suppose was his first observation of sunspots as follows:[13]

11. *Miscellaneous Works and Correspondence of the Rev. James Bradley*, Oxford, 1832, to which was added *Supplement . . . with an account of Harriot's astronomical papers*, Oxford, 1833 (separately paginated), of which see pp. 17–70 on Harriot and pp. 37–8 on sunspots.
12. *Histoire de l'astronomie moderne*, Paris, 1821, p. 631.
13. All references are to MS Leconfield 241, volume 8 as now bound. Speaking generally, folio references are unnecessary insofar as observations are numbered.

1610 Syon
Decemb 8^d mane ♄ [= Saturday]. The altitude of the Sonne being 7 or 8 degrees. It being a frost & a mist. I saw the Sonne in this manner. Instrument. $\frac{10}{1}$.B. I saw it twise or thrise. once with the right ey & other time with the left. In the space of a minutes time. After the Sonne was to cleare.

(See Plate 4.) Little more than a month later, he found the Sun to be clear:

$\frac{1610}{1611}$ Syon

Jannuary. 19.♄ [Saturday] a notable mist. I obserued diligently at sondry times when it was fit. I saw nothing but the cleare Sonne both with right and left ey.

If we are to judge by his surviving papers—and I am not at all sure that we should— nearly a year went by without further observation. On 1 December 1611, in the company of Sir William Lower and Christopher Tooke, he began a systematic study, numbering his observations from this time.

A number of objections may be raised at once to Rigaud's argument that Harriot was at first merely making simple observations of a phenomenon not as yet understood. No astronomer fully appreciated sunspots from the moment he first saw them. The telescopic observer had the inestimable advantage of being able to study carefully sunspot shapes—and hence avoid the planetary interpretation which no less an astronomer than Kepler had previously offered. In Harriot's very first drawings he gave the spots a hard and irregular outline, and on the whole he is unlikely to have thought them planetary or nebulous in character.[14] His second numbered record, although admittedly by this time he must have heard of the speculations of others, refers to the phenomena unambiguously as 'spots in the Sonne', and the appellation never changed.

Our second objection is that Harriot's pages, as they survive, are almost certainly fair copies. We know the appearance of his rough observing notes, from two leaves in the volume of his lunar observations.[15] Are we to suppose that his curiosity during the period from December 1610 to December 1611 never got the better of him? We know that he looked in vain for the spots on 19 January. Are we to suppose that he let matters rest there? He will be shown to have performed a calculation of the sunspot rotation period spanning the interval 8 December 1610 and 28 November 1611, both of which dates are outside the period of the *numbered* observations. No record of any observations of 28 November or of any other date in the interval survive. Scheiner's records subsequently printed and thus available to Harriot,[16] include an observation of 28 November 1611, and we may imagine that Harriot was making use of this in his calculation. But how, without a chain of intervening observations,

---

14. Admittedly there were those who could reconcile an irregular outline with the notion of a stellar constitution. In a debate between a Dominican father and certain Jesuits at the Roman College in September 1612, the Jesuits said that the irregularity was due to the clustering of the stars. See Galileo, *Opere*, ed. Naz., vol. xi, p. 395.
15. MS Leconfield 241, vol. 9, rough draft of the initial leaf *De quadratura Lunae*.
16. See pp. 114-15 above.

did he conclude that there was a connexion between the states of the Sun on the two dates? I shall suggest a solution subsequently. On the whole it is hard to believe that on 1 December 1611 (a highly significant date, as will be explained) Harriot embarked on a systematic programme of recorded observation without due rehearsal.

Rigaud supposed that such of Harriot's sunspot observations as were made with understanding began only on 1 December 1611, and this is very surprising since Rigaud offered what is undoubtedly the true reason for Harriot's great concern over the sunspot configuration on that day.[17] Against the written record and the drawing for 1 December, there are—prominently displayed—the symbols for a conjunction of the Sun and Venus. Maginus in his *Ephemerides* had predicted a superior conjunction,[18] but interest was aroused in what would happen, since there were many who, subscribing to a Ptolemaic view, hoped to see a transit of the planet across the face of the Sun. None was seen, of course. Scheiner's second letter to Welser[19] correctly drew attention to the opposite directions of the apparent motions of the sunspots and Venus, but Galileo took the wind out of Scheiner's sails by pointing out that his calculations proved nothing against those who maintained that Maginus had compiled faulty tables, or that Venus was too small to be seen against the Sun, or that Venus shone by her own light, or that the orbit of Venus was above the Sun. As A.-G. Pingré amusingly commented, 'Galilée étoit un Aristarque un peu trop difficulteux sur les découvertes et observations de ses contemporains'.[20] At all events, there is ample testimony to the general interest stimulated by this important prediction of the *Ephemerides* of Maginus, and that this is why Harriot commenced a careful programme of recorded observations is easy enough to accept, without our supposing at the same time that he made none other in the previous year.

Harriot never published his findings. The first printed account of telescopic sunspot observation was of those by the Frisian Johannes Fabricius, *De maculis in Sole observatis, etc.* (Wittenburg, 1611, dedicated the Ides of June, that is, 13 June). For Fabricius, the spots were certainly on the Sun's surface, whose rotation he observed. He remarked that he had been first led to examine the Sun when looking for an irregularity of the limb, of which his father had informed him by letter (folio C2 verso). Perhaps the first careful modern study of the work of Fabricius was made by Arago, who concluded that the observations date from March 1611.[21] An important document in the form of a rare book, *Prognostication* (1615), by David Fabricius, his father, was reprinted by Gerhard Berthold in 1897. There it is maintained that Johannes, then a student of medicine, in his father's presence first found the spots on

---

17. *Op. cit.*, p. 38.
18. 11 December 1611 (N.S.), $11^h\,40^m\,3^s$. This was in error by less than an hour.
19. See below.
20. *Annales célestes du dix-septième siècle,* ed. M. G. Bigourdan, Paris, 1901, p. 34.
21. *Annuaire du Bureau des Longitudes pour l'an 1842*, pp. 460–76.

27 February 1611, old style.[22] It is now known that Simon Mayr (Marius), in his own *Prognostication* for 1613, and again in his *Mundus Iovialis* (Nurnburg, 1614), corroborates the father's statement, adding in the second work that Fabricius's name had been suppressed on religious grounds.[23]

The East Frisian's book was soon joined by those of Welser's Jesuit correspondent, Scheiner. Although the letters were not printed until 1612, Scheiner long afterwards maintained in his *Rosa Ursina*[24] that he first saw the spots in March 1611,[25] and this confirmed a statement in his letter of 12 November 1611 that seven or eight months previously, with a friend, he had first seen the black smudges. Although Scheiner was to become the principal authority on sunspots during the later seventeenth century, his early writings are not especially good. His first letter shows—and Galileo was quick to point out the error—that he believed the spots to move in a contrary sense to that of the planetary circulation. He also doubted whether the spots adhered to the Sun. In the *Rosa Ursina* he tried to cover up his first error by maintaining that the Sun had a movement about two different axes. The first book was a polemic directed against Galileo. As for the remainder, Scheiner's examples date from 1625, and they are not here of much concern.[26]

Galileo apart, there were other early observers of sunspots. The painter Ludovico Cardi de Cigoli, friend of Galileo, wrote to him on 16 and 23 September 1611 that Cresti da Passigmani had been able to see as many as eight spots, in telescopic observations made at Venice.[27] In November Galileo was informed by Paulo Gualdo that sunspots were under observation in Germany.[28] From this time, over and above the Welser correspondence, references to the subject are frequent. As for interpretations, they span a wide range of specious ingenuity. Jean Tarde, for example, a canon of Sarlat in France, who in visiting Galileo had been irritated by his reticence over the principles underlying the telescope, wrote a treatise (*Borbonia sidera, etc.*, 1620) ridiculing Galileo's views, and contending that the Sun, 'eye of the world', was incapable of experiencing ophthalmia—and hence that sunspots were not a part of the Sun, but planetary. Claimed for the house of Bourbon by Tarde, as his title implied, there was later an attempted *coup* on the part of

22. That is, 9 March 1611, New style. See Berthold's *David Fabricius und Johann Kepler vom neuen Stern*, Norden, 1897; see also *Der Magister Johann Fabricius und die Sonnenflecken*, Leipzig, 1894, pp. 29–38.

23. See Joseph Klug, 'Simon Marius aus Gunzenhausen und Galileo Galilei', *Bayerische Akademie der Wissenschaften, Abhandlungen* (Math.–Phys. K1.) xxxix (1916), pp. 367–8; 403–12, 443–52; 498–503.

24. Printing was begun in 1626 and completed 1630.

25. *Rosa Ursina*, p. 28. It is generally accepted that 'mense Martio anni 1610' on the same page refers to 1610/1611, or is a misprint. The date March 1611 occurs again on the second (unnumbered) page. The title of the book refers to the house of Orsini, the 'rose' being an allusion to the Sun.

26. Scheiner was perhaps the first to notice, and to name, the *faculae*, or 'little torches', the bright areas associated with sunspots (*maculae*).

27. For the letters and replies, see Galileo, *Opere*, ed. Naz., vol. xi, pp. 208–14.

28. *Ibid.*, p. 230.

the Belgian Jesuit Charles Malapert, who wished to have the supposed planets circulate to the glory of the Hapsburgs (*Austriaca sidera, etc.*, 1633).[29]

The evident desire to flatter a potential patron is only one of many reasons for the bitterness of the controversies over priority at this time. Nationalism and a desire for personal fame were equally important motives, the former tending to outlive the latter, and religious dissension cannot be ignored. It has been said that the charge of heresy brought against Galileo, which resulted in his trial and confinement, was in part the result of a deposition made by Scheiner to the Inquisition in Rome.[30] Yet another whose claims to priority have been pressed, this time by Ernst Zinner, was Johannes Baptista Cysat, a friend of whom Scheiner spoke in his letter to Welser of 12 November 1611.[31] The evidence is in the form of a record of a lost manuscript, in which first Cysat and then Scheiner are reported as having seen the spots on a cloudy morning of 6 March 1611.

If the sunspot debate over priorities was acrimonious, this was undoubtedly because large doctrinal issues were at stake, but also because the great Galileo was involved, and Galileo chose quite deliberately to try to capture yet another crown of laurels to add to his collection. His published letters to Welser on sunspots served a double purpose, first as propaganda for the Copernican theory, which he hoped in due course to be compatible with Church dogma, and second, as a defence of his scientific eminence, in the face of what he conceived to be a personal attack by Scheiner. The force of Galileo's scientific arguments in these letters left little to be desired. He argued that by their random formation and inconstancy the sunspots were more like flat and thin clouds than planets, but that they were, if not on the Sun, at least within its atmosphere. He pointed out that the duration might vary from one or two days to thirty or forty, that a monthly rotation might bring a spot back into view, fore-shortening of the shape of the spot occuring near the limb. He explained that the apparent daily motion of each spot diminished as the limb was approached, and that motion was a correct function of angular distance from the solar equator. He saw that the spots favoured a central zone of 28° or 29° width, and he maintained what must have seemed utterly paradoxical to his contemporaries, namely that the intrinsic brightness of the spots was at least that of the brightest parts of the Moon. In all this he proved himself a master of his craft. In dating his first discovery, however, the craft was of another sort.

29. It is amusing to consider the delicacy with which Galileo treated the naming of Jupiter's satellites. He decided at length in favour of 'Medicean' rather than 'Cosmic' that is, in favour of the Grand Duke Cosmo's family rather than simply Cosmo himself, and it was necessary to paste over a correction to the half-title of the first Venice edition of *Sidereus Nuncius* (apud Thomasi Baglionum, 1610), which is uncorrected in the Frankfurt edition of the same year.
30. See Delambre, *Histoire de l'astronomie moderne*, Paris, 1821, p. 681.
31. See Galileo's *Opere*, ed. Naz., vol. v., p. 25. Zinner's arguments were set out in appendix, H (pp. 494–6) of his *Enstehung und Ausbreitung der koppernicanischen Lehre*, Erlangen 1943, Zinner has since repeated his claims elsewhere.

Dame has set forth Galileo's three principal statements on the matter, noting that if we follow the evidence of the *Dialogo sopra i due massimi sistemi* (1632; Imprimatur 1630), Salviati, on the third day, is made to claim Galileo as their discoverer, in July or August 1610, while he was still at Padua:[32]

The original discoverer and observer of the solar spots (as indeed of all the other novelties in the skies) was our Lincean Academician; he discovered them in 1610, while he was still lecturer in mathematics at the University of Padua. He spoke about them to many people here in Venice, some of whom are yet living, and a year later he showed them to many gentlemen at Rome . . .

If, on the other hand, we are to take seriously a letter from Galileo to Maffeo Barberini of 2 June 1612—nearer the occasion by almost twenty years—we conclude in favour of December 1610, since it was then 'about 18 months ago' that, in looking at the setting Sun, he 'perceived several dark spots . . .'. In December he was in Florence. Or again, we might take a letter to Guiliano de'Medici of the 23rd of that same month of June, which places the first observations fifteen months earlier (March 1611), when he is known to have been in Rome. Mention of the subsequent occasion on which he showed the sunspots to a number of important personages in the gardens of the Quirinal is repeated, and attested from other sources.[33]

In a court of law, Galileo's opponents might well have said of the date 'E pur si muove'. They might even have recalled the disparate statements Galileo made in connexion with the re-invention of the telescope, for as Edward Rosen has indicated in an excellent summary of the evidence,[34] Galileo first heard of the Dutch telescopes in or around the end of June 1609 (according to a letter to his brother-in-law),[35] made his first telescope six days before going to Venice, apparently in mid-July,[36] and yet in *Il Saggiatore* (1623) maintained that the interval between hearing of the telescope and making one was no more than a day. In *Sidereus nuncius* (1610), however, he explains how he first heard about the telescope, how the rumour was confirmed a few days later, and how he then gave his attention to the problem, which he solved after a little while (*'paulo post'*). As Rosen very aptly remarked, by the ordinary rules of evidence the early version is preferable to the later.[37]

To complete this Protean chronology, Galileo was in Venice for 'more than a month' showing an improved telescope to the Venetians before presenting it to the Venetian Senate on 25 August 1609, a date confirmed by official docu-

32. Dame, *op. cit.*, pp. 192–4. The quotation which follows is taken from the translation of the *Dialogo* by Stillman Drake, *Dialogue concerning the Two Chief World Systems*, Berkeley and Los Angeles, 1962, p. 345.
33. See Dame, *op. cit.*, p. 194.
34. 'When did Galileo make his first telescope?' *Centaurus*, ii (1951), pp. 44–51.
35. This is reconcilable with an apparently inconsistent statement of *Sidereus Nuncius* (1610). See Rosen, *op. cit.*, pp. 44–5.
36. He arrived there before 25 July. The date depends on the interpretation of the phrase 'more than a month' in *Il Saggiatore*. See Rosen, *op. cit.*, p. 49.
37. *Ibid.*, p. 47.

ments. Accepting with Rosen that 'more than a month' might mean six weeks, we should have Galileo's first demonstration of the (improved) telescope around 10 July, and his first attempt early in July. These dates are all, of course, in the Gregorian calendar. Harriot's first recorded telescopic observation, that in which he made what is the first known telescopic drawing of the lunar surface, is in that calendar to be dated 5 August 1609.[38]

HARRIOT AND GALILEO

Galileo's fame apart, his writings are of such importance to the history of early seventeenth-century cosmology that Harriot's knowledge of them is a matter of great interest. Harriot's earliest known telescopic observation of the Moon, of 26 July 1609 (O.S.) 9 p.m., was made almost three weeks before Galileo presented his telescope to the Venetian Senate, and Harriot must then have been entirely ignorant of Galileo's telescopic activities.[39]

Galileo first saw three of Jupiter's satellites on 7 December 1609 (N.S.). His last observation of the 64 to be recorded in *Sidereus nuncius* is dated March 1610, and the book was dedicated on 12 March and published on 13 March of that year.[40] This book had a number of undated Moon drawings, which in the Frankfurt printing of 1610 were rendered by crude wood-blocks, but in the Venice edition are printed from fine copper plates comparable in quality with Harriot's later drawings. Harriot had no need of the Italian Moon drawings, but we do know that he made ample use of the Jupiter observations, which he himself supplemented by a series commencing 17 October 1610 (O.S.).[41] A letter from William Lower of 11 June 1610 (O.S.)[42] acknowledges a letter from Harriot in which he reports on Galileo's findings, and asks for a copy of the book, meaning *Sidereus nuncius*, of which Harriot certainly then knew and probably possessed. Sir Christopher Heydon had read it by 6 July 1610, as Rigaud has pointed out,[43] and Harriot had probably received a copy at much the same time. Jupiter was then too close to the Sun, conjunction having occurred on 1 June (O.S.), but by 17 October the two were separated by almost exactly 100°.

38. Harriot dates it 26 July 1609.
39. Any doubt which may be thought to attach to the year should be dispelled by the statement that the Moon at the time of observation was five days old. It may be verified that this was so in 1609, but not in 1610, on 26 July.
40. For a manuscript document containing a brief resumé of Galileo's observations of the satellites from 7 January to 15 January 1610 and comparable with the first leaf of his manuscript journal of early observations (*Opere*, ed. Naz., vol. iii. pt. 2, p. 427), see Stillman Drake's 'Galileo Gleanings'—XIII. An unpublished fragment relating to the telescope and the Medicean stars', *Physis*, anno IV (1962), pp. 342–4.
41. Harriot's first observation is explicitly headed 'My first observation of the new planets'. It is copied out again subsequently with a similar heading, as the first of a numbered series of Jupiter observations.
42. British Museum Add. MS 6789, ff. 425–6. See Jean Jacquot, 'Thomas Harriot's reputation for impiety', *Notes and Records of the Royal Society*, xi (1952), pp. 164–87. The letter is dated 'Traventi, the longest day of 1610', and Jacquot's '21 June' is the date in the Gregorian calendar.
43. *Op. cit.*, p. 27.

Harriot's use of Galileo's observations of Jupiter's satellites is manifest in several of his calculations, and in particular in his calendar conversions between the two styles when working out the period of first satellite partly on the strength of what are obviously Galileo's records from *Siderius nuncius*.[44] In its way equally good evidence of the esteem in which he held the Italian astronomer is provided by Harriot's repeated attempts to rearrange the sentence

> Salve umbistineum geminatum Martia prolis.

That this is what Harriot was doing on the verso of a leaf in MS Leconfield 241 (containing on the recto the first rough notes on the observations of the satellites) is obvious from the fact that he has written down an alphabet showing the frequencies of occurrence of different letters, following them by a number of rearrangements in awkward Latin. Thus the first three are:

> Mi tantum Jupiter laus gloria summe brans me,
> Montibus et silva variis martem mage plenum,
> Ignem lunarem in sat. paruum et mobilem visitas.

All of these would be acceptable rearrangements of the original if we were to drop the 'in' from the last sentence. The three arrangements are in a marginal note ascribed to 'M. Thorp', presumably Nathaniel Torporley, while another follows, ascribed to M. W., probably Mr. Warner.[45] The fourth attempt mentions Mars, but then, beginning with the words 'Galilaei Prosopopaea ludibunda', the whole tone of the exercise is lowered with a couple of interpretations, obscene, but in more convincing Latin, the first of them ascribed to an author whose initials are now erased. And so Harriot proceeds down the page with what seem to be his own suggestions, to resume yet again in the same section (page 30, Rigaud's numbering), while scattered through the volumes of his papers in the British Museum there are at least fifty further attempts, all presumably his own, to solve the anagram.[46] These, in fact, are all attempts to solve the anagram which Galileo had communicated to Kepler:

> s/mais/mrm/il/m/epoe/taleum/ibun/enugttauiras.

Kepler's solution was that from which Harriot apparently began ('Salve umbistineum . . .'), a solution which shows that Kepler thought Galileo to have found two satellites to Mars. The correct solution was

> Altissimum planetam tergeminum observavi
> (I have observed the most distant planet to be in three parts),

44. Baron von Zach's very foolish inference from Harriot's use of the date 26 January 1610 to the conclusion that Harriot's first observation was made at that time is disproved by Rigaud, *op. cit.*, pp. 64–7. There is even a reference to Venice, to which Rigaud could have drawn attention, at the side of a calculation of the correction needed for longitude difference of Venice and Syon.
45. I owe the interpretation of these ascriptions to Professor Shirley.
46. Add. MSS 6788, f. 251 (six, with some pencilled attempts); f. 252 (a score or so, with further variants), and 6789, ff. 455–9 (twenty-one in pencil and two in ink).

an allusion to the form of Saturn, which would in due course be seen surrounded by rings, but which to Galileo then seemed like a sphere with two
handles. Kepler published the challenge and his solution in his *Dissertatio
cum nuncio sidereo nuper apud mortales misso a Galilaeo Galilaeo*, published in
Prague, 1610.[47] 'Harriot was obviously working from that source. The date is
uncertain, but it could well have been before 14 December 1610, the date of
observation no. 13. Against his observation of 11 December 1610 we can
perhaps see traces of a sense of humour, if not of a contemplated crypto-
graphic rejoinder:

> Thomas Hariotus
> oho trahit musas
> oho trahis mutas
> oho sum charitas

The self-same phrases are written down on a sheet now among the papers in
the British Museum,[48] and a further nine rearrangements, with variants, are
to be found in the same volume.[49] It is possible that Harriot's 'oho' was in
some way analogous to Galileo's representation of Saturn's telescopic appearance as 'oOo', but if so the Harriot anagram must have been later than
his attempted solutions of Galileo's.

Kepler had owned a copy of *Sidereus nuncius* from 8 April 1610, when one
was sent to him by Galileo through the Tuscan ambassador, with an invitation
to comment on its claims. (He had actually seen the Emperor's copy previously.) His *Dissertatio* was written by 19 April, and was printed in an expanded version by the 3rd May 1610.[50] Knowing that Kepler was in correspondence with Harriot, it is not at all unlikely that Harriot read Kepler's reply
before, or at least simultaneously with, Galileo's original. It is worth observing
that Harriot's efforts at a solution, starting from Kepler's sentence, are on the
reverse of the sheet containing his first rough satellite records—or, conversely,
as the case may be. Harriot has actually at one point added a precise page
reference to Kepler's *Dissertatio*, with the date 6 October 1610.[51]

It has already been made plain that Harriot's first observation of sunspots
owed nothing to Galileo, and even by the end of 1611, knowledge as to the
several observations made earlier in the year does not appear to have been
very rapidly diffused, judging by the apparent self-righteousness of the parties
in the subsequent priority debate. Harriot is not known to have benefited at
any stage, for example, from the solar image-projection technique of European astronomers—advocated in one form or another by Fabricius, Scheiner,

47. Translated by E. Rosen, *Kepler's conversation with Galileo's Sidereal Messenger*, New
York & London, 1965.
48. Add. MS 6789, f. 455r.
49. *Ibid.*, f. 475. It is not unlikely that further examples of this and of the Galilean anagrams
will be found elsewhere in Harriot's papers, which I do not claim to have searched systematically.
50. 5 nonas Maias.
51. Mentioned by Rigaud, *op. cit.*, p. 27.

and Galileo, and already used by Kepler in his unfortunate solar observations of 1607.[52] This does not, of course, mean that Harriot was ignorant of the possibility of producing a solar image by projection, either with or without a lens. Among his papers in the British Museum are some in which he not only considers at great length the mathematics of image formation by a section of a sphere (or, as would now be said, by a thin lens), but actually calculates the image size for a particular lens, taking the solar semi-diameter as 16′.[53] Elsewhere we find related diagrams, some of doubtful significance, but one showing solar image formation on a screen by a hole (*foramen*).[54]

HARRIOT'S METHODS OF OBSERVATION

Harriot has left us two hundred drawings of the solar surface, all but the first in a numbered sequence. Taking into account the fact that he often observed several times in the course of a day, and occasionally omitted to make any drawing for a given day (especially when he was away from Syon) we find that he has left notes of 450 separate observations. Harriot above all else was at the mercy of the weather, which provided him with the means of diminishing the intensity of the Sun's light. He first saw the sunspots 'it being a frost and a mist, (8 December 1610), and on the crucial day, 1 December 1611, he 'observed for half an houre space at which time and all the morning before it was misty'. Clouds were occasionally helpful, but more often not: 'I saw [a cluster of four] twise in a short time through a thinne rag of a cloud, the cloudes quickly obscuring the Sunne' (4 December 1611). By 12 December there was snow, which did not appear to matter much, and on 13 December there was mention of 'thick ayer', a phrase which was subsequently to be often favoured, and of 'coloured glasses', to which we shall later return.

It would be difficult to summarize Harriot's statements of meteorological conditions did he not at length settle for a small number of basic phrases. 'Thick ayer' (by which I understand distant or slight water vapour or dust) might alone be enough, although it might at times leave 'the Sonne . . . somewhat to cleare' (14 December 1611). 'Thin cloudes', often qualified as white, as misty, or occasionally as fast moving, played their part, and mist alone (that is, mist near at hand) often sufficed. Harriot may be said to have recorded observations as made through thin cloud on 114 days, through thick air on 70 days, through mist on 60 days, and under conditions when the Sun was too bright to make observing easy on 22 days. These figures underscore not so much the folly of direct solar observation as the advantages of the Thames valley climate for reducing the light of the Sun to what Harriot called a 'tem-

52. See n. 8. Kepler, and at first Fabricius and Scheiner, projected without lenses. It was Benedetto Castelli, Galileo's pupil, who first suggested to him the use of a telescope for projections.
53. Add. MS 6789, ff. 266–238. See especially f. 230r. On f. 239 he considers a related subject, that of burning glasses.
54. Add. MS 6787, f. 244v. Some of the drawings on the recto are also relevant, but not all are easy to interpret.

perate light'. The Thames vapours were as always a mixed blessing, however, for against the Moon map of 11 September 1610 we find:

I could not set downe the figure of all, nether this but by memory because I was troubled with the reume.

Four days later he was constrained to add to another lunar drawing simply 'The rume'. Harriot generally observed soon after sunrise or soon before sunset, although he was prepared to make his drawing for any hour of the day when conditions were favourable. Occasionally he had only a short time to make his drawing. 'I saw the Sonne thorough misty cloudes for the space of a minute', he wrote on 26 December 1611; and on 25 April 1612, 'the Sonne tarrying but a little while, I set downe there places & number but by coniecture'. On three occasions in the following June he was likewise obliged to conjecture positions, and the fact that he made such notes as 'the distance of the 2 most westerly may be somewhat a misse' (19 August 1612) suggests that he had a certain confidence in the general correctness of his work. That the surviving records are a fair copy of lost originals seems clear enough, both on account of the few alterations needed, and the fact that in the three successive entries for 5, 6, and 8 May 1612 he at first wrote 'April', correcting the mistake later.[55]

Harriot was occasionally careless with his sight. Under 'fayre' conditions, with some mist, he was able to see spots on one occasion with the naked eye: 'Two of the greatest were well seen in the mist (of the cluster [of 10 spots] without instrument)' (23 April 1612). On 1 March 1612 he had looked at the Sun without clouds with his telescope of magnification 20, and twelve days earlier he wrote:

At 12^h all the sky being cleare and the Sonne I saw the great spot with $\frac{10}{1}$ & $\frac{20}{1}$ but no more. My sight was after dim for an houre.

There are several other instances of his having looked into a clear Sun. The hazards of this sort of work are obvious, and the blindness of Galileo and Cassini is often mentioned by way of warning. Rigaud[56] quoted John Greaves, who for some days after measuring the diameter of the Sun thought he saw 'a company of crows flying in the air at a good distance'. Such 'spots before the eyes' surely explain why Harriot was in his first observation anxious to point out that he truly saw what he drew, and with *both* eyes. Rigaud, who adopted an excessive scepticism over the significance of this first observation, has it that Harriot had 'unexpectedly seen what he could hardly persuade himself to be real.[57]

---

55. Drawing no. 146 is repeated on the verso, unnumbered, and this—a note to which mentions much thunder in the west—might be an original sketch. As already mentioned, we have crayon records of some of Harriot's night observations, on the quadrature of the Moon, for example. Against the lunar observations for 23 October [1610] Harriot has added the note 'Looke the papers of obser[vation].'
56. *Op. cit.*, p. 33.
57. *Ibid.*, p. 37.

At only one point does Harriot mention the use of coloured glasses, to diminish the solar brightness (13 December 1611):

I observed thorough the thick ayer and also through my coloured glasses but saw no more: it seemed to me that a greate glasse wold make them appeare devided or more different.

It is natural enough to assume that he here refers to flat pieces of coloured glass, such as were already used by Dutch navigators in taking solar altitudes.[58] Apian had long before recommended coloured glass for eclipse observation in his *Astronomicum Caesareum*.[59] Scheiner, in his letters to Welser, had four different ways of viewing the sunspots, one being to use in addition to his telescope a blue or green plane glass of appropriate thickness, or a thin blue glass alone when the Sun was covered with vapour or thin cloud. Is this the sort of thing Harriot meant? The only reason for doubt is not very conclusive. 'Glass' in Harriot's day commonly meant lens, mirror, or telescope, and in its second occurrence in the quotation the word was certainly used to refer to a telescope. A 'great glasse' capable of improving the resolution of the image is unlikely to have been a thicker slab of coloured glass. It is conceivable that Harriot had telescopes with lenses of coloured glass, but not perhaps very probable, since he never refers to coloured glasses again, and since, as we saw, he hurt his eyes in the following April.

HARRIOT'S INSTRUMENTS

'Glass' was not the word commonly used to designate the telescope in Harriot's writings, when 'instrument' has a clear lead over 'trunk'. The earliest Moon drawing (26 July 1609) has merely a marginal note '$\frac{6}{1}$', meaning that a telescope with a linear (effectively angular) magnification of 6 was used. This instrument was never again mentioned. As for the remaining instruments, we may summarize chronologically the evidence of the entire contents of MS Leconfield 241 as follows:

$\frac{10}{1}$ and $\frac{10}{1}$B: An instrument of this magnification was Harriot's joint favourite, judging by the number of times it was mentioned (thirty-three at least, and perhaps on several occasions by implication). First mentioned 17 July 1610 (Moon). 'Instrument $\frac{10}{1}$B' occurs first on 8 December 1610 (first sunspots record), once again on 11 January 1612 (sunspots) and once by implication in a detailed lunar observation of 9 April 1611, where this note occurs:[60]

58. Delambre, *op. cit.*, p. 681.
59. Ingolstadt, 1540. Tycho later preferred a small pin-hole in a sheet of paper, perhaps because optically sound glass was difficult or impossible to obtain, especially coloured. Robert Hooke is quoted by Rigaud (op. cit., p. 33) as disapproving of 'the coloured glasses' because they 'take off the truth of the appearance as to colour', and detract from the sharpness and distinctness of the image.
60. Here and elsewhere I expand 'Chr' to 'Christopher', and substitute a reasonably intelligible punctuation.

Sr Nicholas Sanders & Christopher were with me & also observed in my garret. Instruments. $\frac{10}{1}$.$\frac{15}{1}$.$\frac{32}{1}$.$\frac{11}{1}$.
          two one one one

Although instrument B was not mentioned again, it was probably often used whilst not distinguished from its partner.

$\frac{20}{1}$: Harriot's other favourite instrument, first mentioned 4 August 1610 (Moon), and on at least thirty-three other occasions, perhaps more implicitly. It was used, like, $\frac{10}{1}$ for all types of observation. On the whole, instruments were not specified in the Jupiter records. That the words 'trunk' and 'instrument' were synonymous is evident from their application to the same telescope, as in Jupiter observations for 14 December 1610 ('my truncks $\frac{10}{1}$.$\frac{20}{1}$') and 25 January 1611 ('my instrument of $\frac{20}{1}$'). Thus also under 3 December 1611 (sunspot record made at Syon):

Sr William Lower & Christopher saw them with me in several trunckes. $\frac{10}{1}$.$\frac{8}{1}$.$\frac{20}{1}$.

$\frac{11}{1}$: Mentioned on only one occasion, in a lunar context, noted above, and certainly then used at Syon (9 April 1611).

$\frac{32}{1}$: Mentioned twice, on 9 April 1611 as in the previous entry, and against a lunar map of the same date on another leaf.

$\frac{15}{1}$: Mentioned on 9 April 1611, as already noted, and again on 9 February 1612 in a sunspot record made in London, where it is possible that the instrument was now kept.

$\frac{8}{1}$: Mentioned only once, in the sunspot record of 3 December 1611 noted above.

$\frac{30}{1}$: The third most frequently mentioned instrument, although only five times noted. It does not appear until 20 January 1612 (sunspot record), unless we are to identify it with $\frac{32}{1}$, which is unlikely.

$\frac{50}{1}$: Mentioned once only, on 27 January 1612, in a sunspot record made at Syon: The southerly [spots] are fayer and great and with $\frac{50}{1}$ seemeth more. He [the Sun] was well seen with $\frac{10}{1}$ but some of the rest not at all, and some difficultely.

Speaking generally, Harriot may be said to have seldom troubled to record the instrument he used, in the sunspot records he made after June 1612. Occasionally he supplied from memory the details of the telescopes he had used—'*memoriter*'.[61]

The greater number of Harriot's observations were made at Syon House,[62] apparently from his garret.[63] Change of residence does not appear to have stemmed the flow of the observations. 'London' was the place of sunspot observations for 7–12 February 1612, and of some unnumbered ones for

61. 10, 12, 13, 15 September 1610, all lunar drawings.
62. About ten miles from the centre of London, and in Syon Park, which now runs along the Thames for about a mile. Across the river is the Old Deer Park containing Kew Observatory. of later fame, which is in fact only half a mile south of Harriot's place of observation.
63. See under instrument $\frac{10}{1}$ above.

12–25 February and 3–7 April 1612, during which interval he was observing from Syon, perhaps suggesting help from a friend. He was in London again between 9 June and 24 July 1612,[64] (a rather scrappy record made at Essex House being included for 25 June), and 6–26 November 1612. His second recorded observation of a satellite of Jupiter (16 November 1610) was 'London at Neales in Black Friers' and he was still in London for the next (19 November), although soon back at Syon, where on 7 December 1610 he twice recorded that Sir William Lower saw two satellites—for the first time, in fact, He made similar observations early in the next year (17 February) 'at D. Turner's house in Little St. Ellens', and on the circumstances of his London observations he is more precise when he four times states (Jupiter, ten observations numbered jointly 81, 25 April 1611 and after) that Christopher [Tooke] observed 'in a gutter', the words 'of a house' being once added.[65] Returning to Syon on 10 May 1611 it was 'we saw . . .', and the phrase is occasionally used in the sunspot records where no other person is specified by name, but where Christopher Tooke is almost certainly always intended, being often also named.[66]

We have no evidence of Harriot's having used a mounting with any of his telescopes. Some drawings with the scarcely legible caption:

in a payre of stone stayres 2 iron barres beare the Instrumentis[67]

in my opinion illustrate some sort of furnace for retorts or glass vessels in certain of the alchemical experiments which come later in the manuscript. There is another deceptive line, in the same section, dated 1601/1602:

New glasses luted with blackest lute.

The glasses were in no sense optical instruments, but chemical retorts, lute being a mud traditionally used to seal (that is, to lute) the same.

It is curious that Harriot does not use the word 'cylinder' for his telescopes, since this is the word used by Lower, replying to Harriot's letters:

I have received the perspective cylinder that you promised me and am sorrie, that my man gave you not more warning, that I might have had also the 2 or 3 more that you mentioned to chuse for me. Hence forward he shall have order to attend you better and to defray the charge of this an others, for he confesseth to me, that he forgot to pay the work man. (6 February 1610)[68]

---

64. Sunspot observations 113 (or 1, in a new numbering sequence) to 139 or 27. Some of these drawings are to a larger radius than usual.
65. Christopher is mentioned in nine of these London observations, which were probably left unnumbered because Harriot had no part in them. All subsequent satellite observation numbers are given in the form '85+9' to allow for the additional nine.
66. On 5, 8, 9, December 1611 it might have included Lower and Tooke, both named on 1 and 3 December.
67. MS Leconfield 241, vol. 12. I owe this difficult reading to Professor Shirley. According to O.E.D. 'pair of stairs' is an accepted usage, known from the sixteenth century, to denote a flight of stairs, or even a floor or storey.
68. Printed by Rigaud (from Baron von Zach), op. cit., p. 42. The original is now lost, but Rigaud identified a later part of the letter.

Lower goes on to describe the appearance of the Moon through what he once calls the 'instrument'. Four months later he writes:

We are here so on fire with thes things that I must render my request and your promise to send mee of all sortes of thes Cylinders. My man shal deliver you monie for anie charge requisite, and contente your man for his paines & skill. Send me so manie as you thinke needful vnto thes obseruations. (11 June 1610)[69]

Lower has in the same letter spoken of observing 'this last winter' with his own cylinder. Writing again on 4 March 1611, he mentions his cylinder, and remarks his inability to see Jupiter's satellites, adding

I impute it to the dullnesse of my sighte, for onlie with your greate glasse I could se them in London.[70]

The 'greate glasse' reminds us of Harriot's use of the phrase, which here apparently refers to the instrument of magnification 20.[71] It seems not unlikely that Lower was simply using a more polite equivalent to Harriot's word 'trunk', which might have referred to a tube of rectangular box-like construction, as a water conduit (*Shorter O.E.D.*, 'Trunk' II. 5), but might equally have had a circular section (ibid., I. 1, 4). Sir Christopher Heydon, in a letter mentioned above,[72] spoke in July 1610 of 'our ordinary trunks', which shows that the word had an early currency.

The word 'glass', as we have already seen, offers great scope for ambiguity. It has been continually used of a refracting telescope,[73] and of its lenses. That it could refer to a mirror is well known from the writings of a number of Harriot's near contemporaries, such as Leonard and Thomas Digges, John Dee, and William Bourne.[74] Harriot was probably using the word in this sense in his *Briefe and True Report* (1588):

Most thinges they [the Indians] sawe with vs, as Mathematicall instruments, sea compasses, the vertue of the loadstone in drawing yron, a perspective glasse whereby was shewed manie strange sightes, burning glasses, wildfire woorkes, gunnes, bookes, writing and reading, spring clocks that seeme to goe of themselues, and manie other thinges that wee had . . .[75]

The variety of English words for the telescope was a reflection of the uncertain terminology throughout Europe. In *Sidereus nuncius* Galileo used the words *organum* and *instrumentum*, but most commonly *perspicillum*. His

69. British Museum Add. MS 6789, f. 425v.
70. *Ibid.*, f. 429v.
71. See the Jupiter record of 7 December cited on p. 117.
72. See p. 118.
73. Robert Burton, for example, in the section of his *Anatomy* from which we quoted at the outset, has this footnote to the third edition (1628): 'Some of those [planets] about *Jupiter* I have seen my selfe by the help of a glasse 8 foot long'. This does not occur in the second edition (1624).
74. Francis R. Johnson, *Astronomical Thought in Renaissance England,* Baltimore, 1937. See esp. Chap. VI.
75. *A briefe and true report of the new found land of Virginia* . . . London, 1588, f. E4r.

favourite Italian word was *occhiale*. Kepler, in his *Dissertatio cum nuncio sidereo*, used *perspicillum* most often, but also *instrumentum*. Harriot, who after the first sunspot record—which was probably copied out a year later—used the word 'instrument' on 7 December 1610 (Jupiter records), and in the context of lunar drawings on 9 April 1611 when he used it in a Latin form (*Instrumento* $\frac{32}{1}$). In all probability he was following the usage of Galileo or Kepler. He apparently never used 'telescope', or its strict Latin equivalent.[76]

Very little is known of the optical performance of Harriot's telescopes, at least by comparison with those of Galileo's which survive. Of Harriot's we know their magnification, and in some instances their field of view. Rigaud was sceptical about the significance of Harriot's field measurements, but does not appear to have correctly understood them. Evaluating them carefully, and making one or two reasonable suppositions about the instruments, we may make a rough comparison with others of the time.

Harriot almost invariably stated the angular separations of Jupiter's satellites from the planet itself, and this usually to a minute of arc, but occasionally to a quarter minute (as for example, on 25 January 1611), or even less (as on 12 December 1610: 'one other westerly, as I thought 10″ or 15″ a [Jupiter]; hardly seen'). We do not know how he gauged these distances, but since, by the time they were recorded, Harriot was in possession of Galileo's work, it is reasonable to suppose that he used Galileo's method. This, which was given near the beginning of the first chapter, involved the observer in the use of a series of object-glass diaphragms of different sizes. Galileo is not very clear on the subject, but it seems that he calibrated his diaphragms empirically, calculating (in terms of its size and distance, using a table of sines) the angle subtended by an object which exactly filled the diameter of the field of view. He first recorded the separation of Jupiter's satellites on 12 January 1610, and subsequently gave 'approximate' separations to an implied accuracy of 20 seconds of arc. Harriot found the semidiameters of the orbits of four satellites to be 14′, 9′, 6′ and 3′ 30″,[77] and Galileo's figures would have been close to these. All these data were probably arrived at by a subjective estimate based on the supposed diameter of the body of Jupiter (taken by Harriot to be

---

76. Edward Rosen argues convincingly that the word *telescopium* was conceived by John Demisiani, and given currency after a banquet given by Cesi on 14 April 1611, at which Galileo was the guest of honour. A report of this occasion dated 16 April 1611 mentions Galileo as a mathematician who

> observed the motion of the stars with the *occhiali*, which he invented, or rather improved. Against the opinion of all the ancient philosophers, he declares that there are four more stars or planets, which are satellites of Jupiter and which he calls the Mediciean bodies, as well as two companions of Saturn.

See Rosen, *The Naming of the Telescope*, 1947, p. 31. Rosen quotes from a letter from Welser of 20 May 1611, also mentioning that Galileo showed to the guests the satellites of Jupiter, 'together with a number of other celestial marvels', *ibid.*, p. 35.

77. Galileo did not analyse his data at this time, but Simon Mayr quoted 13′, 8′, 5′ and 3′. See Rigaud, *op. cit.*, p. 30.

a little under 2′)[78] and the fraction of the supposed field of view occupied by the planets' separation. Both were wrongly estimated, the first no doubt partly by virtue of the indistinctness of the image produced by telescopes of the time; and the second because the conditions determining the field were not properly understood by Galileo and his assumed followers, nor even by Rigaud.

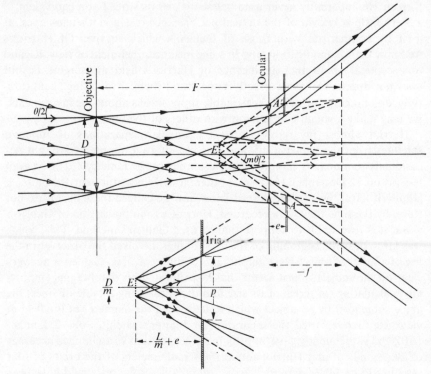

FIG. 1. A diagram of the early Dutch or 'Galilean' telescope.

The diameter of the field of view of a Dutch or 'Galilean' telescope (θ), illustrated in Fig. 1, is shown in Appendix I to be given in minutes of arc by the equation

$$\theta = \frac{10800.(p \pm D/m)}{\pi} \frac{}{(L+em)},$$

where the negative sign gives the clear field, and the positive sign the field with its peripheral dark ring (darkness increasing with radius) formed from light of partly obscured pencils. Here p is the diameter of the pupil of the eye,

---

78. Thus on 16 March 1611 he gives the distance of a satellite as 'about 5 diameters', alternatively '9′ or 10′ '. Earlier, on 28 November 1610, he put a satellite '1½′ a [Jove] or diameter Iovis a [Jove]'.

D is that of the objective aperture, or rather of any diaphragm covering it, *m* is the magnification, *L* is the distance between the centres of the two lenses, taken to be in normal adjustment for a distant object, and *e* is the optical distance of the eye pupil from the centre of the ocular. In writing of the 'field of view' I imagine that all our authors were including the dark peripheral field.

Galileo's method, as explained, took no cognizance of the importance of *p*, or for that matter, *e*. More recent treatises have taken no account of *D* or of *e*. All early observers, following Galileo, however, presumably made estimates of θ by daylight when *p* (for an observer looking into a telescope) is of the order of 5 mm, rather than at night ($p = 8$ mm), when measurements were made. Given a reasonably typical aperture (*D*) of 20 mm, and a magnification of 20, the ratio of an angle supposed correctly measured to the true angle would be 3:2.

We may calculate the true angles subtended at any time in the past in order to discover how close this explanation comes to accounting for the errors of Harriot and others. That the diameters of the orbits subtend different angles at different times follows from the changing geometric distance of Jupiter. On 1 January 1612, for example, Jupiter was 156° removed from the Sun, and at approximately 4·3 astronomical units from the Earth. For the sake of simplicity, consider the case of opposition, a few weeks later, when the distance was 4·202 units. The eccentricities of the orbits of satellites I-IV are all small, and the mean distances from Jupiter are (in astronomical units) 0·00282, 0·00449, 0·00716, and 0·01259. These orbits at Jupiter's opposition then subtended the angles 2'·31, 3'·68, 5'·87, and 10'·31 (approx.)—Increasing then in the ratio 3:2, to follow through our imaginary example, it appears that that erroneous estimates of the angles, using Galileo's method, would have been 3'·47, 5'·52, 8'·80. 15'·46. Comparing these with Harriot's 3'·5, 6', 9', and 14', the explanation offered here for his source of error must be admitted plausible. He was mistaken in adopting Galileo's method, and so, in exactly the same way, was Rigaud mistaken to imply that it was above reproach. Thus, for example, he wrote:

Harriot's quantities were more roughly taken. He gives no reason to think that he had made any previous experiments like Galileo, who determined the value of his apertures by the length which they included of a line at a given distance.[79]

Rigaud is misleading when he says that we have no reason for thinking that Harriot performed the requisite measurements. We know that at least in one instance (see (1) below) he had a preconceived idea of the field of one of his telescopes, and we also know from his papers in the British Museum that he made measurements to calculate the field of view of the human eye.[80] Rigaud goes on to criticize Harriot gratuitously for what he takes to be inconsistent statements as regards the field of view both of the $\frac{20}{1}$ instrument and $\frac{10}{1}$B.

79. *Op. cit.*, p. 24.
80. Add. MS 6786, f. 269r and v. Since he also read Galileo, it would be very surprising if he did not apply Galileo's very obvious method for calculating the field.

The former inconsistency is only apparent, and is very simply explained, for on one occasion the observations were of the Sun, which would have contracted the pupil, whilst the other occasion was by night.[81]

Harriot made four different statements relating to field of view, which we here collect together in order to extract from them what information we may:
(1) Instrument: $\frac{20}{1}$. Date: 7 December 1610. Jupiter record no. 8.

'Two seen on the west side, a little under. Sr W. Lower also saw them here. The nerest fayrest. The farther not well seen within the reach of my instrument of $\frac{20}{1}$ of 14' dyameter.'

The distances marked on the drawing are 7' and 12', −13' being added as an alternative. Compare the record (no. 40) for 25 January 1611, where the outermost satellite, likewise marked as 12' or 13', was 'scarce se[en] within my instrument of $\frac{20}{1}$'.
(2) Instrument: $\frac{10}{1}$B. Date: 11 January 1612. Sunspot record no. 26.

'One of the spots was in the verticall, the semidiameter of my instrument B$\frac{10}{1}$ from the upper edge and the whole diameter from the lower.'

A marginal note adds 'B 30' ', and Rigaud appears to take it that this indicates a field of view (for $\frac{10}{1}$B) of 30'.[82] He writes:

'... we find a notice in the margin, "B 30' ", which, if it is meant to describe the angular value of the field of view, will be another instance, in addition to those already adduced. ... of the little precision with which these estimates were made.'

But the record itself is unambiguous. The field was half as large again as the angular diameter of the Sun, which on the date in question was approximately 32'·5. The marginal note can only have referred to the supposed solar diameter, and the field supposed was thus 20'. Harriot's 'little precision' was not invariably as little as Rigaud's.
(3) Instrument: $\frac{20}{1}$. Date: 16 July 1612. Sunspot record no. 133 (London).

'With $\frac{20}{1}$ I see a little more than $\frac{1}{3}$ of the diameter of the Sonne, that is, 11' or 12'.'

Comparing (1) and (3), it appears that the field of $\frac{20}{1}$ was, to use Rigaud's words 'not very precisely determined'.[83] But only in (1) do Harriot's words suggest a *prior* determination of the field (14'). In (3) he is reporting what he sees, without any preconceived notion of what he *should* see. And what Rigaud overlooked is that, as already explained, Harriot would indeed have seen a smaller field when looking at the Sun than when looking at Jupiter, or at some terrestrial object for calibration purposes. Harriot would not have realized why, but where others would have evaded the apparent inconsistency by omission or distortion of the evidence, he did not, and to accuse him of im-

81. In a bright Sun the pupil might have been 2 mm in diameter, but with a hazy or clouded morning Sun between 3 and 6 mm, according to circumstance.
82. *Op. cit.*, p. 35.
83. *Op. cit.*, p. 24.

precision is to misjudge him. As an observer, the evidence is that he was at least Galileo's equal in the care he took, although perhaps working with inferior instruments[84] and certainly with an inferior night sky.

(4) Instrument:—? Date: 21 October 1612. Lunar observation of the moment of quadrature ('6 houres before at the least by Tycho Brahe').

'At that time the rounde vally (a) was $\frac{1}{3}$ from the lower poynt and $\frac{2}{3}$ from the higher. because *ab* was the diameter of mine instrument, and *ac* semidiameter'.

The lunar marking (a) was on the line of division of the dark and light halves of the Moon. Harriot had earlier said that the line of division was observed to be straight with his 'trunke of $\frac{20}{1}$', 'and if it was not, it rather lacked by some other judgements'. What we know of the field of $\frac{20}{1}$ does not accord with the present statement, which rather suggests $\frac{10}{1}$. We note that when he observed the same phenomenon in April of the previous year he mentioned five instruments, but attempted to judge the line of division by $\frac{10}{1}$.[85] (He did so without instruments on 11 January 1611.) Harriot is unlikely to have calculated or observed the precise angular diameter of the Moon for the day in question, but more probably took a round figure of 30'. The field of view was thus approximately 20', which is to be compared with the field found under heading (2) above for $\frac{10}{1}$B, rather than with any of the data for $\frac{20}{1}$.

After the details of a Jupiter observation of 24 May 1611 (no. 89+9), Harriot made an estimate of the angular separation of Jupiter and Mars, which was nearby: '[Mars] by our instrument was about 31' or 32' of from [Jupiter]'. He comments that they were 30' or 31' apart the night before, and that the emphemerides would have then been 16' or 17' apart. He does not say how he estimated the separations, which were greater than the field of either $\frac{10}{1}$ or $\frac{20}{1}$. It is possible that he used the $\frac{8}{1}$, or made up the distance in two or more steps, using intermediate stars as points of division.

Harriot indicated on several occasions that he was able to see solar detail more easily the larger the magnification used. This is a complex optical problem in the absence of a detailed knowledge of the lens systems used, since considerations of the angular resolving power of the telescope, flare, interference effects, and other aberrations of the image must all enter into any assessment of its distinctness. However, even with a clear aperture of a quarter of an inch, which is as small as any diaphragm Harriot is likely to have used, he should have been able to resolve two stars 18″ apart. The resolution of sunspots is a more difficult problem than that of stars, but the diffraction limits on resolution set by his instrument would not have limited the markings he was considering, when he wrote, for example, (20 January 1612, sunspot record no. 32):

Under the 3 [sunspots] one semed fayer & great & seen alone with the smaller in-

---

84. His circumstances were nevertheless clearly better than Lower's in Traventi. See p. 134 below.
85. See p. 123 above.

strument ($\frac{10}{1}$ but with $\frac{20}{1}$ $\frac{30}{1}$ one more westerly dim yet certayne, & an other smal one easterly & close by'.

Likewise on the following day he could see eight spots, but only seven with less than $\frac{20}{1}$. On 31 January he introduced his largest magnification:

'... the southerly are Fayer & great & with $\frac{50}{1}$ seemeth more. he [the Sun] was well seen with $\frac{10}{1}$ but some of the rest not at all, & some difficultely.'

On further occasions he remarks that a spot was seen 'only with $\frac{20}{1}$','only with $\frac{30}{1}$', or 'only with $\frac{20}{1} . \frac{30}{1}$', and on 25 May 1612, when he could see twelve spots with an unspecified instrument, 'with $\frac{10}{1}$ they seemed to be but 9'[86]. That he could see more with his instruments of greater magnification cannot be shown to have been a consequence of increasing aperture (and hence resolving power) with increasing magnification, although it does seem probable that there was some such connexion.

In order to attempt to derive further detail of Harriot's telescopes (focal lengths, overall length, and aperture) from the fragmentary information already found, we have in effect only three equations in the four variables F, f, L, and D, on the assumption that we know the field (θ), eye distance (e, arbitrarily taken to be 5 mm), and pupil size (p, assumed to be 5 mm, taking a reasonable estimation for a daytime assessment of field). The first equation is that written above, for θ. The second and third are

$$F+f = L,$$
and
$$F = -mf.$$

We obviously cannot solve these equations without making some further assumption, and our safest course is to suppose that the aperture was of the size to be expected of a spectacle lens. All the evidence suggests that the earliest objectives in telescopes of the Dutch type were made on the lathes of the makers of eye-glasses, with standard tools. We know from his will that Harriot possessed such tools. The maximum clear aperture is unlikely, therefore, to have been appreciably less than 20 mm, or much more than about 35 mm, with 25 mm as a likely figure. Taking the above simultaneous equations for each of the three cases, we find these solutions (all in millimetres, and taken to the nearest 5 mm):

Instrument	θ max	$D = 20$ mm			$D = 25$ mm			$D = 35$ mm		
		L	F	$-f$	L	F	$-f$	L	F	$-f$
$\frac{20}{1}$	14′	1370	1445	70	1430	1510	75	1555	1635	80
$\frac{10}{1}$B	20′	1150	1280	130	1240	1375	140	1410	1565	155

Those two of Galileo's telescopes which are still in a complete (if not original) state, have characteristics not vastly different from those of the two Harriot instruments of which the fields of view are known. The following summary table lists some of the parameters of the two complete Galileo instruments, a

86. See for examples the records for 20 February, 27 March, 12 April, 31 March, 7 June, 1612.

further object-glass which he is known to have used, and here conjectured to have been part of an instrument of magnification 30,[87] an extant telescope in a miliary baton once belonging to Gustavus Adolphus of Sweden (of the period 1628–32), and a telescope made by Gascoigne and known from his description (c. 1641).[88]

	F	$-f$	L	θ (actual)	D	m	θ (calc.) $p = 8$ mm	5 mm	3·5 mm
Galileo I	1327	95·2	1232	~15′	26	14	26′·0	18′·1	14′·2
Galileo II	956	48·8	907	~15′	16	20	30′·0	19′·8	14′·7
Galileo III	1689	56·3	1633	—	38	30	17′·9	12′·1	9′·2
Gustavus Ad.	453	60	393	15′·5	15·2	~7·5	—	—	—
Gascoigne	1136	30·5	1035	—	30·5	11·2	33′·8	24′·3	19′·6

From this table and what we have seen of Harriot's notes on the subject, it is easy to see how great were the difficulties under which the earliest telescopic observations were made. To map the Moon, for example, with an instrument capable of revealing only a quarter of its area at any one time (i.e. with field diameter ~15′) required great patience and skill, and Harriot's method of mapping by recording sets of three lunar markings which appeared to be in a straight line was an ingenious—and as it turned out, well executed—solution. (There is evidence that he used the same method with sunspots, since he occasionally mentioned the fact when three or more were in line.) Quite apart from the field of view, few of us now realize how very small was the disk of light seen in the telescope by one looking at the Sun or Moon. To look down Galileo I, for instance, is to see as much light as one sees darkness inside the barrel of a Colt 45 revolver at a distance of two feet. Abetti found that with Galileo I the photosphere of the Sun showed its characteristic granular structure, that little spurious colour was introduced, and that two spot nuclei 17″ apart were well separated. In the Moon he found he could see much more detail than was shown in Galileo's manuscript drawings. Jupiter was not seen well defined, and for some positions of the eye could be seen double. Satellites became difficult to distinguish when within 15″ of Jupiter. With Galileo II, images were less luminous, and clarity was sacrificed to magnification. Jupiter's satellites could now, however, be distinguished when 10″ away from the parent planet. The third object-glass was better than either of the others, and small spot nuclei could be resolved at an approximate separation of 10″. We have at present no idea of the quality of the images formed by Harriot's telescopes. Limited as this was by the properties and quality of the glass, and of the finish of the lenses, it is not impossible that further relevant information

---

87. See Appendix.
88. Essential bibliographical sources for these instruments are given in the Appendix above. All measurements are given in our table in millimetres, as taken or converted from the sources. Note that speaking generally, Galileo stopped down his object lenses to about half their diameters. Figures quoted here are for the clear apertures as stated by Abetti and others.

will come to light in the optical papers at present being studied by J. A. Lohne.

Apart from his telescopes, we know very little of the astronomical instruments available to Harriot. He must have used and might even have designed one of the finest armillary spheres to have survived from Elizabethan times.[89] Passing to Sir Josias Bodley, and in principle presented by him to the Bodleian Library in Oxford in 1601, it reached the library in 1613, and is now to be seen in the Museum of the History of Science nearby.[90] Standing nearly four feet high, and with its three lions supporting an unusually complex system of armillae,[91] the instrument is a fine example of the art of casting in brass, and no doubt in marked contrast with a sphere Lower was obliged to have had made in Traventi out of the hoops of a barrel.[92]

Harriot probably possessed a number of astrolabes. As a device for measuring angles in experiments on optical refraction, the astrolabe is frequently mentioned, and in the lunar records of MS Leconfield 241, for 9 April 1611, we find the entry

    per stellas clock or watch
    $9^h.43'$ $9^h.46'$ Altitudo Canis Minoris $\overline{19}.56'$ by my cathol.
                            Astrolabe, not yet a right line, but almost.

The astrolabe by which he has calculated the time from the altitude of the named star was what Gemma Frisius had called his 'astrolabum [sic] catholicum',[93] which bore a universal astrolabe of a type known in the middle ages as the 'saphea Arzachelis'.[94] This was not an instrument for any but an expert astronomer. It is worth adding, however, that the astrolabe gave the local time much more reliably than did the timepiece. Although Harriot knew, for instance, that longitude could be found by means of 'a perfect watch, or Clock, arteficially made by a Clock-maker',[95] he cannot have held out much hope for an early solution, if we are to judge by his remark made at the foot of the same leaf as the lunar quadrature observations:

'The next morning about $8^h\frac{1}{2}$ my watch was to forward from the Sonne by a $\frac{1}{4}^h$ & somewhat more.'

89. Harriot is known to have been in the employ of Henry Percy at least as early as St. George's Day 1593, when a payment was made to him by the Earl. Percy probably disposed of the sphere in 1600, when he needed money for a military expedition to the Low Countries.
90. See, for an illustration and further details, R. T. Gunther, *Early Science in Oxford*, vol. ii, 1923, pp. 148–50.
91. They show, for instance, the Copernican system for precession.
92. Letter to Harriot, 11 June 1610, Brit. Mus. Add. MS 6789, f. 426r: 'I have supplied verie fitlie my want of a spheare, in the desolation of a hogshead; the hopes thereof have framed me a verie fine one'. In his letter of 6 February 1610 he had asked Harriot to tell the bearer, Vaughan, 'how to provide himselfe of a fit sphere', and how it should be designed (*ibid.* f. 427v).
93. *De astrolabo catholico*, 1550.
94. See my 'Werner, Apian, Blagrave, and the meteoroscope', *British Journal for the History of Science*, iii (1966–7), pp. 57–65, esp. note 8.
95. Quoted by D. B. Quinn and J. W. Shirley, 'A contemporary list of Hariot references', *Renaissance Quarterly*, xxii, No. 1, Spring 1969, p. 13.

THE ROTATION OF THE SUN

Whether or not Harriot used an astrolabe in drafting his sunspot records is a matter for conjecture, although there is one way in which he might well have done so. Each of his drawings, including the very first of 8 December 1610, has superimposed on the solar disk two broken lines as diameters, one invariably vertical, and the other at variable angle to it. The vertical must be along the great circle through the centre of the Sun and the zenith. Were the Sun near the horizon, as was usually the case, this vertical would have been easily judged. The other line undoubtedly represented the plane of the ecliptic. The first intimation of this is in the record of 8 December 1611:

'I saw the cluster of 4 also at 2^h in the afternoone, they with the ecliptick being declined.'

Elsewhere '$\odot$ in 90 gradu' means that the two broken lines are perpendicular, the ecliptic being parallel to the horizon (e.g. at $9^h 50^m$, 18 March 1611). Between the drawings numbered 122 and 155 (inclusive)[96] the lines are not inserted, but on 6 September 1612 we are told that 'the eclipticks were drawn at one time 10 dayes after the observations', and in a similar vein, on the following 6 and 8 November occurs the remark 'eclipt. coniect'. This is an odd remark, since the ecliptic is not in any sense immediately observable. The most natural way of conjecturing the angle between the verticals and the ecliptic would have been to use an ordinary astrolabe, noting the angle between the tangent to the Sun's position on the ecliptic circle and the tangent to the azimuth line through that point, when the instrument was set correctly for the time of day—by solar altitude, in fact. I have verified a score or so of the angles at which the lines are drawn, including those 'conjectured' by Harriot, and find that without exception they are reasonably accurately placed, the error seldom exceeding two degrees. There is thus no doubt that Harriot's use of the word 'ecliptic' is the conventional one.

What did Harriot suppose the significance of the ecliptic line to be? As a Copernican he accepted the fundamental nature of the plane of the ecliptic as that followed reasonably closely by the planets in their orbitings around the central Sun. He saw the rotation of the spots, and readily assumed that they turned about the same axis. Through Salviati in the *Dialogo* (1632), Galileo tells—speaking of himself, of course—how the discoverer and observer of sunspots first took the axis of the sunspots' rotation to be perpendicular to the ecliptic, 'since the arcs described by these spots on the sun's disc appeared to our eyes as straight lines parallel to the plane of the ecliptic'.[97] He went on to explain how the spots were seen to move together and disperse, but how such alterations in their course were accidental, as were the changes in terrestrial clouds, which nevertheless follow the Earth with the general daily rotation'. 'Several years later' than the letters to Welser, the observation of a

96. For 2 July–31 August 1612, London and Syon.
97. *Dialogue Concerning the Two Chief World Systems*, Stillman Drake's translation, p. 345.

slight curvature in the apparent path of a very prominent spot[98] caused Galileo to appreciate the inclination of the Sun's axis to the poles of the ecliptic, and he went on, in the *Dialogo*, to make great philosophical capital out of his discovery.

Harriot's sunspot records do not reveal any such appreciation of the true axis of solar rotation. They show very clearly, however, that the distances of spots from the ecliptic line, although roughly constant were not precisely so. He remarked occasionally on the fact that three spots which had been in line were no longer so, or conversely, and that a spot had moved 'in or nere the ecliptick' (14 February 1612), and even that a southern trio had moved (7 January 1612) so that they were parallel to the ecliptic when they were not so on the previous day. But in none of these cases is there reason to think that Harriot thought the movements to be more than 'accidental', to use Galileo's word. There were, after all, many other ways in which the spots changed in their accidents. They might appear as from nowhere, they might grow 'dimmer' or 'blacker' or 'fayrer', they might grow or diminish in size, and change in shape. They might divide into two or more parts, or merely grow 'ragged'. All these changes Harriot observed and recorded, and he never suggested that the changes in direction in relation to the ecliptic were of any deeper significance than they.

Although it seems that Harriot failed to appreciate the true inclination of the solar axis, yet he paid closer attention to the periodic time of solar revolution than did Galileo. The sidereal period of the rotation as accepted by Carrington[99] is approximately 25·38 mean solar days, making the mean synodic period—namely that which we observe from the Earth—$27^{d.}$ 275. Actually an immense complication is introduced by the longitudinal drift of sunspots, that is, the variation in periodicity with latitude. This we shall overlook, but it must be borne in mind when we are appraising Harriot's measurements that the equatorial sunspot period (synodic) is only $26^{d.}$ 87, and that at latitude 30° this increases to $28^{d.}$ 45. The length of the synodic lunar month (from new moon to new moon) is approximately $29^{d.}$ 53, but the sidereal month (the time to return to a fixed star) is approximately $27^{d.}$ 32. Most early observers and many later ones spoke of the 'monthly' sunspot period, some of the later ones overlooking the fact that they were comparing a sidereal period with a synodic, but others, as for instance Scheiner, were simply mistaken as to the sunspot rotation period. Scheiner in his third letter made this out to be in excess of thirty days, that is appreciably more than the synodic month, and Galileo at one stage put the period at more than twenty-eight

98. It seems probable that it was Passigiani's observations of the elliptic trajectories of sunspots which first led Galileo to realize the inclination of the solar axis. See the letter of Cigoli to Galileo of 23 September 1611, *Opere*, ed. Naz. vol. xi, p. 212.
99. *Observations of the Spots on the Sun*, 1863, pp. 221; 224.

days.[100] Galileo seems to be of the same mind, making the period 'about one lunar month', again almost certainly meaning a synodic month.[101] A difficult problem in accurately establishing the period was that of recognizing a spot, which was bound to have changed after a fortnight's absence from sight. It was possible to follow a spot over the visible hemisphere, but the difficulty here was that its arrival and disappearance were rarely seen, and never clearly seen. Harriot solved this problem in a simple way, and was probably the first astronomer to give a reasonably accurate period for the rotation. When, by 7 April 1612, he had accumulated a collection of 57 drawings of the solar disk, he wrote down without explanation a series of figures beginning thus:

$$\left. \begin{array}{l} \text{Decemb. 1} \\ \text{Decemb. 9} \end{array} \right\} \ 2.8.3 \ \text{II} \ 13.$$

$$\left. \begin{array}{l} \text{Decemb. 13} \\ \text{Decemb. 18} \end{array} \right\} \ 3\tfrac{3}{4}.5.4\tfrac{1}{2} \ \text{II} \ 13\tfrac{1}{4}.$$

Such equations—for the use of the 'Gemini' sign of equality shows them to be such—continued until there were eleven in all, the last ending on 26 February. As Rigaud remarked, the numbers 'seem connected with the time of the Sun's revolution, but it is not easy to understand how they are supposed to contribute to the determination of it'.[102] The points are to be read as addition signs. We notice that the middle number always agrees with the number of days in the interval stated. Clearly Harriot has observed a single spot, or group of spots, or (as in the very first case) a single spot which develops into a group of spots; and he has added on, in proportion to the time it requires to traverse a measured arc, the times it must have required to pass from the eastern solar limb to the first recorded position, and from the second recorded position to the western limb. How he evaluated these proportional times at the limb it is impossible to say, unless further calculations can be found. He might have used a scale drawing, or a table of versed sines (or other trigonometrical tables). In view of the use of fractions, the latter method is unlikely. Taking an average of Harriot's eleven figures for the half-period, we deduce a rotation of approximately $26^{d.}$ 23.

That Harriot later improved on these early estimates is evident from a set of calculations following immediately after the observations, and two auxiliary tables. (The latter are of little moment, having been drawn up to give (i) the number of days from 1 January to the end of each month in a common and a bissextile year, and (ii) the intervals from 1 December 1611 to the end of each month of 1612, and to 31 January 1613.) The new calculations are written in

---

100. See Galileo's first and third letters to Welser (4 May and 1 December 1612), trans. Drake, *Discoveries and Opinions of Galileo*, New York, 1957, pp. 95; 136. All Galileo's and Scheiner's letters are printed in vol. v of the Ed. Naz. of Galileo's *Opere*. Scheiner quoted one period of 15^{d} 2^{h} 22^{m} (i.e. 22 'scruples'), but he also believed that there was a variation in period and distance from the ecliptic, and that this was evidence for the extra-solar character of the spots.
101. *Ibid.*, p. 102, and again in his second letter, of 14 August 1612, p. 106.
102. *Op. cit.*, p. 39, note n.

the manuscript more or less as follows:

$$
\begin{array}{ll}
\text{1610 Dec. 8} \\
\text{1611 Nov. 28}
\end{array}
\quad 355^{\text{d}}\ \text{II}\ \begin{array}{|c} 27 \\ 13 \end{array}\ +\tfrac{4}{13}
$$

$$
\begin{array}{ll}
\text{Dec. 3} \\
\text{Dec. 22}
\end{array}
\quad 385^{\text{d}}\ \text{II}\ \begin{array}{|c} 27 \\ 14 \end{array}\ +\tfrac{7}{14}
$$

$$
\begin{array}{ll}
\text{1610 Dec. 8} \\
\text{1612 Dec. 17}
\end{array}
$$

$$
\begin{array}{ll}
\text{Dec. 21.20} \\
\text{May 30}
\end{array}
\quad 162^{\text{d}}\ \text{II}\ \begin{array}{|c} 27 \\ 6 \end{array}
$$

There are next a large number of supporting calculations, and also, in modern notation, calculations which may be expressed

$$
386/14 = 27\tfrac{8}{14},
$$
$$
739/28 = 26\tfrac{11}{28}.
$$

As to the earlier statements, the first is to be read '355 days separate the two dates, and $355 = 27 \times 13 + 4$', and the others likewise, so far as is possible. In short, the following equations are there to be abstracted:

$$
355/13 = 27\tfrac{4}{13},
$$
$$
385/14 = 27\tfrac{7}{14},
$$
$$
162/6\ = 27.
$$

Were it not for the last, it could be said that Harriot might have been dealing in half-periods. It seems obvious, however, that in some way he believed himself to have done what Rigaud said was impossible by virtue of the 'fleeting nature of the phenomenon',[103] and compared distant observations in the hope of diminishing the error of the inferred synodic period. How else are the calculations to be explained? On the other hand, are we to believe that he compared the Sun on widely separated dates, when between the first pair of those dates we have no evidence that he observed the Sun regularly, and when on some of the terminal dates (28 November 1611; 30 May [1612]; and 22 December 1612) there is no reason for thinking that he observed at all? Harriot was not in the habit of indicating on his drawings those spots which disappeared to be seen again later, but it seems to me certain that he did, at least during 1612, follow spots round full circuit, perhaps helped by the approximate periodicity established earlier. The reason for this claim is that at several points in the records there are numbers which appear to represent counts of sunspots recorded up to that point. Each number is written at the bottom right-hand corner of the page. There are obvious difficulties of interpretation of sunspot counts (unresolved doubles, doubtfully seen spots, and so forth), but making due allowance for these, on referring to the drawings,

103. *Op. cit.*, p. 40.

the numbers seem plainly to be sunspot totals. They are:

Number	43	53	64	65	67	77	84	93	97	98	103	108
Follows obs. number	28	32	36	40	44	48	52	56	60	64	68	72
of date (1612)	14.i	20.i	31.i	5.ii	11.ii	16.ii	20.ii	2.iii	11.iii	17.iii	24.iii	30.iii

There is little doubt that Harriot was the first to attempt this sort of count. Whether he hoped to deduce a periodicity in the count we cannot learn. As Heinrich Schwabe found in the last century, the frequency was to be discovered after years rather than months of counts. Beginning his study of the solar surface in 1826, by 1843 he was able to announce a ten-yearly cycle,[104] subsequently modified to $11 \cdot 11$ years by Rudolf Wolf.[105] This is an involved story which there is no need to recount here, but it is worth noting that Carrington examined Harriot's sunspot records in 1857, made a copy and a duplicate copy of the drawings. One he sent to Wolf, who had earlier placed great reliance on old records, and the other he presented in due course to the Royal Astronomical Society.

If we accept Harriot's five sets of figures as his final estimates of the value of the period of solar rotation, and take their average, we arrive at a period of $27^{d \cdot} 154$. These early sunspot observations of Harriot's were made near the time of a sunspot minimum, when the mean sunspot latitude (regardless of algebraic sign) was in the neighbourhood of $10°$. At $10°$ latitude Carrington found a rotation period of $27^{d \cdot} 06$. I have no wish to obscure the fact that there are difficulties in explaining Harriot's achievement, but under the circumstances I can see no way of denying it. Harriot must be conceded to have interpreted his observations of the solar rotation with a higher degree of accuracy by far than did any of his contemporaries, including Scheiner, who even in *Rosa Ursina* stated the period loosely as 26 or 27 days.[106] All were of course ignorant of the latitude dependence of the periodic time.

If we are to admit that the first of Harriot's deductions of a solar period (8 Dec./28 Nov.) was arrived at from observations sufficiently regular to make the result meaningful, then the price we must pay is that of crediting Harriot not only with the first, but with the first *combined* sunspot observations. These he might naturally have supplemented by the published data of other observers, as he possibly did with Scheiner's for 28 November 1611;[107] but the general conclusion is difficult to avoid, except by taking Rigaud's extreme procedure and simply denying outright that Harriot's calculations could be 'anything like a precise determination of the period of the Sun's rotation'.[108]

104. 'Periodicität der Sonnenflecken', *Astron. Nachrichten*, xxi (1844) p. 234.
105. *Neue Untersuchungen*, 1852, p. 249.
106. *Op. cit.*, p. 215.
107. See p. 113 above.
108. *Op. cit.*, p. 40.

ON THE VALUE OF HARRIOT'S OBSERVATIONS

Harriot's drawings of sunspots are made with pen and ink on ordinary laid paper. Since, as we have already noted, they must be taken to be fair copies, we have no means of knowing whether his originals—perhaps done in pencil— showed any spot penumbrae by shading, but in a small number of cases an effort is made to distinguish the dark from the less dark regions on the ink drawings. Harriot soon realized that spots were changed in appearance when they were at the limb. 'The two occidentall were upon the edge well seen but not black' (14 December 1611; the spots 'generally seem bigger being niere [nigher] the middle than the sides' (15 December 1611); a spot near the limb was 'great long, and somewhat ragged, and not so black as the rest' (5 July 1612); 'the four on the west limb not so black as they were before' (15 August 1612); 'that nerest the limb long and dim, croked and narrower than I have noted' (15 October 1612). These are a few examples in which Harriot recorded not only what he would have expected to see, namely foreshortening of spots at the limb, but also the unexpected change in their intensity. In the last quoted example, he added to his notes a more precise drawing of the spot in question, but neither then, nor on 1 November when he again carefully drew the fine structure of a group of spots near the eastern limb, did he manage to represent the umbral and penumbral differences corresponding to his written descriptions. 'The great long one dim on the ester part', he wrote on 2 October 1612; but his drawing alone would hardly have conveyed the message. Harriot was not here entirely baulked by his direct method of observation, although had he used projection he would have been more at liberty to concentrate on detail, spot positions in that case being automatically registered correctly in the act of drawing.

Harriot recorded changes in the appearances of sunspots of the kinds indicated above—generation, growth and decay, change in intensity, local movements, edge effects, and so on. He left no explanations of the phenomena he so painstakingly recorded, and for this reason he simply cannot be compared in historical terms with Galileo, Scheiner, Fabricius, and the rest. To use a phrase from the end of his third letter to Welser, Galileo's demonstrations of the properties of sunspots set a seal on his other celestial discoveries— those, that is to say, which marked the triumph of Copernicus over the Peripatetics. Whether or not Harriot ever turned his thoughts in the same direction, we are not to know. In considering subjects as diverse as the optimum size for the mast of a ship, the maximum supportable population of the world, or the gaseous yield of a burning candle, Harriot would devise a mathematical theory almost by second nature. Sunspot records of relatively short duration do not lend themselves to mathematics, except in regard to the period of solar revolution, and in this respect it seems that Harriot exploited his work to the full, albeit in a rather mysterious way. It is possible that, perhaps prompted by Kepler, he thought there was a relationship between the planetary periods and that of the solar rotation. But there is no end to possible speculations of this sort, which can lead us nowhere, in the absence of further evidence. No astronomer has a better claim than Harriot to priority in the matter of telescopic observations of sunspots; and yet to make an important issue of this question is to be dragged into the follies of the corresponding seventeenth-century debate. There are no proprietary rights, much less any intrinsic virtue, in an act so obvious as that of raising a telescope to the Sun. When Harriot did so, he undoubtedly showed himself to be an astronomer of a high order of practical and theoretical competence, but we must conclude with regret that in the surviving records of his observations of the Sun we can see little more than his shadow.

Addendum

Since writing this chapter I have learned from Mr J.B. Bamborough that Robert Burton did know of Harriot's work – a fact that makes his failure to refer to Harriot (see p. 109 above) even more melancholy.

APPENDIX

On the early Dutch (so-called 'Galilean') telescope, and its field of view

IN order to discuss the properties of Harriot's telescopes, and to compare them with others of the time, so far as the fragmentary information available allows, a few elementary principles must be set down.

Telescopes first clearly appear in documentary records more or less simultaneously (early in 1609 or thereabouts) in several parts of Europe, with the Low Countries as the most probable source of diffusion. In all instances, where the type of instrument is identifiable, it was that now known as 'Galilean'. This suggests that Harriot's instrument, first used so far as we know in July 1609, was of the same sort. The combination of a converging object-lens with a diverging ocular gives rise to an erect image (as opposed to the inverted image, produced by the so-called 'astronomical' or 'Keplerian' telescope, comprising converging objective and ocular). Harriot's drawings of the Moon and Sun as seen through his telescopes are indeed erect (i.e. looking south, they have east to the left, and often clearly marked as such), and there is thus no reason to suppose that he was using a telescope different in type from Galileo's.

This simple form of telescope has the ocular (of relatively short focal length f) inside the focus of rays from the objective (of focal length F). As in any telescopic system, a focal point of the objective coincides with a focal point of the ocular, in this case behind that second lens. The objective provides an inverted real image, and the ocular an erect and virtual final image. In so-called 'normal adjustment' object and image are at infinity, and prime-foci of the two lenses coincide, their separation being therefore L, where $L = F+f$ (accepting the convention whereby f assumes a negative value). This is the only case to be seriously considered in an astronomical context, although it is admittedly possible to have the instrument out of normal adjustment, the eye being capable of accommodating a wide range of final image positions. The angular magnification (m) for normal adjustment will, for the small angles with which we are concerned, be given by the ratio $-F/f$.

The image of the objective provided by the ocular is virtual, and at a distance L/m in front of the ocular; there is thus no real eye-ring with this instrument, and the eye must be close to the ocular for maximum apparent field of view, the emergent beam having its narrowest cross-section there. (The proximity of the eye to the second lens is responsible for minimizing the chromatism which would be otherwise introduced. That is to say, if the images in light of different wavelengths did not more or less overlap). Unfortunately, the peripheral field of view is seen only by partial pencils of light, and is less bright than the rest (see below). The lack of a clearly defined field cannot be made good by the introduction of stops, as long as the eye is close to the ocular, since there is no real image within the tube of the instrument. Blurring of the peripheral field introduces only very small uncertainties as to the meaning of figures quoted for the field by writers past and present alike.

In calculating the actual field of view (θ), let D be the diameter of the objective diaphragm (assuming it to be in contact with the lens). The size of its (virtual) image as formed by the ocular will be D/m. All rays emerging from the telescope appear to diverge directly from this 'virtual exit-pupil' (E), as we may regard it by analogy with the astronomical telescope. With the eye close to the centre of the ocular, E subtends the same angle (2 arc tan $D/2L$) at the eye as does the objective. It does not follow, however, that the eye is capable of collecting all rays leaving E. Those it collects are those capable of passing through the pupil for any given position of the eye, and what appears to be the standard argument expresses this in terms of the linear diameter of the eye pupil (p). Making use of the accompanying figure, we should on this argument say merely that the eye receives a cone of rays of angle $m\theta$ when p is equal to

AA'. (Actually at the periphery only half of the pencils at A is regarded as collected, but this appears to be one chosen definition of the limit of the visual field.) Then very simply,

$$\tan m\theta/2 = (p/2)/(L/m),$$

and for the small angles involved,

$$\theta = 10800 \, p/(\pi \, L) \quad \text{(minutes of arc)}.$$

Unfortunately, this standard formula may be very far from the truth for a variety of reasons. Not only can the eye pupil never be at the ocular itself, but the eye is a lens system capable of appreciably altering the cross section of a cone of light before it crosses the plane of the iris. Most of the refraction taking place in the eye occurs in fact at the air/cornea surface, before the iris is reached. It is even more mistaken to suppose that the size of the objective, or rather of its diaphragm or mount (namely D) is immaterial to the field of view. If D/m is less than p (see the second figure), the eye can receive no full pencils of light whatsoever. The limit of such full pencils is reached when, taking the pupil to be at the ocular as before,

$$\theta = \frac{10800}{\pi} \cdot \frac{(p - D/m)}{L} \quad \text{(minutes of arc)}.$$

Changing the negative sign for a positive, we have the angle of the field which sends partial pencils into the eye, and beyond which no light is received by the eye. If there is any clear field, the ratio of partially illuminated field to clear field, along any radius of the visual field, is $2D/(pm - D)$.

Supposing, next, that the effective pupil distance from the ocular be e, it will be necessary to replace the distance L in the equation by $(L + em)$. Under the most favourable circumstances, e will be 5 or 6 millimetres, and its introduction into any calculation might affect the result considerably—by 13%, for example, in the case of Galileo's instrument of magnification 20.

Even ignoring the problem of refraction at the eye, we can see that the interpretation of statements concerning field of view—Harriot's in particular—is complicated in several different ways. Since the value of p changes with light intensity, it is meaningless to quote a value for the field of view without some indication of light conditions.

O. Källström and O. Allström have measured the field of an instrument which belonged to Gustavus Aldolphus, a telescope within a military baton, as 4·5 m at 1000 m.[109] This is equivalent to a field of about 15'·5. No comment was made on the clarity of the field selected, although it was pointed out that the field of view of the instrument φ is very limited 'owing to the small objective orifice and because the lenses are placed so far from the ends of the tube'. In fact, given no obstruction, the clear field may be calculated as nearly four times as large as that found (by day, and taking $p = e = 5$ mm.).

In discussing a telescope made by William Gascoigne,[110] S. B. Gaythorpe calculated its field of view as about 25', close to the total field we have calculated with $p = 5$ mm. (day). For other conditions, see our table in our section on Harriot's instruments. Gaythorpe mentions that he takes the pupil aperture of the observer as $\frac{1}{5}$ inch, and makes it clear that he supposes enlargement of the pupil (to $\frac{1}{3}$ inch) with dark-adaptation to increase the field, but gives no indication of his method of calculation. The problem of the Galilean telescopic field was certainly passed over too

109. 'The telescope in the baton', *The Optician*, Nov. 25 and Dec. 2, 1949.

110. 'On a Galilean telescope made about 1640 by William Gascoigne, inventor of the filar micrometer', *Journal of the British Astronomical Assoc.*, xxxix (1929), pp. 238–41.

lightly in earlier writings by G. Abetti and V. Ronchi,[111] a report on whose work was also given in English at the time by D. Baxandall.[112] It is believed that two of Galileo's telescopes (here called I and II) survive, together with an object glass III (broken) of a third. (The ocular of II is a replacement, and the precise details of the original are not known. If we accept a magnification of 30 for the complete instrument, III, we may deduce a focal length of 5·63 cm for its missing ocular.)[113] At the suggestion of George Hale, of Mount Wilson, Abetti and he initiated the study, and 'ha portato a concludere che il cannocciale I ha un potere risolutivo di 20″ e un campo di 15′, il II ha 10″ e 15′ rispettivamente'.[114] As may be seen from our table, the term $\pm D/m\,\varphi$ is of much less importance than is the aperture of the eye pupil. Telescope II, with the eye adapted to darkness ($p = 8$ mm, we suppose), might, it seems, provide a total field of over 30′ (taking $e = 5$ mm), or 24·6 clear. Bringing the effective pupil aperture down, however, to 3·5 mm—a reasonable figure for bright Italian day-light—we find for Galileo I a total field of 14′·2 (clear 4′·3), and for Galileo II a total field of 14′·7 (clear 9′·2). These figures suggest that the 1923 tests were made under the untypical conditions of bright daylight—not even relevant to Galileo's most important sunspot observations, which were made by projection. The figures also give incidentally some idea of the limitations of the Dutch type of instrument, with which the *area* of the unobstructed field is in one case less than a tenth of the total area seen. (As we know, the unobstructed field might in principle be entirely non-existent.)

111. G. Abetti, 'I cannocchiali di Galileo e dei suoi discepoli', *L'Universo*, Anno IV (Sept. 1923) no. 9; V. Ronchi, 'Sopra i cannocchiali di Galileo', ibid., (Oct. 1923) no. 10, summarized in 'Le caratteristiche dei cannocchiali "di Galileo" e sulla loro autenticita', *Rendiconti della R. Accad. Naz. dei Lincei*, xxxii (Nov. 1923) pp. 339–43. A French translation of Abetti's paper is in *Ciel et Terre*, Anee XL (Jan. 1924) no. 1, p. 12.
112. 'Replicas of two Galileo telescopes', *Trans. of the Optical Society*, xxv (1923–4), pp. 141–4. Note that he there warns against some errors in an earlier paper.
113. The fact that the Dutch type of telescope is inherently inexpensive explains the relatively large numbers of such instruments mentioned in the writings of Harriot and Lower, Galileo and others. We know, for example, that Galileo's first instrument was of magnification about 3; that on returning to Padua he made one of magnification 9 (*Sidereus nuncius*) and another of 20; and that he observed Jupiter on 7 January 1610 with one of magnification 30. He made many others, for mention of which see Ronchi's article in *L'Universo*. It seems that Galileo's highest magnification was 33. Objective III is supposedly that with which Jupiter's satellites were discovered. It was presented by Viviani to Leopoldo de' Medici, who had it mounted in its present ornamental frame.
114. Ronchi, in the second article cited, p. 339.

7

A POST-COPERNICAN EQUATORIUM

Although usually conceived against a background in which one theory predominates, scientific instruments are for the most part neutral with regard to their theoretical — one might even say historical — setting. If the hand which held an astrolabe was a little less steady after Copernicus than it had been before, this was not because he had altered in any way the mathematical principles on which the astrolabe depends. The equatorium, however, is another matter entirely; and this, together with the fact that until a year ago only two complete equatoria in metal were known, should be enough to justify the following somewhat lengthy account of the fine post-Copernican example recently brought to light in the Liverpool Museum.

An equatorium is an instrument designed to lighten the task of computing the planetary positions corresponding to any epoch. To 'equate' the motion of a planet — in medieval Latin 'equare' — is to combine its mean motion, easily calculated, with certain correction terms known as 'equations' ('equationes'). The whole calculation, even with the help of astronomical tables, was lengthy and — for all but the best astronomers of the time — difficult. The equatorium was an arrangement of movable disks simulating the circles of planetary theory. Since on both Ptolemaic and Copernican theories movements around these circles were uniform with respect to suitably chosen centres, the disks could be set correctly knowing only *mean* motions. The most common sorts of equatoria therefore showed the positions of the planets in the zodiac (usually in longitude only) without the explicit evaluation of the equations. Needless to say, the positions of the planets were to be referred to the Earth, what-

PLATE I. — Side (i) of the equatorium, showing in particular the solar and lunar epicycles, the dragon plate, and the lunar latitude plate. (Photograph by the courtesy of the City of Liverpool Museums).

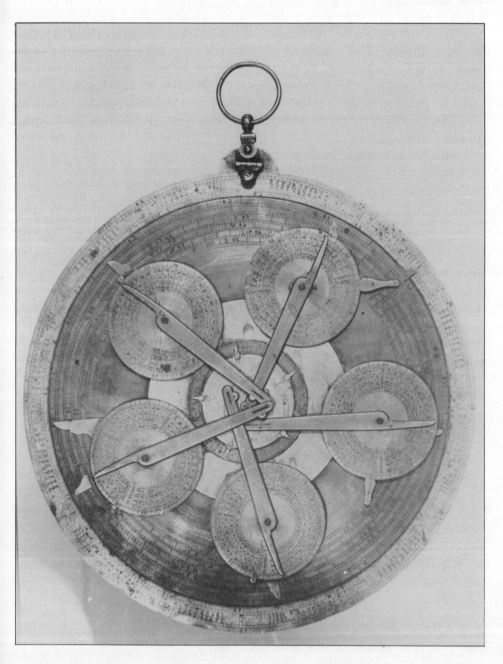

PLATE II. — Side (ii) of the equatorium, with the five planetary epicycles. (Starting at the upper left hand side, and working in a clockwise direction, these are for Mercury, Venus, Mars, Jupiter, and Saturn). Notice the large index (Z) following the main Saturn index. (Photograph by the courtesy of the City of Liverpool Museums).

ever the theory accepted, whether it be geocentric or heliocentric, or a mixture of the two. The astronomer who designed a Copernican equatorium was thus committed to performing a geometrical transformation of Copernican theory, with results which need not be discussed at greater length here. It is not necessary to discuss here the history of the equatorium, but some familiarity with the astronomical theories on which it was based must be presumed. An introduction to the history of the equatorium — as yet only partly written — can be obtained with the help of Emmanuel Poulle's bibliographical notes to articles in « Physis », III (1961), pp. 223-251 and (with Francis Maddison) V (1963), pp. 43-64. A survey by Poulle which has since appeared under the title *L'équatoire de la renaissance*[1], although dealing with relatively late examples of what is essentially a medieval instrument, nevertheless provides an excellent account of the context of the instrument described here.

Very many equatoria survive, made of parchment or paper, in the pages of manuscripts and printed books. (The art of the printed volvelle may be said to have culminated in Apian's *Astronomicum Caesareum* of 1540, although this is a highly derivative work). Only a few brass equatoria, or fragments thereof, survive. The reasons for this are clear enough. First, they were never so common as astrolabes, playing no part, for instance, in the curricula of the schools. Second, equatoria became outmoded as the parameters of planetary theory, and even the theories themselves, were revised. Thirdly, as a purely calculating device (in contrast to the astrolabe, which was used both as an observing instrument and as a computer), the equatorium did not require great durability. Admittedly there were equatoria designed as both observing and calculating instruments, but they were relatively uncommon. In fact, apart from the ' sexagenarium ', the only complete metal equatoria of any complexity which now remain are, it seems, the Sarzosius-type equatorium now in the Museum of the History of Science in Oxford[2], and that described here, which has been in the possession of the City of Liverpool Museum for about a century, without having been appreciated for what it was[3]. The Sarzosius-type equatorium was seemingly made in about 1530, the text on which it was based having first been published in 1526. It is strictly Ptolemaic in conception. The instrument presently described was most probably

---

[1] In *Le Soleil à la renaissance* (colloque, A. Birkenmajer et al., Brussels and Paris, 1965), pp. 129-148.

[2] This is described and illustrated in the article by Poulle and Maddison, *op. cit.*. On the sexagenarium, see Emmanuel Poulle, *Théorie des planètes et trigonométrie au XVe siècle, d'après un équatoire inédit, le sexagenarium*, « Journal des Savants », 1966, pp. 129-61.

[3] Accession number: 7.1.69.39. Mr Alan Smith, Keeper of Ceramics and Applied Art at the City of Liverpool Museums, first drew our attention at Oxford to this instrument, and we are grateful to him for this, for lending us the instrument for inspection, and also for providing two photographs and allowing their reproduction here. The equatorium was presented to his museum in 1869 by a well known collector of the time, Mr Joseph Mayer of Newcastle, Staffs. Its earlier history is unknown.

made in, or around, 1600. It incorporates, as we shall see, sixteenth-century astronomical data, and is in a sense a compromise between the Ptolemaic and Copernican theories. (Throughout the following description, references to sides (i) and (ii) are to Plates I and II respectively).

Since an observer wishes ultimately to know the *geocentric* longitudes of the Sun, Moon, planets, caput and cauda draconis (the ascending and descending nodes of the Moon's orbit on the ecliptic), the instrument is itself geocentric. Parallax due to the finite distance of the observer from the centre of the Earth is, for the planets, ignored. There are pairs of zodiac scales on each side, their significance being explained below. Both sets are described in an anti-clockwise sense, which is customary on the back of an astrolabe, that is, they are described as viewed from the north celestial pole. There is no standard astrolabe convention for placing the initial point (the first point of Aries) on the calendar scale (where present) on the back, and for the principal zodiacs here we find it placed at ' 3 o'clock ' on one face and at roughly ' 2 o'clock ' on the other. In addition to the fixed zodiac scales, and still on the main disk of the instrument, we find (side i) a scale giving the motion of caput and cauda draconis, and (side ii) a set of tables giving the motions of the planetary apogees. This mixture of scales with tables is symptomatic of the instrument as a whole, and is not a characteristic of the Sarzosius equatorium. In describing the instrument we shall therefore distinguish between the information which can be gleaned, first — especially as to provenance and date — from the general appearance of the instrument; second, from the tables engraved on it; third, from the scales; and fourth, from the character and dimensions of the epicycles, eccentricities, and so on.

GENERAL APPEARANCE, PROVENANCE, AND DATE.

As is evident from the plates, the engraving is accurately and skilfully done. The zodiac illustrations in particular (side ii) show that the maker was not new to his craft. Since the auges are acknowledged to be changing position (not all astronomers admitted as much), and since the aux-positions marked on the instrument correspond, according to the accompanying tables, to the year 1600, there can be little doubt that the instrument was made within a few years of that time. (Further confirmation of the date will be given in due course.) We are not aware of any astrolabe maker whose style resembles that of the equatorium. In searching for comparable astrolabes, there are obvious points for comparison: the equatorium uses the flat-topped figure 8 (8), a cursive form of the figure 1 (*1*) and a very unusual letter F (Ɛ). This form of 8 was not unusual in Italy, France and England, at the very least, in the sixteenth and seventeenth-century, and yet Henry Sutton and Thomas Gemini are the only late sixteenth-century instrument makers we have found using the cursive ɪ. We have found no instrument maker using this form of F. The use of the letter D to indicate degrees suggests an English or French

PLATE III. — The two solar epicycles. Notice the scale (one of two called here scale (8)) on the rule to the left.

origin ('degrees', 'degrés'). The usual symbol at this time, even in those two countries, is of course G, the initial letter of the Latin 'gradus'. On the whole, the D-abbreviation was perhaps more usual in France, where it was in common use until the nineteenth century. One point of resemblance between ours and the Sarzosius equatorium which is not apparent in photographs, lies in the fact that both make use of radially sliding joints to vary the distances of certain of the epicycles from the centre of the instrument. (It will be appreciated that the points about which the centres of the epicycles move with constant angular speed, namely the

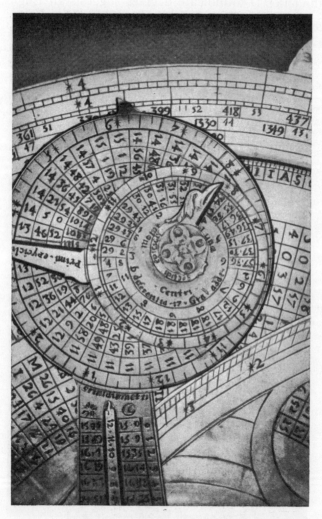

PLATE IV. — The two lunar epicycles.

equants, do not coincide with the centres of the deferent circles on which the epicycles are required by Ptolemaic theory to move — hence the need to alter their radial distances within certain limits.) This common character of the two designs makes a French origin seem slightly more probable than an English, although the evidence is far from conclusive.

The instrument is substantially complete, the only missing part being whatever was attached to the hinge near the ring and shackle (side ii). This hinge might have carried a hinged rule, or cover-plate, or even — more trivially — an arm or thread carrying a slender pin to be used to

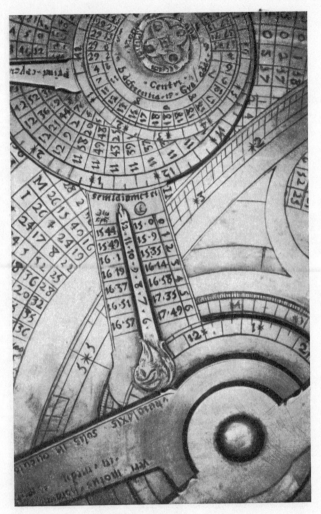

PLATE V. — The central scales of side (i), with a detail of the engraving on the lunar arm.

locate the centres and equants of the planets in turn. Different parts of the instrument are admittedly of different sorts of brass[4], but this was quite common, and is of no significance. The whole is assembled with great

---

[4] The only iron is in the form of pins securing the epicycles to the sliding joints (side (ii)).

skill, with most of the pivots carefully concealed, and with the main pin hammered down so that the instrument cannot be dismantled without first cutting away the head of the pin. This feature has undoubtedly helped to preserve the instrument in its virtually complete state.

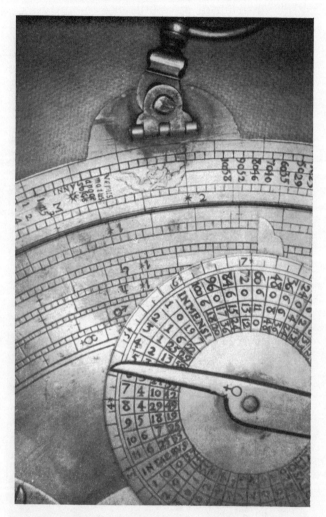

PLATE VI. — The Venus epicycle, rule, and index. Notice the style of engraving of Gemini, and the upturned hinge.

The equatorium is assembled for the most part from plates presumably cut from sheet brass. The disk of equants (side ii, centre) with its five pointers, has the appearance of a casting, but in fact the pointers are cut from sheet brass and brazed to the disk.

Such asterisks as mark the planets on the epicyclic disks (side ii) were occasionally used in printed volvelles for the same purpose. They were, for example, used by Apian in his *Astronomicum Caesareum* (1540). The lunar latitude scale (side i - see below) also resembles that used by Apian, although it did not originate with him. On the whole these are trivial resemblances, and the instrument owes little else to Apian.

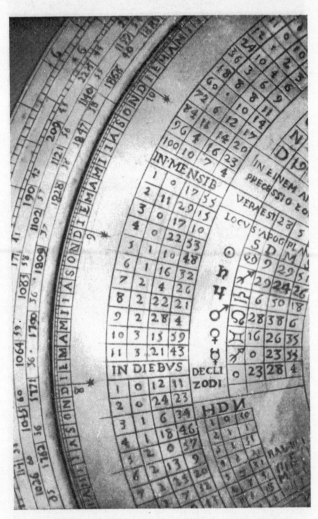

PLATE VII. — A detail from side (i), showing the curious F.

The dimensions of the several parts of the instrument may be evaluated from the photograph, knowing that the largest diameter is $10\frac{1}{2}$ English inches (= 26.6 cm). It is perhaps no more than a coincidence that this diameter is close to 10 pouces (10.66 inches) in French measure.

THE TABLES. SIDE (i).

Beginning at the limb, there are no tables until we reach the large moving disk, which contains tables abstracted (without regard to order) as follows:

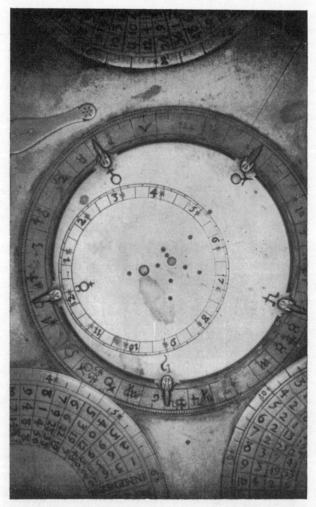

PLATE VIII. — The movable plate of equants and centres, side (ii). The scale below is fixed. Compare this photograph with Fig. 2. Note the asterisk on the (Venus) epicyclic plate above.

(I) MOTUS SIMPLEX ☉

The mean motion of the Sun in signs (S), degrees (D) and minutes (M); for individual years (1, 2, 3... × 1 to 12, × 12 to 96, and 100); for days (1, 2, 3... × 1 to 10, and 20) and for hours (1, 2,... × 1 to 8). At the head of the table is the radix, indicated by the letter ' R ' and presumed to be for the end of 1600 or 1599. If so, and if this is not in error (for any year it corresponds to the mean motus on a November date), then it must be referred to an equinox not adjusted for precession. See table (7) etc. The present tables include these entries:

Radix	8ˢ 11° 39′ (for 9ˢ 21° 39′ ?)
1 year	11ˢ 29° 45ˢ
1 month	1ˢ 0′ 33′
1 day	0ˢ 59° 8′
1 hour	2ˢ 28″

The figures are not precise enogh for us to pronounce on their source. The phrase ' motus simplex ' (as opposed to ' motus medius ') gained currency with Copernicus. He used a sign of 60°, rather than the common sign of 30° used here.

(2) ELONGATIO ☽ A ☉

The elongation of the first epicycle of the Moon from the *mean* Sun for individual years (as in table (1)), for months (1, 2, 3... × 1 to 11); for days (1, 2, 3... × 1 to 10, and 20) and for hours (1, 2, 3... × 1 to 8). As before, there is a radix which is now confirmed to be for 1 January 1600. Selected entries are as follows:

Radix	10ˢ 15° 44′
1 year	4ˢ 9° 37′
1 month	0ˢ 17° 55′
1 day	0ˢ 12° 11′
1 hour	0° 30′

As before, the figures could originate with any author of the time. We observe that the Radix would have been accurate during the afternoon of 1600 January 2 had the *true* Sun and Moon been intended. We shall later explain how the (mean) elongation enters into the determination of the Moon's true position.

(3) TABULA MEDIAR' ☌ ET ☍ ☉ ET ☽

Table of mean conjunction and opposition of the Sun and Moon for years (1, 2, 3... × 1 to 20 and × 20 to 100); and for months (given by initial - I, F, M,A, etc.); specimen entries:

1 year	18ᵈ 21ʰ 32ᵐ 41ˢ
January	29ᵈ 12ʰ 44ᵐ 3ˢ
February	28ᵈ 1ʰ 28ᵐ 6ˢ
April	28ᵈ 2ʰ 56ᵐ 13ˢ

To be used together with the following table.

(4) RADICES ♂ ET ☌ MEDIAR' ☉ ET ☽ A NATIVI D̄N̄I

Radices of mean conjunction and opposition of the Sun and Moon for centuries from the birth of Christ, from 00 to 2000. Specimen entries:

00 years	14^d 5^h 0^m 38^s
1500 years	20^d 6^h 54^m 58^s
1600 years	5^d 2^h 24^m 25^s

This means, for instance, that the first conjunction of the Sun and Moon after the beginning of 1600 was on January 5, at the hour, minute and second stated. The error in this example, which does not seem to be Alfonsine, is difficult to assess since we are dealing with the mean Sun and Moon, and since we do not know whether the day was taken to begin at midday or midnight. At the most, the error for a longitude appropriate to western Europe was of the order of a few hours only.

(5) ARGUM' MEDII ☽

Argument of the mean Moon, with radix, the argument being for years (1, 2, 3... × 1 to 12, and × 12 to 96, and 100); for months (1, 2, 3... × 1 to 11); for days (1, 2, 3... × 1 to 10, and 20) and for hours (1, 2, 3... × 1 to 8). This argument gives the movement in the first (larger) lunar epicycle of a Copernican model. Specimen entries:

Radix	0^s 7° $11'$
1 year	2^s 28° $43'$
1 month	1^s 15° $1'$
1 day	0^s 13° $4'$
1 hour	0^s 0° $33'$

Apart from the radix, all are commonplace, as before.

(6) DIMID' MENSIS

Figures for one half the synodic month, that is, the time from conjunction to opposition; and the time to quadrature, that is, one half the first figure. The half month quoted is:

$$14^d \ 18^h \ 22^m \ 2^s$$

Copernicus quotes this parameter in terms of minutes (i.e. sixtieths), seconds, and thirds, of a day — rather than in hours, etc. as here.

(7) IN FINEM ANNI 1600 PRECESSIO EQUINOC' VERA EST 28° $5'$ $34''$

The designer's views on precession are discussed in the comments on scale (10) and in a subsequent section. It is clear from scales (9) and (10) that table (7) is evidence for a date of design near 1600. The words of (7) suggest that a radix was taken for the *end* of 1600, and yet with other radices the beginning of that year seems to have been accepted. We are reminded of the contrc-versy as to when a new century begins.

(8) LOCUS APOG PLANET

The apogees of the planets are listed with the symbols for the zodiacal signs-altered here to numerals:

Sun	3ˢ 9° 29′ 51″
Saturn	8ˢ 29° 24′ 26″
Jupiter	6ˢ 6° 50′ 18″
Mars	4ˢ 28° 38′ 6″
Venus	2ˢ 16° 26′ 35″
Mercury	8ˢ 0° 23′ 55″

The fact that Venus is listed separately from the Sun gives us an intimation that the instrument is not designed in accordance with the usual medieval version of Ptolemaic astronomy, in which the auges of the Sun and Venus are identical. The figures are not Ptolemy's, with a constant added for precession, nor are they Alfonsine. Copernicus had made known the intrinsic movements of the lines of apsides with respect to the stars, in his *De Revolutionibus*. With the benefit of this knowledge, Longomontanus and Lansberg both subsequently compiled lists of auges for 1600, but our data are not from either source. Nor are the data Copernicus's, modified by the addition of terms determined by the velocities of apogee as given in table (15). They are nevertheless all within half a degree of such modified Copernican figures, which suggests that the in-strument did not involve any new observations of much merit. (Some interesting remarks on the difficulty of finding the solar apsides are to be found in *De Rev.*, III, 20). Copernicus, incidentally, does refer apogees to the head of Aries rather than to the star γ Arietis, to which he often refers other longitudes.
The planetary apogees listed above are engraved on scale (21), and discussed below.

(9) DECLI ZODI

The obliquity of the ecliptic:

$$23° \ 28′ \ 4″$$

This is not a constant quoted by any well-known astronomer although it might have been calculated in accordance with data provided by an astronomer who thought the obliquity to vary. Tycho's figure was generally available in 1600, but was not being followed here, for Tycho had proposed 23° 31′ 30″. Copernicus believed in a changing obliquity, proposing as he did his own brand of trepidation theory (*De Rev.*, III, 2 sqq). He believed that the obliquity varied between 23° 28′ and 23′ 52° (*De Rev.*, III, 10) and though it to be 23° 28′ 24″ at the time of writing (c. 1524). It seems probable that the figure engraved on the instrument was calculated on Copernican principles (*De Rev.* III, 12), but if so it was erroneously calculated, the correct Copernican figure for 1600 turning out to be 23° 28′ 24″. We notice, in passing, that the Copernican William Gilbert, in chapter 7. Book 6 of his *De Magnete*, first published in 1600, makes the obliquity 23° 28′, adding that others put it at 23° 29′.

The only tables remaining on side (i) are those on the epicyclic volvelles, which are as follows:

(10) *Semidiametri* ☉ [et] ☽

The table, on the stem of the lunar volvelles, has only six entries for each luminary, i.e. for the first point of each zodiacal sign, taking each entry twice over. Specimen entries are:

		Sun	Moon
12ˢ	0ˢ	15′ 44″	15′ 0″
9ˢ	3ˢ	16′ 19″	16′ 14″
6ˢ	6ˢ	16′ 57″	17′ 49″

These are Copernican data for the Moon (cf. *De Rev.*, IV, 24), giving semidiameters at conjunction or opposition, the left hand columns giving position in the epicycle. The data were also accepted by Reinhold, Maginus (occasionally), and Lansberg. Of these, only Copernicus and Reinhold[1] accepted the last solar figure, and both differed from the first, Copernicus making it 16′ 22″ and Rheinhold 16′ 18″. It may well be that the originator of the above figure of 15′ 44″ obtained it by wrongly dividing the diameter quoted by Copernicus.

(11) *Semidiamet. Umbr.* [et] *Motus diurn. solis*

Each of the tables occupies about half of the centre solar disk. The first gives the angular semidiameter of the Earth's shadow at the lunar distance, and the second gives the diurnal motion of the same, both at syzygy. Speciment entries:

		Semidiameter of shadow	Solar motion per diem
12ˢ	0ˢ	39′ 58″	57′ 1″
9ˢ	3ˢ	44′ 18″	59′ 5″
6ˢ	6ˢ	49′ 49″	61′ 26″

The shadow sizes are not Copernican (*De Rev.*, Tables IV, 24), and the same is true of most of the tables which follow. It is perhaps worth remarking on the fact that the *ratio* of the angular size of the shadow to that of the Moon at maximum distance (cf. (10)) is almost 400/150, as against the 403/150 of Copernicus (*De Rev.*, IV, 23) and 390/150 of Ptolemy.

(12) *Veri motus horarii solis centri in medii eccentrotetis* [et] *Paralaxis solis in circulo altitudiniis*

 Each tables occupies half the lower solar epicyclic disk, the titles being on the arm which carries the disk. Specimen entries:

		Motion per hour	Solar parallax in altitude	
12ˢ	0ˢ	142″	Zenit.	0 [5]
9ˢ	3ˢ	148″	40	114
6ˢ	6ˢ	154″	85	78

The first table conforms exactly with the second of (11).

(13) *Motus horarii Lunae Centri b* (?) *diferentia 17 Gra adde*

 Each table occupies half of the upper lunar disk:

		Hourly motion	
12ˢ	0ˢ	29′	2″
9ˢ	3ˢ	32′	36″
6ˢ	6ˢ	37′	36″

 Common differences between entries are also engraved on the instrument, but it would be meaningless to give them in our excerpt. Was the ' b ' an engraver's mistake for the following ' d '?

(14) An unnamed table occupying the whole of the lower (fixed) lunar disk.

 Specimen entries:

0ˢ	11	37	17
3ˢ	13	10	0
6ˢ	15	1	12
9ˢ	12	54	6
12ˢ	11	37	17

The second table of differences is omitted here.
These are clearly *daily motions* of the Moon for different positions on the cycle, with the latter at apogee, and are an extension of (13).

---

[5] Adjacent marks: 5/19.

The tables. Side (ii).

Tables (15)-(19) are each engraved on the appropriate epicyclic disk.

(15) MOTUS ☿ IN EPICICLI

(16) MOTUS ☿ IN EPICICLI

Following a radix, motions are given for years (1, 2, 3,... × 1 to 12, and × 12 to 96, and 100), months, and days (1, 2, 3... × 1 to 10, and 20 or 28). The following are excerpts:

	Mercury	Venus
R	3ˢ 18° 35′	1ˢ 23° 46′
1 year	1ˢ 23° 57′	7ˢ 15° 2′
100 years	2ˢ 13° 23′	6ˢ 18° 20′
1 day	0ˢ 3° 6′	0ˢ 0° 37′

(17) MOTUS ♂ IN ECCENTRI

(18) MOTUS ♃ IN ECCENT

(19) MOTUS ♄ IN ECCENT

Excerpts from the tables, which in arrangement resemble those above, are:

	Mars	Jupiter	Saturn
R	5ˢ 2° 35′	11ˢ 2° 47′	9ˢ 28° 53′
1 year	6ˢ 11° 16′	1ˢ 0° 20′	0ˢ 12° 12′
100 years	1ˢ 29° 29′	5ˢ 4° 35′	4ˢ 21° 6′
1 day	0ˢ 0° 31′	0ˢ 0° 5′	0ˢ 0° 2′

The figures of tables (15)-(19) have a more or less anonymous appearance, expressed as they are only as far as minutes. In medieval terms, the first two gave the *mean argument* for the inferior planets, and the last three the *mean motus* (apart from the radices) for the superior. That the tables are Copernican is vaguely suggested by the annual (365 days) motion of Mars, which in Ptolemy's *Almagest* is given as 191° 16′ 54″, in the Alfonsine tables as 191° 17′ 5″, but in *De Revolutionibus*, V, 1 as 191° 16′ 20″ (all approximate). The years are not here the Egyptian years which Copernicus favoured, but are calendar years of which the years numbered 4,8, and so on are bissextile (a fact not evident from the excerpts above). This is what we should expect where the radix is for the end of a bissextile year. But what is the year in use? The last three quoted radices, if they are of mean motus, could not belong to any date between the thirteenth and seventeenth centuries, let alone to the beginning of the Christian era, or to any of the standard eras before it. In fact they prove to be radices of elongation from aux (apogee), as will be shown, when we have discussed scales (9)-(5) of the instrument, and are almost certainly for the end of A.D. 1600.

(20)-(25) *Verus motus apog. Solis ... Saturni... [etc.]...*
           Verus motus prece. equi. Vern.

Movements of precession and of the apogees of all planets, excepting the
Moon and Venus, in minutes of arc for individual years (1, 2, 3... × 1 to 10 and
× 10 to 100). We excerpt the movements per century. The tables occupy the
space between the two zodiacal scales.

Sun	191'
Saturn	120'
Jupiter	77'
Mars	107'
Mercury	155'
Precession	58'

These tables supplement table (8). They are not strictly Copernican, although — precession apart — they stem from an essentially Copernican source.
At the time of writing the *Commentariolus* Copernicus believed, as Ptolemy
had done, that the apogees of the planets were fixed with regard to the fixed
stars. He later decided on the basis of new observations that the longitudes of
apogee were slowly increasing. His figures — explicit and otherwise — for the
mean secular motions of the apogees, which did not (as those above must be
presumed to do) take precession into account, were, to the nearest minute:

Sun	137' per century [6]		*De Rev.*, III, 16
Saturn	60' » »		V, 7
Jupiter	20' » »		V, 12
Mars	47' » »		V, 16
Mercury	95' » »		V, 30

In *De Revolutionibus*, V, 22 it is hinted that the apogee of Venus has not
changed, and this might account for the omission of a table for the apogee of
Venus on the instrument: « Inveniuntur autem haec omnia, quae hactenus de
Venere demonstrata sunt, etiam nostris consentanea temporibus, nisi quod ec-
centrotes 1/6 fere parte decreverit... ». In the case of Mercury Copernicus sti-
pulates 1° per 63 years only if the motion is uniform (« si modo aequalis fuerit »).
To the movement of the Sun's apogee he most emphatically denied regularity,
putting rather too much weight on astronomical observations intermediate between Hipparchus and himself (See *De Rev.*, III, 16).

Now three of the equatorium's apogee movements differ by 60' per century from those taken from *De Revolutionibus*, the remaining two differing by
57' per century, adjusting the Sun's rate to that appropriate in 1600. These
differences are so close to the designer's figure for precession (which is not of
course Copernican) that there can be no doubt of the ultimate influence of Copernicus. It is possible to say this dogmatically since the likelihood of another
astronomer arriving at them independently is too remote to be entertained

---

[6] I calculate 134' for the year 1600. See below.

(Thus Tycho decided that the Sun's apogee moved 75′ per century, without precession).

Finally, there is a possibility that the figure for the *rate* of precession marked on the instrument, which figure is again very poor, and markedly inferior to medieval determinations, arose from a misunderstanding or an over-zealous simplification of the theory of trepidation in the *De Revolutionibus*, III, 2 sqq.

THE SCALES. SIDE (i).

(1) The outermost scale is the true zodiac and is graduated concentrically. It is used in conjunction with the pivoted rule or 'ostensor'.

(2) The scale within it is not uniformly divided, and is presumably analogous to the equant scales (12)-(15) discussed below. It is graduated with respect to a centre which lies at a small distance (approximately 3.1 mm) from the true centre, and its significance, used as it obviously was in connexion with both lunar and solar epicycles, will be considered in a later section. This scale should, of course, agree with scale (1) at only two points: apogee and perigee. Again it is not possible to be precise, but the scales seem to agree at longitudes of 100° and 280°. This is very little different from the information given in Table (8), where the longitude of the Sun's apogee in 99° 29′, and tends to compensate for the slow change in the direction of the apseline. As it stands, the instrument would, theoretically speaking, improve in accuracy for 30 years or so, but lose in accuracy thereafter. This fault of manufacture need not have been a fault of design, for the scale (2) might have been designed to move relatively

(3) Verus motus ♌ a nativi Domi' ad 2000 annorum (sic)

(3) Verus motus ♌ a nativi domi' ad 2000 annorun (sic)

This scale of the dragon (lunar nodes) still on the fixed outer limb, carries divisions, in three banks, oo, 19, 38... × 19 to 1956 and 1600 × 19 to 1999 (erroneously written 2999). The break, involving a shift of a little less than a quadrant, is merely to allow the use of 1600 as a convenient standard epoch. Interspersed between these numbers is another sequence, in a smaller script, running 35, 35, 36, 36, 37, 37, 38, 39... 66 (maximum)... 45. These are explained in an inscription:

numerorum minores anuum equinoctiorum motum significant qui toties e motu draconis subducendus est quoties qui ex 19 annus instat unitatem complectitur.

The lunar nodes move round the zodiac once in about 18.6 years. The interval between divisions on scale (3) represents the excess over 360° moved by the dragon in 19 years. Careful measurement of the scale shows that this excess was on

---

average close to 7° 27′.5, which implies an annual movement of the nodes of 19° 20′ 24″. This differs by only a few seconds of arc from the Alfonsine figure (approx. 19° 20′ 29″) in common use during the late middle ages and after. Copernicus appears to have had nothing new to say on the values of this parameter. Tycho, it is true, had appreciated that the nodes did not move uniformly (see J. L. E. Dreyer's *Tycho Brahe*, reprinted 1963, p. 343), but the instrument does not take his ideas into account.

The scale under discussion was, of course, to be used with the scale on the limb of the main disk entitled:

REVOLUTIO CAPITE DRACONISS (*sic*)

On this are the years of a 19 year cycle, divided into months and thirds of a month, and going a little way beyond the full circle in order to accommodate the last few months of the cycle. (These graduations are near the index.) The whole style of these two scales (required, needless to say, for eclipse calculation, especially when taken in conjunction with scale (4)), is very reminiscent of a volvelle in Apian's *Astronomicum Caesareum*.

(4) The outer scales on the movable and pierced lozenge-shaped plate are scales of lunar latitude, the indexes at its extremes serving to mark the nodes of the lunar orbit. These indexes are positioned correctly when the index on the main disk is set to the beginning of the appropriate 19 year cycle, and when they are subsequently placed over the position of the Moon and the zodiac, its edge crosses scale (4) at the appropriate latitude.

The scales bear no inscription. The author did not make use of Tycho's data on the lunar inclination (Dreyer, *loc. cit.*), but used the round figure of 5°, current throughout the middle ages and renaissance. It would, of course, have been very difficult to take into account the changing inclination which Tycho found.

(5) Scales of the two solar epicycles.

(6) Scales of the two lunar epicycles.

There are two epicyclic volvelles for the Sun, and two for the Moon, which suggests acceptance of the Copernican representation — inverted, of course, in the case of the Sun, so that the observer may be at the centre. This inversion will be discussed below. The scales are uniformly graduated 0ˢ-12ˢ, both clockwise for the Sun, and the upper anticlockwise, the lower clockwise, for the Moon. The indexes are in the case of the Moon labelled *primi epicycli* and *secundi epicycli*. They move, respectively, clockwise and counterclockwise as required by Copernican theory (*De Rev.*, IV, 3). Only the index traversing the smaller of the Sun's disks is labelled (*eccentrotetis*, a word used by Copernicus in place of the more usual *eccentricitas*).

(7) Two zodiac scales, each used in conjunction with the radial edge of one of the two arms (solar and lunar) carrying the epicyclic disks.

The upper scale may be used to assess the mean elongation of the Sun and Moon.

The lower scale, being on the lozenge-shaped plate carrying scale (4), may be used to give directly the distance of the (mean) Moon from the lunar nodes.

(8) There are two short and ungraduated scales on the diametral rule. One, in five divisions covering less than 1 cm, shows the distance of the Sun (on the second solar epicycle) from its positions of minimum and maximum distance from the centre; while the other, in 16 (nine plus seven) divisions, shows the analogous lunar position. These distances, according to Copernicus, are recpectively 95 and 18 Earth radii in reality.

THE SCALES. SIDE (ii).

(9) The outermost is a uniform scale, beginning, like scale (1), from ' 3 o'clock ' of the instrument. It is shown below to be the ecliptic of date.

(10) Immediately within it (disregarding tables (20) to (25)) is a concentric and uniform scale, graduated like (9) with the numbers of the signs. Since the zodiacal images drawn between (9) and (10) are placed opposite the correct divisions of scale (10) (but not of 9), which is out of phase with (10)), we presume that it is this which was intended as the zodiac scale, that is, the one graduated in keeping with the rough positions of the constellations at an earlier date (exactly which is a matter discussed shortly). Scale (9) will be the ecliptic or zodiac of date. In fact the angular separation of (9) from (10) is on average constant at a few minutes of arc over 28°. This is surely conclusive evidence for a date near 1600 as that of design and — presumably — manufacture of the instrument; for as we saw from table (7), precession for the end of 1600 was believed to be 28° 5 ′ 34″. Remarks made earlier in connexion with scales (1) and (2) apply here with equal force: since no provision is made for the relative movement of (9) and (10), the instrument will soon become out of date.

(11) A uniform scale of signs and degrees on the large moving plate, the graduations beginning from the protruding index on the plate, to be known henceforth as index T, and itself moving over scale (10) only.

(12)-(15) A series of four concentric scales each divided in accordance
with a rule illustrated in Fig. 2, so that the scales may serve to position the epicyclic disks of the planets — which all revolve about a
fixed centre — correctly with respect to different equant points. The
outer scale is for Saturn, the next for Jupiter, the next for Mars, and

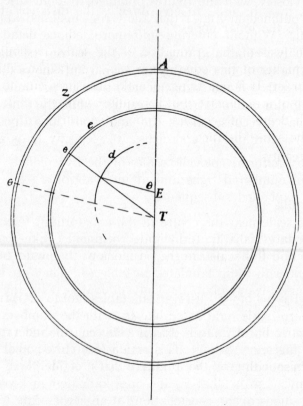

FIG. 2. — The scale *z* represents the zodiac or ecliptic, while
e is the equant scale. The angle at *E* (the equant point)
marked θ is represented by the two graduations θ on
e and *z* · *A* is the aux of date, *d* the deferent circle,
and *T* the Earth.

the innermost for Mercury and Venus. All, like (11), are graduated,
beginning from the protruding index *T*, which therefore marks the
apogee of whatever planet is the subject of calculation. Measurements made of the equatorium suggest that these scales were graduated for the following distances (mm) between equant and centre:

Saturn	Jupiter	Mars	Venus	Mercury
8.2	6.9	15.0	3.1	2.7
(8.2)	(6.5)	(13.7)	(3.0)	(2.8)

The probable error in these measurements is difficult to discuss, and on *a priori* grounds one would put it at approximately ±0.5 mm or better. In fact the bracketed figures are taken from *direct* measurement of the spacing of the Earth (*T*), centre deferent (*C*), and equant (*E*) marks on the central disk (see below), and they conform so closely with the figures obtained from the use of the construction outlined in Fig. 1 that the error suggested appears unduly pessimistic. Without entering into more tedious detail with regard to the analysis of the graduation of the scale, it should be evident that the maker of this equatorium was a craftsman with skill of the highest order. It is one thing to make measurements to an accuracy of a tenth of a millimetre, but it is quite another to make graduations of a high degree of accuracy from an eccentric centre positioned to within the same distance.

(16)-(20) The (uniform) epicyclic scales of the five planets. Each is graduated in numbered signs and at intervals of 5°. The use of these scales is explained subsequently.

(21) A zodiac scale near the centre of the equatorium, fixed to the main plate (the largest of all), and aligned with scale (9), the ecliptic of date. Engraved on this scale are the positions of the planetary apogees, in agreement with those tabulated in table (8). They are each signified by a fleur-de-lys symbol, the conventional sign for the planet being placed to each side of it.

(22) An eccentric but uniformly divided scale on the innermost plate, the plate of auges (see below). This scale is centred on what is clearly the hole corresponding to the auxiliary circle of Mercury, in Ptolemaic astronomy.

THE PLANETARY RADICES

It is convenient here to assign letters to the indexes, beginning with that for Saturn (*U*), continuing with Jupiter (*V*), Mars (*W*), Venus (*X*), Mercury (*Y*), and ending with the index (*Z*) attached to the epicyclic plate at the side of the Saturn index *U*. The purpose of index *Z*, as Dr Emmanuel Poulle has pointed out to me, is to allow the use of scales (11)-(15), which are obscured by the epicycles in such a way that indexes *U*—*Y* can only have been intended for use with scale (11). It is tempting from this to conclude that the 'motions on the eccentric' given in the tables on some of the disks were to be set on scales (12)-(15) with the help of *Z*, in which case the radices (see the notes on tables (17) to (19)) would be elongations from aux suitably displaced by about 22° (the angle between Z and U), 94° (Z and V), and 166° (Z and W). This hypothesis raises an insuperable difficulty: there is no obvious date at which the radices hold good. (There is no point in going into the details of our fruitless search). We may therefore simply suppose that index Z is for converting a true elongation from aux (to be known here as γ — measured along scale (11), and usually known as 'true centre') to a mean ($\bar{\gamma}$ measured along scales (12)-(15) and usually known as 'mean centre'), or conversely. In short, it provides us with the 'equated' elongation from aux (or 'centre'), the 'equation of the centre' being the name usually given to the difference between true and mean. We are still left with the problem of the radices which, since they are not of mean motus for any reasonable date, might prove to be radices of mean centre, $\bar{\gamma}$. Let us consider the predicted mean motus for the planets Saturn, Jupiter and Mars, on the assumption that the radices are for the end of the year 1600 A.D.:

	Saturn	Jupiter	Mars
Quoted radix	9ˢ 28° 53′	11ˢ 2° 47′	5ˢ 2ᵈ 35′
Aux positions for end of 1600 from table (8)	8 29 24	6 6 50	4 28 38
(Sun) Resulting mean motus according to the instrument	6 28 17	5 9 37	10 1 13
Alfonsine radices of mean motus for the same date	6 29 34	5 10 6	10 6 42

The proximity of the two sets of radices, one based on our hypothesis that the instrument's radices are of mean centre, the other taken from manuscripts of the Alfonsine tables[7], leaves us in no doubt that the hypothesis is correct, and confirms the year of the radices[8]. When we examine the remaining radices of mean argument for Venus and Mercury, we find that they are respectively 7° 22′ and 13° 34′ in excess of the Alfonsine figures. These differences are not substantial, but they confirm us in the belief that the planetary part of the instrument is not strictly Alfonsine.

---

[7] We could more conveniently have used the Alfonsine figures in Schoener's, *Equatorii astronomici omnium fere uranicarum theorematum explanatorii canones*, Nürnberg, 1522, which differ at most by 1′ arc from these.

[8] The differences are such that Saturn would cover the interval in 38 days, Jupiter in 6 days, and Mars in 9 days.

THE THEORY OF PRECESSION ACCEPTED ON THE INSTRUMENT

Since we have seen the equatorium to comply with Copernican astronomy in certain respects, it is worth our while to calculate precession for the end of 1600 A.D. in accordance with the tables of *De Revolutionibus*, III, 6:

i) 1600 Julian years = 1601 Egyptian years and 35 days
$$= 26 \times 60^y + 41^y + 35^d$$

	Aequalis motus	Anomaliae motus
ii) 26×60ᵛ	21° 45′ 14″ 0‴	163° 32′ 27″ 0‴
41ᵛ	34′ 18″ 15‴	4° 17′ 53″ 30‴
35ᵈ	4″ 48‴	36″ 11‴
	22° 19′ 37″ 3‴	167° 50′ 56″ 41‴
		(×2)
		335° 41′ 53″ 22‴
Radix (AD)	5° 32′ 0″ 0‴	
Prosthaphaireseon	28′ 0″ 0‴	
	28° 19′ 37″ 3‴	

This is not close enough to the figure of 28° 5′ 34″ quoted on the instrument to allow us to say that the latter was calculated on Copernican principles, but it suggests at least that the same standard radix has been accepted as that of the *De Revolutionibus*, namely one according to which precession at the death of Alexander (a little under 452 years after the first Olympiad, which Copernicus favours as a time origin) is only 1° 2′. Consider, hower, the statement made on the instrument that precession is 58′ per century. Extrapolating backwards in time we find that at this steady rate the movement of over 28° quoted takes us back to 1310 B.C., an unlikely date. Turning, however, to Copernicus's own calculation for 16 Kal. May 1525, yielding a precession of 27° 21′, we find to our astonishment that the difference between this and the quoted figure would be covered *at a constant rate of 58′ per century* in a time of a day or two over 76 years 9 months. This brings us so close to the date quoted (the end of 1600) that we are driven to suppose this monstrous mixture of two different astronomical theories of precession to have been used in arriving at the constant engraved on the instrument, subsequently used in positioning the scales on side (ii). And here again is one of many instances from medieval and renaissance astronomy proving that there was a certain cachet attaching to any precession figure which had been derived using a trepidation calculation, perhaps simply because the calculation was not really understood.

ON THE THEORY AND USE OF THE EQUATORIUM. SIDE (ii)

We begin with side (ii), which is purely Ptolemaic in form, if not
in the parameters it embodies. It does not involve a reinversion of Copernican principles, for this would require it to have small secondary epicycles, among other things.

Although they are now too stiff to move, it is clear that there are
sliding joints on which the epicyclic disks may both revolve and move
radially, to allow for the varying geocentric distance required of them

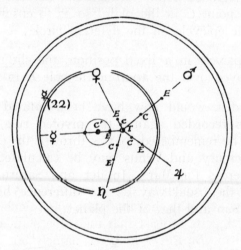

FIG. 3. — The arrangement of centres of deferent circles (*C*) and equant centres (*E*)
on the central plate of side (ii).

by Ptolemaic theory. The fact that the central disk is pierced with holes
(as illustrated in Plate VIII and Fig. 3), and that the rules pivoted at
the centres of the epicycles are pierced each with a hole and a slit,
suggest that

i) the aux of date (calculated from Tables (8) and one of (21)-(25))
   would have been correctly set with respect to the outer zodiac[9];

ii) the central plate, carrying pointers which mark the directions of
    the auges (i.e. lines of centres), would have been set for the planet
    in question, scale (21) being engraved for this purpose, and actually
    having the 1600 aux positions marked on it;

------

[9] We observe that although the index or auges only reaches as far as zodiac scale (10), whereas
the auges cited in table (8) are with respect to the equinox of 1600, the central scale (21), used in
conjunction with either a missing radial rule or even the thin radially engraved line (see Plate II)
on the movable plate of epicycles, is all that is needed.

iii) as explained earlier, the mean centres would have been set on the scales (12)-(15) using index Z to discover the true centre (scale (11)) corresponding to the desired mean centre (scales (12)-(15)). The appropriate index (U, V, X or Y) could then have been set on scale (11), the concealed scales of mean centre being thus correctly adjusted. It is worth observing that the mean motus of Mercury and Venus was presumably taken in the usual way as equal to the mean motus of the Sun;

iv) a pin would have been put through the hole in the rule and the corresponding point C in the plate, so as to ensure the correct positioning of the epicycle on the deferent circle;

v) holding the epicycle in a fixed position, the slit in the rule would have been moved over the appropriate hole E in the central plate;

vi) the epicyclic disk would now have been rotated until the correct argument was recorded against the pivoted rule, the direction of which, it will be remembered, passes through the equant E. The argument of Mercury and Venus may be calculated from table (15) and (16), whereas for Mars, Jupiter and Saturn the argument is calculated in the usual way as the difference between the mean motus of the Sun and that of the planet;

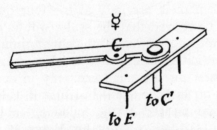

Fig. 4. — A device which might have been used as an accessory, in determining the centre of the deferent circle of Mercury. Cf. Fig. 3.

vii) finally, a rule (perhaps one was originally attached to the hinge near the shackle, in some form or another) would have been placed so that its edge passed through the centre T and the asterisk (*) on the epicyclic disk. The true position of the planet in longitude would then be where the rule crossed the ecliptic scale of date. It is always possible that a thread was used, rather than a rule, but the presence of the hinge remains to be explained. The procedure is to be varied only in the case of Mercury. The centre of Mercury's deferent circle lies on an auxiliary circle drawn lightly on Fig. 2, but not on the equatorium itself. The actual position of the deferent centre is decided using scale (22) with the well known Ptolemaic principle (the radius of the auxiliary circle through the deferent centre makes the same angle with the aux line as the line from the equant to the epicycle, but the two are on opposite sides of the aux line). It is conceivable that the instrument was originally used with an attachment of the form shown in Fig. 4, serving both as a moving centre (C) and an index for scale (22).

Certain small problems remain concerning side (ii) of the equatorium. Each epicycle has a small hole drilled into it in the direction of the planet's position (*), but at a radial distance having no obvious significance. It is probable that these holes were to aid the user in rotating the disks, with the help of a small pin. Finally, there is the question as to why Mercury and Venus share an equant scale, when each of the other planets has its own. That this is, strictly speaking, geometrically impossible, even had Mercury a fixed centre deferent, should be evident from a study of Fig. 1. The eccentricity is all important in evaluating the position of the graduation θ as a function of θ itself. By suitably choosing the deferent radii (i.e. the lengths of the pivoted rules) for the two planets, however, the graduations on the equant scale may be theoretically made to suit the eccentricities of both planets *precisely* at the four quadrants; and when this is done, it may be proved that the scales will be correct at all points, to well within the limits of accuracy of the equatorium as a whole. (This is so because the eccentricities of Mercury and Venus are so little different.) This is the justification for the small economy in design.

The parameters of this side of the instrument, calculated after careful measurement, are summarized in the following table. The Earth-equant distance (*TE* in Fig. 5) is $2e$, (except for Mercury, when $TE = e$); and the radius of the epicycle is r, both e and r being expressed in the usual sexagesimal parts. Figures in brackets refer to parameters obtained from measurement of the equant scales. The probable error in measurement of all but the bracketed figures is of the order of $\pm$ 5′ arc at the most.

	$2e$	Ptolemy g.e.c. (see below)	r	r, Ptolemy
Saturn	6ᵖ 46′ (6ᵖ 46′)	6° 31′	6ᵖ 45′	6ᵖ 30′
Jupiter	5ᵖ 16′ (5ᵖ 36′)	5° 15′	11ᵖ 28′	11ᵖ 30′
Mars	11ᵖ 42′ (12ᵖ 49′)	11° 25′	39ᵖ 40′	39ᵖ 30′
Venus	2ᵖ 23′ (2ᵖ 25′)	2° 24′	43ᵖ 8′	43ᵖ 10′
Mercury	$e =$ 2ᵖ 31′ (2ᵖ 25′)	3° 1′	23ᵖ 24′	22ᵖ 30′

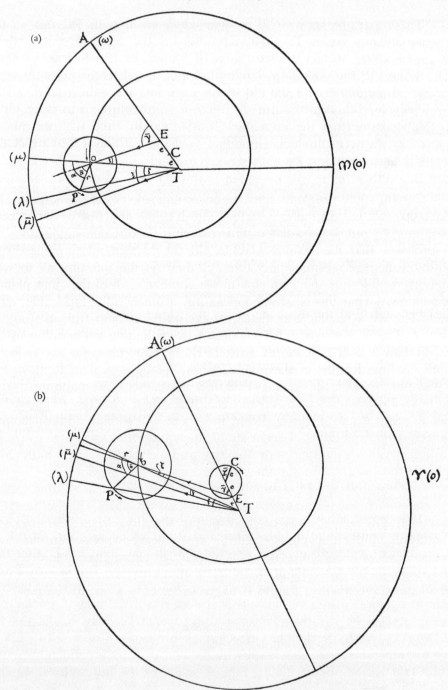

FIG. 5. — The Ptolemaic constructions for planetary longitude. Bracketed figures around the outer (ecliptic) circles are longitudes. Barred quantities represent mean motions. *A* is the aux, *E* the equant centre, *C* the centre of the deferent circle, *T* the Earth's centre, and *P* the planet. γ is the first point in Aries, from which longitudes are measured. $\bar{\alpha}$ and $\bar{\gamma}$ are respectively mean argument and mean centre, with ξ and ζ the so-called equations of centre and argument. (a) shows the construction for Venus, Mars, Jupiter, and Saturn, while (b) shows the construction for Mercury.

These parameters are all in good agreement with Ptolemy's, but it goes without saying that they are also in good agreement with those of almost all the standard astronomers of the middle ages and renaissance, who altered Ptolemy's parameters by amounts less then the probable error in our measurements. (They did so for no worse a reason than that Ptolemy was very close to the truth.) Copernicus, of course, is to be included in this remark. Once again, we are entitled to point out the extreme accuracy of the workmanship embodied in the equatorium. The small differences between r and Ptolemy's r, for example (supposing that Ptolemy was the author followed), reflect in part the errors of our measurement, and in part the errors of manufacture, and show that the latter are laudably small. Had I listed Ptolemy's eccentricities, the accuracy would have seemed appreciably less; but it is known that eccentricity was often regarded (then, as now, unfortunately) as identical with the greatest equation of centre (g.e.c.). The two differ by between 7' (Venus) and 28' (Mars). Comparing Ptolemy's greatest equations (the equation for an argument of 90°, in Mercury's case) with the first column, I think it is clear enough that the designer of the instrument used them, rather than strict eccentricities, in drilling the plate of centres. One might imagine that it is possible to decide on the authorship of the design by searching for a writer who made the greatest equation of centre of Mars more nearly equal to ours than Ptolemy's; and yet almost all writers after Ptolemy actually *reduced* the figure. (An exception was Lansberg, who made it 12° 9'; but there are many reasons why the instrument cannot be in a Lansbergian tradition.) [10].

THE THEORY AND USE OF THE EQUATORIUM. SIDE (i).

It is immediately apparent on examining the instrument that in a European context the double solar and lunar epicycles mark it out as a product of sixteenth or early seventeenth century thought. During that period there were far more variants of the relatively well-known Ptolemaic, Copernican, and Tychonic systems than is generally realized, and this fact makes it imperative that every aspect of the accepted solar and lunar models be carefully considered. It would be easy to dismiss the instrument, with its double epicycles, as obviously Copernican; but if the Copernican system was used, why the non-uniformly divided 'equant' scale (2)? By analogy with the equant scales of the last section, we find

---

[10] As it happens there was an equatorium designed according to Lansberg's astronomy, and this as late as 1646: Silvius Philomantius (= Bonaventura Cavalieri), *Trattato della ruota planetaria perpetua e dell'uso di quella, ecc.*, Bologna, 1646.

that scale (2) was divided for an eccentric equant point 4.1 mm from the centre of the uniform zodiac scale (1), suggesting an eccentric of about 0·0332 (or 2ᵖ 0′ to the nearest minute). The main epicycle centres are now pivoted, however, at constant distances from the *true* centre of the instrument, and not as before at constant distances from suitable eccentric centres. In Ptolemaic theory the Sun's eccentric equant is identical with

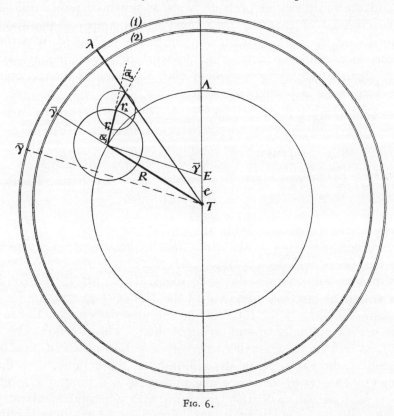

FIG. 6.

the centre deferent, and this theory is therefore — even overlooking the epicycles — ruled out. The same goes for Tycho's theory, for he required one lunar epicycle to be roughly twice the size of the other, whereas here the ratio is more like five to one. In Kepler's famous amendment (c. July 1600) of Tycho's solar theory, when — to use a well-known phrase — he « bisected the eccentricity », the equant and centre deferent were separated in the same way as in the Ptolemaic theory of the superior planets and Venus; which means that yet another possible source for the design is closed to us. Here we have simply a concentric deferent with an eccentric equant point, and for both Sun and Moon the theory on which the design is based appears to be as illustrated in Fig. 6.

The parameters of the equatorium, deduced from measurements made to an accuracy of a little less than a tenth of a millimetre, are summarized in the table below. These are quoted for a radius of 10^4 units simply because — following Copernicus — such was the most common convention of the time. The figures quoted have not been rounded off, but the radii are nevertheless susceptible to an error of the order of 15 units, and the eccentricities perhaps twice as much or more, quite apart from inaccuracies of manufacture. I may say at once that although I have not found the precise source for the theory embodied in the instrument, certain parameters taken from the work of Copernicus are so close to those deduced from measurement that once again the accuracy of manufacture must be remarked upon.

$(R = 10^4)$	r_1	r_2	e
Sun - instrument	379	110	332
Sun - *De Rev.*	369	48	—
Moon - instrument	1105	221	332
Moon - *De Rev.*	1097	237	—

Copernicus, when explaining the use of the double epicycle for the Sun and Moon, does not *at the same time* make use of an eccentric deferent. His ideas will be explained, nevertheless, since it is clear that the instrument has a close affinity to them. There are in the *De Revolutionibus* two geometrically equivalent theories of solar motion (*De Rev.*, III passim, and especially III, 20). In one description, the Earth (*T* in Fig. 6) moves annually round a large circle, whose centre, the mean Sun (*C*), moves round a small circle with a period of about 3434 years, the period of the variation in the obliquity of the ecliptic, according to Copernicus. The centre of this circle moves in turn around a somewhat larger circle (centred at *S*, the true Sun) with a period of about 53242 years, the regular component of the movement of the Earth's line of apsides, precession excluded. The relative sizes of the circles are given in the last table. Copernicus decided that in 1515 A.D. the distance *SC* was 323 units, *C* being only 14° 21′ along its circle from the nearest approach to S. The second scheme which he proposed while acknowledging that he was not able to pronounce on which was right[11], is geometrically equivalent to the first — ignoring stellar parallax — and is indicated by the broken-line diagram in Fig. 7. This is the system which makes the *true* Sun the ' centre ' of the Earth's motion, by the expedient of a double

---

[11] « Cumque tot modi ad eumdem numerum sese conferunt, quis locum habeat, haut facile dixerim, nisi quod illa numerorum ac apparentium perpetua consonantia credere cogit eorum esse aliquem », *De Rev.*, III, 20.

epicycle. It will be seen that the periods in the epicycles correspond to the very slow periods prescribed for the motions which previously together altered the position A' of the aphelion. Ignoring these motions entirely, it is easily seen that the two models reduce to a simple eccentric of the

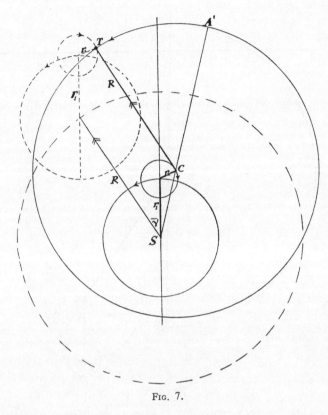

FIG. 7.

familiar Ptolemaic type, whether it be the Earth or the Sun which is taken as stationary, and that its eccentricity according to Copernicus was (for 1515 A.D.) 1^p 56′, compared with Ptolemy's 2^p 30′.

If we translate the centre of motion to the Earth (T), we find the equivalent scheme of Fig. 8. This, with its double epicycle, closely resembles that of the instrument (Fig. 6), the only obvious difference being that the one deferent is homocentric, the other eccentric. The diameter drawn through T is, be it noted, *not* the apse-line of date, but the line through the centre of the circle on which C moves, last reached by C, according to Copernicus, in 64 B.C. It is not even itself fixed in direction, for as explained, it revolves with a period of 53242 years (about 40′ 36″ per century). If, on the other hand, we know its position for a particular time, this position changes so slowly that the motion in the larger epicycle may be taken as equal to that in the deferent circle, namely the 'motus simplex' or mean motion of the Sun. What is more, the motion in the second epicycle on this theory is so very slow — about $10°\ 30'$ per century with respect to the radius of the first epicycle — that its position may be

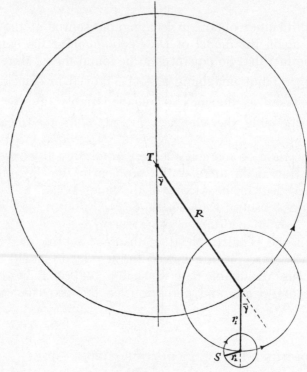

Fig. 8.

left unaltered, to all intents and purposes, for months at a time. Is this, then, the model on which the equatorium is based?

There is no doubt but that the equatorium *could* have been used in this 'Copernican' way, even though the effective radius of the second epicycle is apparently too large by a factor of two. (The pivoting 'error' involved is even so only about 1 mm, and I have certainly no wish to read the maker's mind on the strength of such a small discrepancy.) But why, then, the equant scale (2)? The eccentricity of this is such that it could have been used *alone* to give reasonable solar longitudes, just as the essentially similar calendar scale on many astrolabes is used for this purpose. However, the edge of the diametral rule on the equatorium is without doubt meant to pass through the point which is the 'nose' of the Sun-symbol on the second solar epicycle, touching only the *outer*

of the two zodiacs, whereas the epicycle index touches only the *inner* of
the two scales. I have not found a single author of the time advocating,
for the Sun, a double epicycle with eccentric equant, let alone an author
who also accepted Copernican parameters. There is little point in listing
negative evidence, but it is perhaps worth noting Tycho's use of a simple
eccentric.

Similar difficulties obtain in locating the source of the lunar equatorium. Here again the model of *De Revolutionibus* fits extremely well,
apart from the fact that no equant is to be found there. More specifically,
turning to Fig. 5, but imagining E to be moved to coincide with T, $\bar{\gamma}$
is now to be taken as the mean elongation (table (2)), while $\bar{\alpha}_2$ is the
supplement of double the elongation, and $\bar{\alpha}_1$ is the mean argument
(table (5)).

Again, the radii of the epicycles are admirably fitted for the Copernican model; but again there is the problem of the equant, for which
I know no precedent in the writing of the time. Tycho, to be sure, made
use of a moving eccentric with double epicycle, but in its details his model bears no resemblance to the equatorium.

Finally, if the peculiar blend of theoretical principles embodied in
our instrument is rare, we are indeed fortunate, for should they ever be
found in a treatise of suitable date we shall in all probability be very close
to the source of design, and perhaps even the provenance, of the instrument.

ON THE USE OF THE INSTRUMENT IN COMPUTING ECLIPSES.

From what has been said of the way in which the Moon's elongation
enters into the Copernican model it should be clear that during an eclipse
— which must be very near opposition or conjunction — all three radii,
R, r_1 and r_2 *will* be approximately in a straight line. (They will fail to
be collinear only because the elongation is measured from the mean rather
than the true Sun.) The only factor of importance in deciding the lunar
diameter (see table (10)) during an eclipse is therefore the mean argument,
which is either $0°$ or $180°$. The angular diameter of the Sun and of the
Earth's shadow at the lunar distance offer essentially the same problems
as in Ptolemaic astronomy. The calculation of them, and of their relative
velocities during conjunction and opposition, is explained in the final
chapters of *De Revolutionibus*, Book IV. Copernicus has virtually nothing to add to Ptolemaic eclipse techniques, although he paid more attention to the importance of parallax than medieval astronomers were
wont to do.

The equatorium makes no provision for the final calculation of the circumstances of an eclipse (its time, its duration, its magnitude and so on), but it does provide a means of obtaining all the necessary ancillary information for a graphical solution or a short calculation. The table (3) allows an estimation of all mean oppositions and conjunctions, and from a knowledge of eclipse limits and the use of the scales (3) for the lunar nodes, most of the oppositions and conjunctions may be rejected. Using the solar and lunar equatoria, the moment of true opposition and conjunction may be found, together with the angular distance of the eclipsed object from the node at the instant the eclipsing object passes the node. Parallax in latitude and longitude having been obtained, the usual calculations may be made. Copernicus is certainly not the best guide in this task, which he admits has been treated more fully by others. As he explains, he is in a hurry: « festinamus enim ad reliquorum quinque syderum revolutiones, quae in sequentibus dicentur » (*De Rev.*, IV, 32).

SOME FURTHER NON-PTOLEMAIC EQUATORIA AND PLANETARIA.

It was a common complaint of the sixteenth century that the Copernican system could not be easily visualised. The extent to which equatoria were made would clearly be dependent on the ease with which astronomers could calculate in the new system rather than the ease with which it could be visualised; but there is no doubt that many mechanical models were soon to be made to help lesser mortals to understand the new cosmology more clearly. The eighteenth-century examples of such models — ‘orreries’, as they are often called when driven by wheelwork — are widely known. By such makers as Stephen Hales, John Rowley, Thomas Wright, and Benjamin Cole, they were notable only for the skill with which they were made and the number of phenomena which they exhibited[12]. Within a decade of the publication of *De Revolutionibus*, Girolamo Fracastoro had made a model of the new system[13]. Such models were clearly expensive, and seem to have been more often the property of princes than of their subjects. Thus Riccioli in his *Almagestum Novum* mentions Giannello Torriano da Cremona, « commonly known as Iancellus Cremonensis », instrument maker to Charles V, as

---

[12] Orreries were so named by John Rowley after Charles Boyle, Earl of Orrery. A well-illustrated account of the English and American tradition of making such instruments, together with a useful bibliography, is to be found in H. C. Rice, *The Rittenhouse Orrery* (Princeton, 1954). This eighteenth-century tradition might be said to have begun with Stephen Hales, maker in about 1705 of a machine now lost, but sketched by William Stukeley in a work in the Bodleian Library (shelf mark: MS Gen. Top d. 14).

[13] See Fiorini, « Rivista geografica italiana », 1900, pp. 438-45.

making planetary automata, and he goes on to mention devices owned by Christian, King of Norway and Denmark, Maximilian, and Ferdinand[14]. It is unlikely that such automata were anything but Ptolemaic. J. L. E. Dreyer, in his biography of Tycho Brahe, refers to the fact that Raimarus (Nicholas Reymers Bär) sent the geo-heliocentric hypothesis — which Tycho accused him of plagiarizing — to the Landgrave of Hesse, who was so delighted with it that he had the instrument maker Bürgi make him a model[15]. The reference is to the fifth chapter of Reymers' *Fundamentum Astronomicum* (1588); and the very last diagram in this work, which as a matter of interest was that cited by Tycho as evidence of the plagiarism[16], actually shows a suggested arrangement for the wheels of a geo-heliocentric planetarium, dedicated to the Englishman John Dee[17].

In the seventeenth century the planetarium tradition was enlarged by the discovery of new phenomena. Thus Olaus Roemer, long after Galileo's discoveries, made an instrument (c. 1679) for the king of France, showing the movements of the satellites of Jupiter[18]. In mid-century A. M. Schyrleus de Rheita had described what he called his « trochlea », a planetary device with a rope-drive, incorporated in his « planetologium », partly driven by water using a mixture of gearwark, endless screws, rope and pulleys. This Heath Robinson affair incorporated among other things a revolving crystalline sphere covered in stars and decorated with gold and other colours, being half-filled with liquid to show horizon effects[19]. Schyrleus de Rheita based his work on both Tychonic and Copernican principles. Later in the century Christian Huygens designed a planetarium more notable for its ingenious construction than its showmanship[20]. Roemer made a complete planetary machine (c. 1697), borrowing, as it turns out, his wheel ratios from Huygens, and making some serious mistakes in the process[21].

---

[14] Riccioli, *Almagestum Novum*, 1651, p. 595. On the emperor's clocks see *Los relojes del Emperador* and *Los relojes de la exposicion « Carlos V y su ambiente »*, « Cuadernos de Relojeria », Madrid, 1959, no. 18.

[15] *Tycho Brahe*, 1890, p. 184.

[16] At the time the idea was stolen, according to Tycho, he had put the orbit of Mars outside the Earth, as Reymers did in the diagram.

[17] Dreyer's remarks about it make it clear that he cannot have seen the offending diagram.

[18] He sent a sketch of the device in a letter to Flamsteed (1679), and this is redrawn and printed with a description in William Derham's The *Artificial Clockmaker*, 4 th ed., 1759, pp. 130 sqq. This parallels an equatorium for the satellites, known as the « Giovilabio », for which see 'Galileo and the history of scientific instruments', forthcoming in *Saggi su Galileo Galilei*, Pisa, by F. R. Maddison.

[19] *Oculus Enoch et Eliae, sive Sidereomysticus*, Antwerp, 1645. For the trochlea see pp. 132-9, and for the planetarium, pp. 325-35.

[20] There is an English outline and illustration of this mechanism by G. H. Baillie, in « Clocks and Watches », 1951, pp. 129-30. The planetarium was made by Van Ceulen in 1682 and is now in the Ryksmuseum voor der Geschiedenis der Natuurwetenschappen in Leyden. A valuable account of this and other comparable instruments, although to be used with caution, is J. H. Van Swinden, *Beschrijving van het Rijks-Planetarium te Franeker, van 1773 tot 1780 uitgedacht en vervaardigr door Eise Eisinga*, Schoonhoven, 1851.

[21] See his *Basis astronomiae*, 1735.

Two much earlier mechanical devices, which deserve to be mentioned separately before we turn to the equatoria proper, are Jobst Bürgi's clock of 1591 and Edward Wright's « celestial automaton ». The clock by Bürgi, one of several produced by the man who was undoubtedly the finest maker in Europe in his time, is a table clock, now in the Kassel museum [22]. The lunar dial is, from our point of view, its most notable feature, for it incorporates primary and secondary epicycles. Stylistically, it may be added, the clock has nothing in common with the Liverpool equatorium.

Edward Wright's device was perhaps a similar example of a post-Ptolemaic planetarium, for it was driven apparently by clockwork [23]. It was first brought to light by E. G. R. Taylor, who unfortunately made several misleading remarks about it [24]. The treatise *The Description and Use of this Coelestiall Automaton more particularly* undoubtedly deals with an instrument which was indeed made, for, quite apart from its tone, there is a specific reference to the inscribed 'letter *E* in Cancer'. It is equally certainly not to be identified with the Liverpool equatorium. Both, however, are geocentric in construction, altough whereas our equatorium is in some respects 'Copernican', Wright's Automaton — so far as I can follow his description — seems to resemble basically nothing so much as the Tychonic device drawn by Reymers. There are, to be sure, primary and secondary epicycles on each planetary eccentric, and all were seemingly geared — suggesting a *tour de force* on the part of whoever made the automaton. The Moon's orb is so small that no involved machinery existed, apparently, to represent the complexities of its movement. Whether Wright meant his device to be Tychonic is not evident, for although he speaks of the Earth as centre of the world « according to these Hypotheses » (that is, those involved in the instrument), he cites a few parameters which are roughly Copernican, and refers the reader to « Astronomical tables made according to Copernicus hypotheses », going on to mention Reinholdt and Magini. We are not told what trains of gears were used, but they cannot have been very accurate for the text includes a table of annual adjustments to the angular displacements of the orbs. It also includes the positions of the planetary apogees which

---

[22] See, for a brief description, H. A. Lloyd, *Some Outstanding Clocks*, 1958, pp. 67-9.

[23] There is mention of « the Ballace, that continually swingeth to and fro on the backside of the automaton ». There was a handle enabling the device to be used for demonstration, if required.

[24] *The Mathematical Practitioners of Tudor and Stuart England*, 1954, p. 341. The shelf mark of the manuscript is not as given there, but rather (British Museum) MS. Sloane 651, item 4 (ff. 39v - 50v). Her first title is merely that added by a later librarian. The date « 1610 ? » should read « c. 1620 », the year for which the planetary apogees are given on the last verso. There is no evidence of which I am aware that « it was one of a number of instruments which Wright made for the instruction of Prince Henry, who died in 1612 », and it is impossible to equate the Automaton with Jonas Wright's Ptolemaic sphere. The work which Wright did for the prince is itemized in his own hand on f. 331r of (British Museum) MS. Cotton Titus B. 8, and there is no mention of the Automaton there. Had there been such a reference, the dates quoted would have presented something of a problem.

are very close to Copernican values, and hence are quite consistent with those on the Liverpool instrument.

Equatoria were less likely to attract the attention of princes and common men than were automata [25], and astronomers who made equatoria usually demanded something better than a crude approximative mechanism for the mean motions. Not all astronomers, of course, sympathized even with equatoria, as Rheticus's reference to the « art of threads » and the exasperated Kepler's invective against Apian and makers of automata « with 600 or even 1200 wheels » remind us [26]. But equatoria continued to be designed long after Kepler's time, and on principles other than Ptolemaic. Apart from the contemporaneous Liverpool equatorium, there are a number of comparable paper instruments, a few of which deserve to be mentioned. They range in value from that of the extremely crude Copernican volvelle printed at the end of *Astronomia Crystallina* (by J. H., London, 1670, in English), to that of the superbly executed Tychonic volvelles printed in Robert Dudley's *Dell'Arcano del Mare* of 1646 [27]. This was the golden age — if we except our own century — of the multiplication of cosmological species, and Philip Lansberg's was a system which obtained a certain success after tables (1632) based on it gave quite fortuitously a good account of the Venus transit of 1639. It comes as no surprise, therefore, to find a treatise « of the perpetual planetary wheels ... being how to find the places of the planets according to the manner of Lansberg », by « Silvius Philomantius », better known as Bonaventura Cavalieri [28]. Even Kepler's system attracted attention, but his ellipses saved him at least from the ignominy of having one instrument maker attempt a Keplerian equatorium. For in the words of Samuel Forster, « Kepler makes the Orbits of the Planets to be Ellipses, which

[25] Thus one wonders what Henry VIII would have made of Apian's book of 1540: « Petrus Apianus, " a man of great name in the arts mathematicals ", has shown Wotton that hearing of Henry's delight in liberal sciences, he meant to present a book of his own, named Astronomicum Caesareum, containing divers new things. He has printed it himself, as he does all his own books, and not above sixteen or seventeen copies, and, albeit it is dedicated to the Emperor and his brother, he would send it because otherwise Henry could not come by it... ». From *Letters and papers, foreign and domestic, of the reign of Henry VIII*, vol. XIX, pt. I, p. 423.

[26] « Iam quis mihi fontem porriget lacrymarum, quibus ex merito suo deplorem miserabilem APIANI industriam, qui in suo OPERE CAESAREO Ptolomaei fidem secutus tot bonas horas impendit, tot ingeniosissimas meditationes perdidit, ut spiris et corollis et helicibus et volutis et universo illo intricatissimorum flexuum labyrintho figmenta hominum exprimeret, quae natura rerum pro suis plane non agnoscit? Sed ostendit nobis vir ille, se divinis ingenii perspicacissimi dotibus facile naturae parem esse potuisse de caetero animum oblectavit suum praestigiis hisce (in quibus naturam ipsam provocaverat) fortissime superatis et in schemata conjectis, palamque inde famae perennis est adeptus, quicquid Operibus ipsis fortuna ista detrimenti attulerit. De Automato poeorum vero κενοτεχνία. quid dicemus, qui sexcentas imo milleducentas fabricant rotulas, ut de latitudinibus (hoc est de figmentis humanis) in Operibus suis expressis triumphare preciumque eorum intendere possent? ». Kepler, *Astronomia Nova, seu Physica Coelestis, tradita commentariis De Motibus Stellae Martis, etc...*, 1609, p. 82 (end of Pt. 2, ch. xiv).

[27] Published in Florence. The plates were engraved by F. A. Lucini.

[28] *Trattato della ruota planetaria perpetua e dell'uso di quella, ec.*, Bologna, 1646.

is the better way and I here doe make them perfect circles, which is the easier way » [29]. Curiously enough, Forster's instrument does incorporate elliptical curves for computing planetary latitude, and it is by no means as despicable as the apology just quoted might suggest; for Foster actually evaluates some of the errors implied by his design, which errors « may well be endured in these mannuary theories ». John Twysden, who in 1685 published this for Foster, wrote a book of his own, in the second part of which was yet another « planetary instrument », designed by the clergyman John Palmer and printed for Walter Hayes, who engraved the plates and made and sold the instrument in question [30]. (Anthony Thompson had made Foster's original instrument, judging by an advertisment in the *Miscellanies*.) It seems that in the latter half of the seventeenth century some of the best English instrument makers and designers were still unprepared to deem the equatorium passé. And here again, in Palmer's case, we find an attempt to reflect what he wrongly took to be the best modern theory: the aphelia and nodes were fixed 'according to Mr Street's hypothesis', although precession was allowed for.

So much for the printed sources of non-Ptolemaic equatoria. There must be many comprable examples, just as there are many manuscript instruments of the same period. But in the seventeenth century there is a steady decline in the ingenuity brought to bear on the many practical difficulties of design. The Liverpool equatorium leaves us with one or two unsolved problems, in particular as to its provenance; but nevertheless, as the finest surviving example of a type of instrument which was gradually passing into obscurity, it provides an invaluable point of contact with a tradition which, although reaching back to antiquity, belongs essentially to the middle ages.

---

[29] Samuel Foster, *Miscellanies, or Mathematical Lucubrations*, published posthumously ' by the care and industry ' of John Twysden, who added some things of his own. Printed by R. and W. Leybourn, 1659.

[30] *The use of the General Planisphere, called the Analemma...*, London, 1685. Another tract, *The Planetary Instrument, etc.*, to be sold with Hayes' production, was published earlier in 1672.

8

THE SATELLITES OF JUPITER, FROM GALILEO TO BRADLEY

The satellites of Jupiter are hardly to the forefront of the typical astronomer's consciousness, despite the remarkable findings of the Voyager mission. Even historians of astronomy tend to pass over the satellites in silence, perhaps thinking of them as nothing more than a trivial extension of the solar system—and paralleled by roughly comparable systems of moons around Mars, Saturn, Uranus, and Neptune. In the seventeenth century, however, they carried a cosmological message of great importance, for they were first seen at a time when the old and the new world systems were contending for the favor, not merely of astronomers, but of a significant fraction of the educated world. They were seen by Galileo in 1610. In 1676 Ole Rømer made use of them to show that light takes time to travel. My account runs for roughly half a century beyond this date, stopping at Bradley because, as I hope to explain, he marks the end of the first, largely empirical, phase of investigation of the satellites. I do not mean by this that no further empirical work was done—on the contrary, the most dedicated work of this sort was still in the future. Bradley's proof of the aberration of light nevertheless clinched the argument for the finite velocity of light, at least in the eyes of reasonable men. He was one of the first to allow for the velocity of light in tables of the four satellites then recognized, and he it was who first saw that the inequalities in their motions are interconnected—and thus possibly a consequence of gravitational interactions. In a sense, therefore, he opened the way to the theoretical studies of this problem by Euler, Bailly, Lagrange, Laplace, and others.

[AUTHOR'S NOTE: For the sake of those who are not familiar with the Jovian system, so far as it is known at present, it should be pointed out here that the satellites are numbered in the order of their discovery. They fall into three groups. The inner satellites (V,I,II,III,IV) revolve in much the same plane, with a direct motion. The next three (VI,VII,X) revolve in the same sense, but in three different planes; but then we have VIII, IX,XI, and XII, all in different planes but with retrograde motion. I am mostly concerned with I,II,III, and IV.]

THE DISCOVERY OF THE SATELLITES

The effect of the discovery of the satellites by Galileo on his own intellectual development was itself dramatic. The story is well known: by the beginning of 1610 he had made telescopes in the Dutch manner, the most powerful of them giving a magnification of about thirty. This he turned to the skies in January 1610, to discover that the Moon was clearly mountainous, that the Milky Way is a collection of individual stars, and that Jupiter is accompanied by lesser planets. On 7 January he saw three, but took them to be merely bright stars, close to Jupiter, and in a straight line.* The following day they were, to his surprise, so arranged that it seemed to him that Jupiter must have a *direct* motion, whereas all the standard ephemerides, the "computations of the astronomers," in Galileo's words, suggested a *retrograde* motion for the day in question. On 9 January the sky was cloudy, but on the 10 January one of the "stars" had disappeared while the other two again seemed to have changed sides. (Similar phenomena had shown themselves a few days earlier to Simon Mayr, if we are to believe his story. His findings however, which were not to be published until 1614,† had a much slighter cosmological impact than Galileo's.) Galileo now decided that Jupiter's motion could not be responsible for this strange behavior, and by 11 January he had decided that Jupiter must be accompanied by three wandering stars, comparable, as he says in the *Sidereus nuncius*, to Venus and Mercury who revolve around the Sun.‡ In this book, the *Sidereal Messenger*, he published his new telescopic discoveries. The book appeared at Venice in March 1610, and a day-by-day account of Galileo's observations of the satellites was there published, ending only ten days before the publication.§ The satellites proved to follow Jupiter in his retrograde and his direct motions, said Galileo, adding that they settled the doubts of those who, while accepting the Copernican system as a whole, were disturbed at the thought that our own Moon was unusual in revolving around the Earth, which in turn moved round the Sun. (One consequence of this was that the Moon indulged in a double motion of a sort thought contrary to the canons of Aristotelianism.) But our own eyes show us, he said, four stars which wander round Jupiter as does the Moon around the Earth,

---

*For the page of the ms. describing the following sequence of observations, see Galileo[1] [3 (2), p. 427, and 3 (1), pp. 35–37, 80–81], and Drake.[2]

†References to his works are given below. Mayr (Mayer, or Marius) is said to have made a partial announcement of his discovery in an astrological calendar of 1612, but an even earlier printed reference is in the introduction to Kepler.[3]

‡Galileo.[1]

§Galileo[1] [3 (1), p. 81].

the five bodies together circuiting the Sun in about twelve years. His parting words in the book amount to a hastily formed hypothesis to explain the appreciable fluctuation in the brightness of the satellites. This fluctuation he put down to the interposition of Jupiter's atmosphere.* This hypothesis, which seems to us so innocuous, must have been profoundly irritating to the Aristotelians, who would have seen here yet another threat to their system in the rival "elmental" sphere of Jupiter's "envelope of vapors."

The *Sidereus nuncius* was Galileo's first printed testimony to his belief in the Copernican hypothesis.† The first printing of 500 copies sold out immediately, and he achieved instant fame. A second edition was published in 1610 in Frankfurt. By the summer of 1610 he had resigned his chair at Padua and returned to Florence as mathematician and philosopher to the Grand Duke of Tuscany, in honor of whose family, the Medici, Galileo had, with great foresight, named the new satellites of Jupiter.‡ He was also made chief mathematician at the University of Pisa, without teaching duties. Galileo was a maestro at the art of orchestrating his own fame, but the excitement engendered by his discovery of the satellites was as spontaneous as such things can be. Within months, Kepler had published two short works on the subject, the *Dissertatio cum Nuncio Sidereo* (Prague, 1610),§ written before he had seen the new planets himself, and the *Narratio de observatis a se quatuor Jovis satellitibus* (Frankfurt, 1611), after he had seen them (between 30 August and 9 September). ‖ The second book, or rather pamphlet, was at once reprinted in Florence. It was, after all, by the Imperial Mathematician, and was useful ammunition in the battle for the respect of the Jesuits of the Collegio Romano, a battle which Galileo won on his visit to Rome early in 1611.¶

For two years, Galileo steered a reasonably safe theological course. His troubles began in 1613, with the publication of his letters on sunspots.** It would take me too

---

*Galileo[1] [3 (1), pp. 95–96].

†See, however, the earlier *Lettera a Jacopo Mazzani* (30 May 1597), Galileo[1] (2, pp. 197–202). This received a small circulation, hence my word "printed" rather than "published."

‡Following his example, Cassini named the satellites of Saturn the "Ludovician" planets, in honor of his own patron, *Le Roi Soleil*. The individual names still used for the first four of Jupiter's satellites were provided by Simon Mayr. See P. Humberd.[4] See also¶ note on p. 191 below, and Galileo[1] [3 (1), pp. 94, 136].

§Why, asked Kepler, should Jupiter be circled by four moons, if there is no one on the planet to admire the spectacle? He concluded that both our own moon and Jupiter may be inhabited, and that we may some day fly to them. See Kepler[5] (2, pp. 497, 502), and Kepler.[4] Antonio Maria Schyrlaeus de Rheita[7] indulged in a similar conjecture. If Jupiter has inhabitants, he said, they must be bigger and more handsome than we, and in the shape of a pair of spheres. He considered whether they might still be in a state of moral innocence. Huygens in his *Cosmotheoros*[8] describes Jupiter and his satellites by analogy with the Earth and Moon, in ways reminiscent of de Rheita. Galileo[1] (5, p. 220), on the other hand, violently rejected the notion that there are living creatures on the planets.

‖Kepler[5] (2 pp. 510–513).

¶Christoforo Clavius to Galileo in Florence, 17 December, 1610, Galileo[1] (10, pp. 484–485). Clavius says that the Medicean planets had been observed at Rome, congratulates Galileo, and provides some data. There are various letters concerning Galileo's triumphant visit to Rome in Galileo[1] (11, pp. 78–84).

**Galileo, 10 reprinted in Galileo[1] (5, pp. 71–245).

far afield to consider either this controversy, or Galileo's personal fortunes, but I must say more about his work on the satellites of Jupiter, if only as a corrective to the view that they represented to him no more than a qualitative cosmological spectacle, a Copernican system in miniature, as it were.

Already, whilst he was in Rome to show off his astronomical findings to the cardinals, to the Jesuits, and to the members of the Academia dei Lincei, Galileo wrote to Belisario Vinta that just as God had let him discover them, so He would let Galileo determine the laws of motion of the new planets, even though this required nightly observation.* His observations of them were indeed carried on for many years, but his tables of their motion, which were to have been published by a follower (the Olivetan and Professor at Pisa, Vincenzo Renieri), were lost to the sight of his contemporaries. Renieri actually completed the tables after Galileo's death in 1642, but was unable to publish them before he himself died in 1647, and not until the nineteenth century were the tables found—in a private library in Rome.† Accurate as even Galileo's early tables were, for their time, all were thus without influence, although many of Galileo's observations were used by later astronomers to assist in the derivation of the periodic times of the satellites.‡

Galileo's observations were made from the time of the discovery until the end of 1619—first at the Villa delle Selve, near Florence, then in Rome, and finally at the Villa Segni, in Bellosguardo, from whence he moved to Arcetri.§ He followed the movements of the Medicean planets, therefore, for somewhat less than one cycle of Jupiter.‖ His first technique of angular measurement is one which I have discussed previously, in connection with Thomas Harriot's work on sunspots.¶ It involved gauging angular distance in terms of the apparent field of view. Quite apart from the fact that it is unsuitable when the field is dark, it is unreliable to the extent that the field of view is a function of the size of the eye pupil. In 1612 Galileo made himself a crude micrometer.** A scale was placed *outside* the telescope, and was viewed with one eye, whilst the other was applied to the telescope. This crude device worked moderately well in expert hands.††

One of the principal attractions of the Jupiter system was that it presented a

*Galileo[1] (11, pp. 79–80).

[1]Galileo[1] [3 (2), pp. 403–424; 18, passim].

‡Some early data were published in the "Discourse Concerning Things Which Float on Water," for example; cf. Galileo.[1] Galileo communicated more accurate information than that in the *Sidereus nuncius* to G. B. Agucchi. See Galileo[1] (11, p. 205, 9 September 1611, pp. 214–215, 7 October, 1611, pp. 225–227, 29, October 1611).

§Galileo[1] [3 (2): *I Pianeti Medicei*]; S. Drake.[11]

‖One condition for satisfactory value for the periodic times is that averages are taken over long intervals between moments when Jupiter is at an identifiable point of its orbit.

¶(pp. 145–146); see above, chapter 6, pp. 109–44.

**Galileo[1] [3 (2), p. 446].

††See letters, 548 and 561 of G. L. Ramponi to Galileo in Galileo[1] (11, pp. 133–136 and 159–162) where such measurements are discussed. R. M. McKeon[13] argues that such a scale was used by Galileo. See also A. Borelli.[14]

challenge to astronomers. Here was a restricted system with fast-moving planets for which the earlier history of astronomy might be regarded as a sort of rehearsal. Over the next two centuries the truth was slowly appreciated: the satellites presented all the problems of conventional planetary astronomy, and some others besides. I will consider the complications as they were appreciated historically, beginning with the most obvious, namely, that we observe the system from a moving Earth.

Making the approximation that the orbits of the Earth, Jupiter, and those of his satellites that were then known all lie in the same plane, a reasonable idea of what can be expected of a single satellite can be had from a drawing in correct proportion; and if you graduate this with scales (of angle or of time) of the sort used on medieval planetary equatoria, you will have a useful nomogram for evaluating actual and expected phenomena. Galileo designed such a device, called in Italian "giovilabio." More precisely, I should say that there are several related nomograms in the mss., reproduced in facsimile in the National Edition of Galileo's writings.* The principles implicit in the nomograms were incorporated into a brass *giovilabio* which is now in the Museum of the History of Science at Florence, and which has been described by Maria Luisa Righini Bonelli.† It is certainly Galilean in general conception, although it is probably by a late contemporary. I shall have more to say of its design and its engraved tables later.

THE SATELLITES AND THE LONGITUDE PROBLEM

The brass *giovilabio* was meant not only to allow the expert astronomer to arrive rapidly at the probable disposition of the satellites, but to assist the mariner who wished to determine longitude at sea. This was a particularly vital subject for most of those who investigated the satellite orbits during the seventeenth and early eighteenth centuries. There is even reason for supposing that Galileo was driven to develop the pendulum-controlled clock by his ambition to solve the longitude problem through the use of Jupiter's satellites.‡ This method was first mentioned in a letter dated 11 November 1635, from Galileo to Jean de Beaugrand.§ Nine months later Galileo outlined the method, and offered it *gratis* to the States General of the United Provinces of the Netherlands.‖ The States General had followed the example of Philip III of Spain, who in 1598 offered large sums of money to the first who should solve the problem of longitude. Huygens speaks of the "long promised reward," but the details

---

*Galileo[1] [3 (2), pp. 486–487; cf. partial diagrams at pp. 477, 479, 481, 483].
†Bonelli.[15]
‡The first use of the pendulum for timekeeping was apparently a medical one, namely, for taking the patient's pulse (published in 1602 by Sanctorius, generally attributed to Galileo). See especially Ariotti.[16]
§Galileo[1] (16, p. 344).
‖Galileo[1] (16, pp. 463–468, letter 3337).

are obscure. The governments of Venice (?),* Britain (1714), and France (1716) all offered similar rewards, although the British were the only government ever to pay, when a satisfactory solution was forthcoming.†

By the mid-eighteenth century Jupiter's satellites were out of favor for this purpose, chiefly because it was so difficult to make precise telescopic observations from a rolling and pitching ship. Throughout the previous century, however, astronomers had lived in the hope that they would solve the practical problems entailed, especially that of making a compact and powerful enough telescope. In a letter to the States General, dated 13 August 1636,‡ Galileo had given a masterly summary of the problem as a whole, together with an indication that he knew very little about the practical realities of seafaring. His clock was for keeping time only from midday to the sightings—but of course it would not have been accurate without a stable platform, any more than could the mariner have viewed the planets without some leveling device. (Remember that good telescopes were around two meters or more in length!) One could not risk losing sight of Jupiter for any length of time, since eclipses of the satellites lasted for only a minute or so. It was these eclipses, of course, which were to be used as the phenomena to fix a universal time, so to speak, and it was the times of their occurrences that Galileo now felt he was in a position to tabulate with reasonable accuracy.

The Dutch admiral Laurens Reael showed interest, but asked about the stabilizing of the telescope.§ In March 1637 Galileo begged Elia Diodati to assure Constantijn Huygens that he had a solution to this problem, and he himself sent it to Reael in June.‖ It involved a massive universal joint, with one hemispherical component moving inside a second, fixed to the ship. The two were to be separated by water or oil, the gap between the hemispheres being preserved by some eight or ten springs.¶ The arrangement simply could not have been made to work effectively. Perhaps I should add here that, despite numerous attempts to solve this technolgically difficult problem, a solution became feasible only in our own century, with the possibility of large, high-speed, gyroscopes.** In a final letter to Galileo, Reael tells him that his method is implausible, and impracticable for his sailors, "who are rude people, men only superficially acquainted with mathematics and astronomy . . . and who still find insuperable the problem of using your discovery on a moving ship, continually tossed about."†† For his work, the States General nevertheless offered Galileo the gift of a gold necklace, but he declined the offer.‡‡ According to Gould, the method of finding longitude

---

*I have included Venice, following Gould,[17] but have so far been unable to confirm the claim elsewhere.
†Gould.[17] The British Board of Longitude disbursed more than £100,000 before 1828, when the Board was finally abolished.
‡Galileo[1] (16, pp. 464–467).
§Galileo[1] (17, pp. 39–41).
‖Galileo[1] (17, pp. 46–49, 96–105).
¶Ariotti[16] (p. 368); Galileo[1] (16, pp. 96–99).
**Gould[17] (p. 7) lists eight later attempts to design a stabilized observing point, between 1719 and 1858.
††Galileo[1] (17, pp. 116–117, 22 June 1647); Ariotti[16] (p. 369).
‡‡Galileo[1] (17, p. 371).

from observations of Jupiter's satellites has never been successfully applied at sea.*
This should not obscure the fact, however, that the method was frequently used in
the seventeenth and eighteenth centuries on *land*; nor should it obscure the fact that
the intense study of the problem of Jupiter's satellites had motives which were largely
utilitarian, that is, had to do with the longitude problem. For this, moreover, they
were by many thought to offer the *best* solution.†

PRIORITY DISPUTES

As Galileo and others found, the motions of the satellites were more complex
than had at first been suspected. Galileo's figures for the mean motions are much more
accurate than is generally recognized, although it is a matter of taste, whether or not
they justify the energy he threw into the project. Simon Mayr's tables of the satellite
motions have occasionally been claimed as more accurate, which as a whole they were
not, although it would be quite wrong to go to the extremes of Robert Grant, who
wrote that Mayr

> ... who contended for the independent discovery of the satellites, resolved to
> strengthen his claims by the construction of tables of their motions. The crude
> labours of this impudent pretender were, however, no sooner given to the world
> than they fell into deserved oblivion.‡

Grant was here falling into line with propaganda of Galileo's making. Galileo
and Mayr were old antagonists, but not until 1623, in *Il Saggiatore*, did Galileo go
into print maintaining that Mayr's *Mundus Jovialis* of 1614§ was a complete
plagiarism.‖ Galileo was of course stung by Mayr's claim to priority, but Mayr had
also tried to name the satellites the "Brandenburg stars,"¶ after his own patrons, thus
jeopardizing Galileo's standing vis-à-vis the Medici. On Mayr's side, one might point
out that he claimed to have begun to observe the satellites one day after Galileo,**
and that the *tables* in his book antedate anything comparable by Galileo; they incorporated more accurate motions than Galileo's first published motions, as will be seen
from my table of comparison. I shall raise other points of comparison as the opportunity arises.

---

*Gould[17] (p. 7).
†For this note see the account in Halley[18] of his reduction of Cassini's tables, discussed again here below.
‡Grant[19] (p. 79).
§Mayr.[20]
‖Galileo[1] (6, pp. 197–372).
¶He also called them for the first time by their customary names, Io (I), Europea (II), Ganymede (III),
and Callisto (IV), as well as "the Mercury," "the Venus," "the Jupiter," and "the Saturn" of Jupiter.
He acknowledges that Kepler first suggested the now customary names to him. Hevelius used the names
"Saturn" etc.
**Mayr[20]; see the *Observatory* reference at p. 372.

The Galileo/Mayr controversy is one capable of arousing strong emotions, or so it seems, for a prize essay on the subject by Josef Klug (the only entrant), written at the end of the last century,* was so hostile to Mayr that one of the judges appointed by the Dutch Academy of Sciences, namely J. A. C. Oudemans, joined with Johannes Bosscha to defend Mayr.† The controversy continued for some time, and more recent reference to it may be found in a summary article by J. H. Johnson, and an excellent article, introducing a number of contemporary documents for the first time, by Pietro Pagnini.‡

The controversy raises three separate questions: Was Mayr capable of observing Jupiter's satellites early in December 1609(O.S.), the date at which he claims to have seen three of them? Second, did he draw up his tables from Galileo's published data? Third, did he falsify the dates assigned to his own observations, assuming that we suppose them to be genuine? Priority disputes may be tedious, but the noise created by them is a part of history. Unfortunately, this example cannot be settled except in terms of the characters of the disputants and the potential interdependence of Mayr's data and Galileo's relevant publications.§ Klug argued that Mayr could have relied on measurements taken from the plates of the third *Solar letter* of 1613, but Bosscha showed that had he done so he would have found results much less accurate than those he published.‖ It appears that the chief inaccuracy in Mayr's tables was due to badly chosen *radices* for the motions. His estimate of the diameters of the satellites, of the radii of their orbits, and of the mean motions, are good and on the whole better than any Galileo had then made public. (A cut-and-dried comparison is impossible, since Galileo changed his parameters from time to time, and there is little to be gained by chronicling the changes here.)

Mayr's book does not read like an exercise in studied deception, and a reasonable conclusion might be that he it was who first saw three of the satellites, and who first realized that they were moving, but that Galileo first saw four satellites, and did so independently. That Mayr is telling an honest tale in the *Mundus Jovialis* seems all the more probable since a letter of his, reproduced in the preface of Kepler's *Dioptrice* of 1611, tells, in outline, much the same story.¶ (Why did this not irritate Galileo? Did he not see it until after 1614?) At this level of naive discovery, their findings were virtually simultaneous. What is not at all obvious—and here I follow Pagnini**—is that Mayr was able to *explain* the observed phenomena before he had read the *Sidereus nuncius* in 1610. Mayr probably reached a full understanding of what he saw

*Klug.[21]
†Oudemans and Bosscha[22] and Bosscha.[23]
‡Johnson[24] and Pagnini.[25]
§These are: *Sidereus nuncius* (1610), *Discorso interno alle cose che stanno in su l'acqua . . .* (1612), the plates of the third *solar letter* (1613)—see Galileo[1] (5, pp. 241–245). Of less significance is the earlier correspondence between Agucchi and Galileo, for which see Klug.[21]
‖See Oudemans and Bosscha[22] and Bosscha.[23]
¶Kepler[3] (pp. 27–28).
**The crux of the argument is the absence of Mayr's name in certain key letters dating from 1610.

only towards the end of 1610. This is compatible with his good faith as regards the wording of the preface to the *Mundus Jovialis*, but it means that he is to be placed at a lower level of originality than Galileo, however accurate an observer he might have been.

In addition to Galileo's and Mayr's, another important series of observations of Jupiter's satellites was Thomas Harriot's. Harriot was in possession of a copy of the *Sidereus nuncius* by June 1610, and he began his systematic observation of the satellites on 17 October 1610(O.S.).* At first he could see only one, and not until December 14(O.S.) could he see all four of the satellites known to Galileo.† It is necessary here to remind ourselves that in 1610 there were telescopes and telescopes. William Lower observed with a friend, and could see not one of the satellites, as he tells Harriot in a letter of 4 March 1611.‡ Peiresc similarly found it impossible to obtain a telescope equal to the task before November 1610.§ Harriot has left us records of his own observations of the satellites made between October 1610 and February 1612, and calculations of their motions. Dr. John Roche has made an analysis of these unpublished ms. records. He has found that in 1611‖ Harriot made numerous calculations of the periods of the satellites, sometimes using only Galileo's published data, sometimes only his own, and sometimes comparing his own with Galileo's data. Making such a comparison he found the sidereal period of the first satellite to be 42.4353^h (correct value: 42.4582^h). In preferring to establish sidereal rather than syndoic periods, Harriot was wiser than Galileo and Mayr. For some reason he failed to use the same methods with the other satellites, and indeed did not persist in his satellite observations after February 1612—perhaps because he was disappointed with his observing conditions.¶ Much of what Rigaud has to say about Harriot's Jupiter observations is unfortunately incorrect.** He was wrong, for example, to say that Harriot gave no indication that the satellites could be off the ecliptic.†† Harriot called attention to the fact on several occasions,‡‡ although he made no attempt to lay down a theory of latitude.

There is ample testimony to the excitement caused by Galileo's announcement of his discovery of the moons of Jupiter,§§ and it is hardly surprising that others—including Mayr, Harriot, the Jesuits at the Collegio Romano,‖‖ and Agucchi—imme-

---

*Harriot[26] (ff. 1–14); article by J. D. North in Shirley[12] (pp. 136–137). On 1 June 1610 (O.S.) Jupiter was in conjunction with the Sun, and the satellites would have been difficult to see then and for several weeks thereafter. See above, pp. 118-19.

†Harriot[26] (f. 2).

‡Halliwell.[27]

§Rigaud.[28] See the "Supplement . . . with an Account of Harriot's astronomical Papers," p. 28.

‖Roche.[29]

¶Roche.[29]

** See above, pp. 129-30.

††Rigaud[28] (p. 25, supplement).

‡‡For example Harriot[26] (f. 4).

§§See, for example, Kepler[6] (pp. 9–10).

‖‖Galileo[1] [3(2), pp. 861–864]; Christoph Scheiner;[30] de Rheita[7] (p. 282); P. Gassendi;[31] Delambre;[32] Riccioli;[33] Hérigone (Cyriaque de Mangin);[34] Jacques Godefroy, see Brown;[35] Fontana;[36] for Hevelius see Delambre[32] (2, p. 437) and Hevelius.[37]

diately followed him in recording similar observations. I could refer also to the announcement of observations by such men as Giuseppe Bianchini, Christoph Scheiner, Schyrlaeus de Rheita, Pierre Gassendi, Pierre Hérigone, Jacques Godefroy, or Francesco Fontana, but this would only be to pile the obvious on the obvious. It seems to me that the study of the satellites took a more serious turn with the writings of Gioanbatista Odierna* and Giovanni Alfonso Borelli.

ODIERNA'S THREE INEQUALITIES AND BORELLI'S CELESTIAL MECHANICS

Odierna's *Medicaeorum ephemerides* ... was published at Palermo in his native Sicily in 1656. It was dedicated to the Grand Duke Ferdinand II of Tuscany, and indeed Odierna makes an abortive attempt to rename the satellites after members of the Medici house—Ferdnipharus (IV), Cosmipharus (III, the brightest), Victripharus (II, after Victoria, wife of Ferdinand), and Principharus (I, after the heir-apparent). "Florence" was to be the face of Jupiter, and "the Arno" the bands across the face.† What makes Odierna's work noteworthy, however, was neither such window-dressing as this, nor his data for the mean motions of the satellites,‡ but rather the introduction of new parameters into the theory of the satellite motions. He enlarged upon a theme first published by Mayr, namely, that of motion in latitude. Galileo had expressed the opinion (in his polemic against Mayr in the introduction to *Il Saggiatore*)§ that the satellites are always in a plane parallel to the ecliptic. When Jupiter is in the ecliptic they will therefore appear to us to be in a straight line, on this view; and when, for example, *Jupiter is north of the ecliptic*, the satellites will seem to be *north* of the main planet when in the *inferior* position. They will be *south* when in the *superior*. When Jupiter has a southerly latitude, the appearances will be reversed. Mayr took a different view, namely, that the satellites were in orbits permanently inclined to the ecliptic. He gave tables for the satellites' latitudes based on estimated inclinations of the orbits (10, 10, 12, and 15 seconds of arc for I–IV, respectively), but in his explanation of the tables failed to take into account the movement of the parent planet in latitude. Neither Galileo nor Mayr therefore gave an adequate account of movements in latitude. (And for the record, I should here note that Galileo's attempt to discredit Mayr fails when he argues that he could not have seen the movements in latitude he claimed to have seen in the first half of 1611.) As a matter of fact, the satellites are in orbits virtually coincident with Jupiter's equator, and the latter is inclined at about $3°04'$ to Jupiter's orbit. Unless we know the *direction* of Jupiter's axis, or alternatively, the nodes of the orbits, we cannot make any progress with the determination of lati-

---

*"Hodierna" in most early printed sources.

†He used alternative names for the satellites: Alphipharus (I), Betipharus (II), Cappipharus (III), and Deltipharus (IV). These are a Graeco-Latin alphabetical compromise.

‡His figures were inferior to those of the best of his predecessors.

§Galileo.[38]

tudinal movements. Odierna was a careful observer, capable of dismissing on empirical grounds the opinions of Galileo and Mayr, if not of reaching a completely satisfactory theoretical interpretation himself. Against Galileo he noted that conjunctions may take place more than half a diameter from Jupiter's center. Against Mayr and Galileo he found from observation that the satellites are *always north* of Jupiter when *superior*, and *south* when *inferior*.

Odierna set forth three sorts of periodic inequality. The first was a consequence of changes in satellite latitude. Delambre seems to believe that he began by taking the plane of the satellites to be inclined at 45° to that of Jupiter's orbit, and that he then expressed the maximum latitudes in tenths of a disk and found for the four satellites 1°59′, 3°7′, 5°6′, 8°29′—which he later found inexact.* What Delambre does not seem to have realized is that these figures fit together with Odierna's maximum distances for the satellites in such a way as to suggest that he believed them all to lie in a plane inclined (on my calculation) at about 3°21′ to the ecliptic. This is by no means a despicable result.

Odierna's second inequality, according to Delambre, was the annual parallax, and the third was due to Jupiter's movement round the Sun.† From Delambre's account it is hard to decide exactly how these were calculated, but it is doubtful whether together they gave more than that correction which is a function of the Jupiter–Sun–Earth angle, and which will be discussed later. Odierna was perhaps the first to note that the outermost of the four satellites then known is (unlike I–III) not invariably eclipsed at superior conjunction with the shadow cone.‡ He was also aware that with I, and usually also with II, only immersion or emersion is observed. (The unobserved phenomenon is hidden by Jupiter.)§ He listed past eclipses, some of which were spurious, but whether because they were imagined, whilst he observed under poor conditions, or whether because erroneously retrodicted, it is impossible to say. Odierna is the first to stress the importance of eclipse observations, and his tables bring eclipse calculations to the fore. He failed to account accurately for eclipse durations, but those who know the corresponding difficulties in connection with eclipses of our own Moon will forgive him his shortcomings. His ephemerides of the satellites, running from 1656 to 1676, are not as accurate or as thorough as the few who possessed telescopes might have liked, but they did help to prepare the ground for later ephemerides.

The ephemerides of 1656, with a memorable terminal date of 1676, were just half-expended, so to speak, when Borelli published his work *Theoricae mediceorum planetarum ex causis physicis deductae* (Florence, 1666). Borelli's strategy was to describe the inequalitites in the motion of our own Moon and then to transfer them to the

---

*Delambre[32] (2, p. 329). The original is rare, and I have not seen a copy.
†Delambre[32] (2, p. 329).
‡We know that the eclipse fails to occur when the orbital plane (of IV) is inclined at more than about 2° to the Sun–Jupiter line.
§For III and IV, both phases are usually visible. In general, eclipses are seen on the western side of Jupiter before opposition and on eastern side after opposition. Near opposition the shadows may be behind the disk, as seen from the Earth, so that eclipses are invisible.

theory of the motion of Jupiter's satellites. His book is perhaps best known for its celestial mechanics, that is, for its explanation of the elliptical orbits of planetary bodies* in terms of three types of forces: (1) a central force; (2) rays of force acting like spokes from the central body, i.e., rays which, with the rotation of that central body on its axis, move the planet round; (3) a centrifugal force. The satellites of Jupiter were supposed to be simultaneously acted upon by Jupiter's rays [as under (2)], and the rays of the Sun, and these influences were thought to combine to give the satellites their complex motions.† Borelli's planetary mechanics is not a subject for a brief aside, and I shall leave it well and truly alone, merely pointing out that it follows upon a general description of the satellite motions.‡ In his preface, Borelli tells the reader the circumstances under which the book was written. Campani sent a fine telescope to the Grand Duke, who then commanded Borelli to use it to study Saturn and Jupiter and to investigate the accuracy of Galileo's tables of the satellites' motions. Accordingly, in the summer of 1665 Borelli began his work, in an observatory in the castle of San Miniato, near Florence.

Borelli was one of those who finds it easier to lay down the lines of a bold program than to fill in the details. Working by analogy with Keplerian cosmology, he postulated ellipitical orbits for the satellites, with moving lines of apsides, inclined orbits with moving nodes (retrograde, of course), evection and variation—all as in the case of our own Moon. There were the three inequalities reckoned by Odierna, and in principle many others besides. But when it came to the extraction of new empirical parameters, all of Borelli's grandiose theory came to nothing. He made age and ill-health his excuse, but it is quite clear that the analysis was as far beyond his talents as were the necessary observations beyond the reach of seventeenth-century instruments.

It is difficult to decide how important were good telescopes to progress in this part of astronomy. If we consider the satellite observations of Hevelius, as recorded in *Selenographia* (Danzig, 1647),§ we find that the derived motions were decidedly inferior to those by Galileo, Mayr, and Harriot well over thirty years previously. Hevelius complains of his telescopes, but it is hard to believe that they were no better than those of the 1610 era. Nevertheless, whilst Hevelius could see the great spot on Jupiter, he could not see the bands which Odierna had named "the Arno." Hevelius gave almost the same figures for the maximum elongations of the satellites from Jupiter as his predecessors had done. All were estimates made in terms of the diameter of Jupiter's visible disk, usually loosely set at one minute of arc. Needless to say, the crudity of such an estimate made nonsense of Borelli's ambitious scheme, although it must be granted that he had recognized the virtue of a form of micrometer (here essentially a

---

*Delambre[32] (p. 333) says incorrectly that Borelli indicates no physical cause for the motions of the satellites.

†Borelli[14] (pp. 45–81).

‡Koyré[39] and other works cited there.

§See §§ note, p. 193 above.

scale in the focal plane of the eyepiece), which he says was communicated to him by Buono.*

GIAN DOMENICO CASSINI

If one may talk of a "heroic" period in the history of Jupiter's satellites, it must be taken to begin in or around 1664, when Gian Domenico Cassini,† still in Italy, obtained fine telescopes of long focal length from his acquaintances in Rome, the great telescope makers Giuseppe Campani and Eustachio Divini. I shall consider Cassini's observations as a natural prelude to Rømer's, and those which followed in their wake in England.

Within weeks of obtaining the new telescopes, and whilst using one of Campani's with a length of about 5 meters, Cassini was observing the shadows of satellites II and III on Jupiter's disk when he noticed an uncharted spot. Some days later he saw two or three movable dark spots which he assumed to be clouds, and bright marks, which he took to be volcanoes. He explained what he saw, when in 1665 he published *Tabulae quotidianae revolutionis macularum Jovis . . .* ‡ He was a lover of analogy, and his analogies with known phenomena did not stop with clouds and volcanoes, but extended to the theory of the satellite motions. He observed a notably large spot on the face of the planet, and was at length able to use it to settle the planet's period of rotation at 9^h56^m—say 10 hours.§ Now the Sun rotates once in about 28 days, and it provides Mercury—assuming the solar-ray theory of celestial motion—with a period of revolution of about 88 days. Should the innermost satellite not therefore have a periodic time of about 31.4 hours? In fact satellite I has a period of about 42.5 hours, a third as much again. (Cassini knew nothing, of course, of satellite V, the nearest now known to Jupiter, with a period of nearly 12 hours.) Cassini, who is here following Kepler's own reasoning, but who lacked Kepler's analytical skills, decided that Jupiter was less efficacious than the Sun, each judged vis-à-vis his planets. As others before him had done, Cassini alluded to the possible habitation of Jupiter.

The tables of Jupiter's rotation were published on 12 June 1665. On 22 July of the same year there appeared from Cassini's pen *Lettere astronomiche al signor ab. Falconieri spora l'ombre de pianetini di Giove*. This work reveals that Cassini, through a study especially of shadow transits, was aware of inequalities of only a few minutes of time. He studied many shadow phenomena, and made a restricted attempt to correlate them with variations in Jupiter's ecliptic latitude. He also made a half-

---

*Hevelius[37] (p. 145).
†Cassini[40] and Taton.[41]
‡Cassini.[42]
§A modern figure for the rotation period is $9^h 50^m 30^s$ (equatorial), increasing to the figure quoted by Cassini, in the temperate zones.

hearted attempt to find algebraic relationships between central distances, periodic times, and transmit times, for the satellites.

Dissatisfied with existing tables of the motions of the satellites, Cassini observed them carefully for more than two years, before publishing his *Ephemerides Bononienses mediceorum syderum* ... in 1668. Apart from further descriptive material concerning Jupiter's disk, the work was not unusual in character, and yet the sheer accuracy of predictions made from the tables was impressive, to contemporary eyes. Picard, in particular, lavished praise on them, and almost simultaneously with their publication Cassini received Colbert's invitation to come to the then recently founded Académie Royale des Sciences, in Paris. Cassini's reputation throughout Europe was considerable, and although in retrospect he appears to have had a knack of making wrong conjectures, his observational work was already both extensive and accurate.

Cassini regarded his visit to Paris as a temporary affair, but in fact he was never to return permanently to Italy, so generous were the terms of his stay. His work on the satellites of Jupiter continued, although naturally enough he had—as always—many other irons in the fire. He apologized later for the inaccuracies in his 1668 tables, due, he maintained, to the haste with which he was pressed to publish them. The apology is in the introduction to *Les Hypothèses et les tables des satellites de Jupiter*, published in Paris in 1693, the culmination of many years of careful work. In the same year, Cassini published his *De l'Origine et du progrès de l'astronomie*, in which some pertinent remarks are made about the satellite problem. Seventeen years had passed since Rømer's greatest achievement, and still Cassini was reluctant to accept the idea that light travels with a finite velocity. Edmond Halley, who reduced Cassini's new tables to the meridian of London and the Julian calendar,[*] could not understand Cassini's reluctance to utilize the "equation of light."[†] If we look at *De l'Origine*, however, we find evidence that Cassini had himself considered the possibility of such an equation, and had rejected the idea. This evidence supplements a number of indirect statements by Montucla,[43] Du Hamel,[44] the *Mémoires* of the Academy,[45] and others, quoted by I. B. Cohen in his fine memoir on Rømer.[‡] Here, writing in 1693, Cassini gives us the reasons for his decision, reasons which we shall find far from cogent. The passage is interesting enough to be quoted at length:[§]

> The observations which the Academy made of the satellites of Jupiter provided an occasion to examine one of the most beautiful problems of physics, which is, to know whether the movement of light is successive, or whether it is effected in an instant. The times of two successive emersions of the first satellite during one of Jupiter's quadratures were compared with the times of two successive immer-

[*]Halley.[18]

[†]Halley[18] (p. 239).

[‡]Cohen.[46] Note a small error in regard to the date given in the *Histoire*[45] for Cassini's publication of his conjecture, August 1675, not 1674.

[§]Cassini.[47] The methods which Galileo proposed, and which are called "useless," involve signaling with widely separated lanterns. See Galileo[1] (8, p. 88).

sions of the same satellite during the opposite quadrature; and although the light of the satellite during the first quadrature covers less distance to reach the Earth when it approaches Jupiter than in the second quadrature when Jupiter is separating from the Earth, and although this difference amounts to 60,000 leagues at the very least as between the two occasions, nevertheless, scarcely any appreciable difference was found between the two intervals of time. This gave occasion to believe that the observations one can make at the surface of the Earth, or even throughout space as far as the Moon, do not suffice to form a definite conclusion on the problem, and that, as a result, the methods which Galileo proposed for this effect in his mechanics, are useless. It is not that the Academy, in the series of observations, failed to notice that the time of a considerable number of immersions of the same satellite is appreciably shorter than a similar number of emersions, which may be explained by the hypothesis of the successive movement of light; but that this did not seem enough to convince that the movement is really successive in fact, for one is not certain that this inequality of times may not be produced either by the eccentricity of the satellite, or by the irregularity of its movement, or by some other cause at present unknown, which in time one might be able to clarify.

Delambre was perhaps the first to draw attention to the fallacy in Cassini's procedure.* I will rework the problem and assume the modern value of c, the velocity of light. We may reasonably take the Earth's approach at quadrature to be directly towards or away from Jupiter throughout the interval of time between successive observations of the satellite's reaching a particular recognizable position (e.g., maximum elongation). The technique is to measure this interval when the Earth is at opposite quadratures. If T is the true period of satellite I, and t the time for light to travel between the two points on the Earth's orbit from which the satellite is successfully observed, then the periods measured at opposite quadratures will be $T - t$ and $T + t$. The difference, $2t$, was what Cassini wished to establish. To a reasonable approximation, taking the Earth to move uniformly round a circular orbit, $2t = 4\pi aT/cY$, where a is the radius of the orbit and Y is the length of the year. We find that $2t$ is a little more than 30 sec, when $T = 42.5$ hr. For the record, the corresponding times for the other three known satellites are 61 (II), 123 (III), and 287 (IV) sec (all to the nearest second, in keeping with the above formula, although the implied accuracy is somewhat spurious, in fact, since the approximations made are invalid for large intervals).† Had he observed the outer satellites, or taken, as did Rømer, the time for a large number of revolutions of the satellites, he would have found the appearances much harder to explain. To be fair to Cassini, he did believe that Rømer's hypothesis was inconsistent with what was known of the outer satellites, and this was ostensibly the reason why, after Rømer had proved his point to the satisfaction of most of the

*Delambre[32] (2, p. 736).

†Grant[19] (p. 80) repeats a remark by Maraldi, without correcting it, to the effect that if errors in the eclipse times depend on a finite velocity of light then "they should be equal for all the satellites when the earth was in the same part of her orbit." My rough figures show clearly that if this means what it seems to mean, then it is mistaken.

best astronomers in Europe, Cassini nevertheless continued to ascribe the observed inequality to an unknown cause. A few followed his example—notably Giacomo Maraldi, his nephew, who was congratulated by Fontenelle for doing so.*

Since it seems that Cassini thought of the equation of light before Rømer, it is worth asking whether he found the idea elsewhere. The astronomical test for a finite velocity of light which Descartes proposed, in terms of the concomitant delay in the perceived eclipse of the Moon, had been mentioned at intervals ever since Descartes explained it in a letter to Beeckman in 1634.† It is possible that Descartes' proposal put the idea into Cassini's mind, but it seems to me not improbable that Huygens, whom Colbert had also brought to Paris, had considered the test carefully. It is well known that he discussed the Cartesian test at length, together with Rømer's "ingenious proof," in the first chapter of his *Traité de la lumière*.‡ This was first published in 1690, but we have evidence that the lunar eclipse test was "in the air" at least in 1665, when Hooke published *Micrographia*, for he there refers to it.§ Descartes' ideas on the velocity of light were constantly under discussion in Paris, especially in connection with the "proof" he gave of the law of sines, and I cannot help feeling that the general principles of eclipse tests must have been actively discussed publicly in the period terminating in Rømer's discovery.

How severely should we judge Cassini for the stand he took against Rømer's interpretation of the satellite observations? One has to remember that Cassini was the first to bring tables of satellite motions to such a state of perfection that they were reasonably reliable for years on end. For the first satellite, this was especially true; and since the same techniques and care had been applied to the others, was it unreasonable for Cassini to postulate unknown causes for the inequalities? We know that the irregularities in the outer motions are due to mutual perturbations, and these should indeed be identified with a part of Cassini's "unknown causes." But of course they do not cover the equation of light, and Cassini never really faced up to Halley's simple remark, made when he reduced Cassini's tables to the London meridian. It is hard, said Halley, "to imagine how the Earth's Position in respect of *Jupiter* should any way affect the Motion of the *Satellites*."‖

I will return shortly to the English treatment of Cassini's work, but first I should note some of his findings in regard to the satellite orbits. In his tables of 1693 they were all placed in a plane inclined at 2°55′ to the ecliptic, with nodes fixed at within two degrees or so of 10ˢ 14°30′ longitude.¶ His nephew Maraldi claimed to find from a study of eclipses that the orbital inclination of satellite II was changing. (He made

*For reference to the often-quoted letter see Grant[19] (p. 81), and Cohen[46] (pp. 348–349).
†Descartes.[48]
‡Huygens.[49]
§Hooke,[50] quoted by Cohen[46] (p. 337). Strangely enough, Hooke does not appear to have been convinced by Rømer. See Cohen[46] (p. 354, note 84). Hooke's argument can seemingly be paraphrased: "If light is as fast as *that*, then I don't see why its travel should not as easily be considered instantaneous"!
‖Halley[18] (p. 239).
¶Cassini[66] (pp. 43, 52, 99). For some of Cassini's predictions, which would disprove the claims of Galileo, Mayr, and Odierna concerning satellite latitudes, see Cassini.[51]

it 3°33' in 1707.) Both men were often working beyond the limits of meaningful observations, and the wonder is that they came so near to the truth. (Their only way of
finding the nodes, for example, was to search for the eclipses of longest duration, and
there were many obvious uncertainties even so. As already explained, one never sees
both phases of an eclipse of satellite I, and rarely both of II.)

To give an idea of the accuracy attainable with Cassini's tables of 1693, he listed
four Paris observations (three immersions and one emersion) of the first satellite, all
timed to the nearest second. The errors involved in calculating the phenomena were
+2 sec, +58 sec, +53 sec, −9 sec.* It will later be shown that the errors are much
greater with the other satellites; but in the *first* the astronomer was seemingly provided with a celestial timepiece accurate to better than a minute, and the longitude
problem for places on land was solved, in the sense that a single observation of an
eclipse of Jupiter's first satellite, combined with a star transit (or its equivalent), should
give the terrestrial longitude within 15'(arc).† A whole series of observations should
obviously give much better results, and especially if they were made simultaneously
from some central observatory. The tables of 1668 had already led the French to
launch a program of longitude determination the world over—along the French coast,
in Cayenne, Egypt, the Cape Verde Islands, and the Antilles, for example.‡ Cassini's
Paris observations served to control and coordinate the rest. Well into the eighteenth
century the method was a standard one for the longitude of land bases—witness
Lacaille's determination of the longitude of the Cape in 1750.§ The Cayenne expedition led by Richer in 1672–1673 had, as is well known, a more serious object than
that of merely determining the longitude of the place. One of its prime purposes was
to determine the parallax of Mars during the opposition of 1672,‖ when the planet
was simultaneously observed by Richer at Cayenne, and Cassini and Picard in Paris.
(The measured solar parallax of 9".5 was 8% high, and inevitably affected estimates
of the velocity of light made on the basis of Rømer's method, but most writers were
wise enough to state only that light took such and such a time to transverse the diameter of the Earth's orbit.)¶ Needless to say, one of the most significant of all the
French longitude expeditions was that which brought Picard, Bartholin, and Rømer

---

*Flamsteed put the maximum error at about 4 min. He explained at some length the method of finding
longitude on land.
†Note Halley's enthusiasm for the technique: "[Cassini's] Account of the *Longitudes* observed, has put
it past doubt that this is the very best way, could portable telescopes suffice for the work." Halley[18] (p.
237). He looked forward to the invention of shorter telescopes, manageable on board ship. Note that
Newton observed Jupiter's satellites with his first reflector—a short telescope. See Newton.[52]
‡Cassini[47] (pp. 40–42) and passim.
§Lacaille.[53,54] Note Bradley's determination of the longitudes of Lisbon and New York by the same
method—from observations made by others and compared with his own at Wanstead. In his mss. (e.g.,
Bradley 35, Bodleian Library) he lists longitudes for places the world over—from Peking to Paraguay.
‖"Observations astronomiques et physiques faites en l'Isle de Cayenne," in Cassini[47] (p. 2).
¶Rømer (according to Huygens) 22^m; Horrebow 28^{m}20^s; Halley 17^m; Wargentin 16^{m}26^s; Delambre 16^m
26^s.4. The last is only about 10^s too low. Cohen[46] (pp. 351–353) gives some calculations which show that
Rømer's data could have been used to give a more accurate value than that actually found.

to the island of Hven, to measure the longitude of Tycho's observatory, Uraniborg.*
Again, Cassini's observations, made in Paris, served as a control. Picard returned to
Paris with the information he had sought, with the original records of Tycho's obser-
vations, and with Rømer, whom he had persuaded to work at the Academy.

RØMER AND THE FINITE VELOCITY OF LIGHT

Rømer's discovery—or "hypothesis", if you accept Cassini's attitude to it—is not
well documented,† and we have no certain way of knowing how he calculated the
time for the travel of light, although what we might call "Cassini's method" (already
sketched), using eclipses of satellite I, was undoubtedly the framework of his calculation. In September 1676 he announced to the Academy that the eclipse predicted for
9 November would be ten minutes later than "naive" calculations based on previous
eclipses would seem to suggest. When observations confirmed his prediction, he could
state with some confidence that the cause of the delay was the finite velocity of light,
and that its value was such that light would take 22 minutes to cross the diameter of
the Earth's annual orbit. The published accounts are relatively brief. A translation of
that in the *Journal des Scavans* appeared in the *Philosophical Transactions*,‡ and
indeed it was this version that Huygens first saw. As explained, there is nothing there
to augment Cassini's analysis of the satellite motions from a theoretical point of view.
What Rømer managed to do was to dismiss the doubts that had prevented Cassini
from reaching the same conclusion, and that Maraldi used as ammunition in a polemic
against Rømer—doubts, that is to say, as to the simplicity of the problem. Was there
not potentially a *multiplicity* of interconnected sources of irregularity in the motions?
Rømer answered Maraldi's criticisms in letters to Huygens,§ showing that he regarded
the first satellite as a simple case (with the light equation alone significant), and the
outer satellites as presenting all the problems that Cassini had suggested. He never
published anything to suggest that he had made a systematic investigation of the satellites II, III, and IV. It is possible that he failed to obtain inconsistent results, and that
his intuition told him to abandon the subject with right on his side.

After Rømer, astronomers were left in a curious situation. The best available
"calculus" of the Jupiter satellites was drawn up according to principles involving a
mysterious "second equation" which no one appeared to be in any hurry to rationalize
by the application of Rømer's principle. I have mentioned Halley's skepticism in
response to Cassini's 1693 tables. Halley was content to make some small emendations, and merely raise his eyebrows at Cassini's techniques. The first complete systems to make consistent use of the equations of light were by Pound[58] and Bradley. I

*"Voyage d'Uraniborg . . ." with Cassini.[47]
†Most of his mss. were lost in the great fire that destroyed much of Copenhagen in 1728. See Eibe and
  Meyer,[55] containing a hitherto lost work.
‡See Rømer.[56]
§Huygens.[57]

will try to illustrate the main differences between the early systems I have mentioned, and this by reference to that class of instruments which goes under the generic name of Jovilabe.*

SOME INEQUALITIES—THE JOVILABE

The first and greatest inequality in the apparent motions of the satellites about Jupiter as viewed from the Earth is that produced by the annual motion of the Earth. In Fig. 9 (taken for convenience from Lalande's *Astronomie*, Vol. III), angle *SIT* is the annual parallax of Jupiter, which may be as much as 12° or so, and the interval between conjunction with the Sun and conjunction with Earth, i.e., the time the satellite takes to cover 12°, is appreciable [1^h25^m (I), 2^h50^m (II), 5^h44^m (III) and 13^h24^m (IV)]. This inequality was, however, of slight interest to those concerned with the prediction and use of eclipses.

Eclipse phenomena would not recur at equal intervals of time, even if light traveled instantaneously and the angular motion of the satellite about the planet were constant, since the planet itself moves with varying velocity in its orbit around the Sun. (One might explain this simply by saying that the "starting line", i.e., the shadow, is moving with variable velocity.) The satellite must perform its own (sidereal) revolution, so to speak, and then move further through the angle traversed by the shadow, before it is eclipsed. Cassini,[51] in 1676, was critical of Galileo and other astronomers who had ignored this inequality. As he said, they took as simple a movement com-

---

*This word, as Delambre points out with a trace of disdain, joins a Latin word with a Greek: Delambre[32] (2, p. 331). We have met it already in Galileo's Italian "*Giovilabio.*" The Latin form is "*Jovilabium.*"

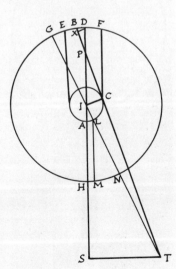

Fig. 9.   The effect of the annual motion of the Earth (T) around the Sun (S) in producing the first inequality in the apparent motion of a satellite (moving on the larger circle about Jupiter, the smaller circle—all on a very distorted scale). SIT is the annual parallax of Jupiter. The figure is redrawn (in its entirety, for completeness) from Lalande's *Astronomie*, Vol. 3, Fig. 244.

pounded from a uniform and a nonuniform motion. If this factor is not taken into account, the only way of obtaining reasonably accurate data for the mean motions is by observing Jupiter at the beginning and end of complete revolutions round the Sun, i.e., at intervals of about 12 years. This is more than most early observers were prepared to do, before publishing their findings. Galileo's manuscripts nevertheless show that he was aware of the problem. On one page, for example, he had drawn an eccentric orbit for Jupiter and marked part of it with what seem to be Jupiter's positions at ten-day intervals. Whereas the observations of 15 June 1610 to 17 March 1611 are given without correction for Jupiter's prosthaphairesis, those of 17 March to 16 July 1612 do have this equation incorporated.* As mentioned earlier, Harriot in his best calculations also took this inequality into account.

The simpler "Galilean" scheme (adopted by his contemporaries, of course, by reason of its simplicity) is well illustrated in the *Jovilabe* now in Florence (Fig. 10).The lower circle represents the Earth's orbit, with the Sun at its center, and the long (slotted) arm the line of sight to Jupiter. The four circles within the large upper scale represent the orbits of the satellites, and the transversals their separation from Jupiter, to a unit of the radius of Jupiter's disk. (The advantage of this unit is that it changes— as Jupiter's distance changes—in the same ratio as the apparent sizes of the orbits.) The parallactic angle ($\pm 12°$) I mentioned a moment ago is measured on the small

*Galileo[1] [3 (2), pp. 489–542].

Fig. 10. *Jovilabe* attributed to Galileo. (Courtesy of Museo di Storia della Scienza, Florence.)

TABLE 1.   Table of Mean Motions of Satellites I–IV

Source	I d	h	m	s	II d	h	m	s	III d	h	m	s	IV d	h	m	s
Berberich, calculated for 1612–14	1	18	28	34	3	13	17	42	7	3	58	48	16	18	0	0
Galileo, published 1612	1	18¼			3	13¼			7	4			16	18		
Galileo, Bellosguardo tabs	1	18	28	28	3	13	17	28	7	3	59	50	16	17	51	38
Mayr, published 1614	1	18	28	30	3	13	18		7	3	56	34	16	18	9	15
Harriot MS	1	18	28	48	3	12	33	36	7	7	33	36	16	16	0	0
Odierna	1	18	28	44	3	13	18	15	7	4	1	26	16	18	14	33
Florence giovilabio, deduced	1	18	28	31	3	13	17	40	7	3	58	49	16	17	59	57
Cassini	1	18	28	36	3	13	17	54 3	7	3	59	39 22	16	18	56	50
Hevelius	1	18	28		3	13	18		7	3	57		16	18	9	
Rømer's mechanism (Derham)	1	18	28	58	3	13	20	0	7	4	0	0	16	18	0	0
Newton, *Principia*	1	18	27	34	3	13	13	42	7	3	42	36	16	16	32	9
Bradley	1	18	28	36	3	13	17	54	7	3	59	36	16	18	5	8
Delambre (seconds quoted to nine places of decimals)	1	18	28	35.9	3	13	17	53.7	7	3	59	35.8	16	18	5	7.0

scale at the top. The sizes of the circles for the satellite orbits are to scale among themselves, but there is no need for them to be drawn to the same scale as the *Earth's* orbit, although the distance between Jupiter and the Sun must be to scale with this.

The significant failing of the instrument is not that the distance between Jupiter and the Sun is constant (in fact it is exactly 5 times the Sun–Earth distance), but that there is no means of compensating the satellite motions (in the tables engraved on the plate) for the varying velocity of the shadow-line. One other notable deficiency is less interesting: there are no radices for eclipse times marked on the plate, and auxiliary tables would have been needed. There are also no tables for yearly motions. The instrument does, nevertheless, have one unexpected virtue. When we come to calculate the parameters on which the existing tables are based (see Table 1), they seem to be of extraordinary accuracy, even better than Cassini's. This is something which can hardly be properly discussed, since we do not know the date of the instrument, but some comment should be made on the remark by Maria Luisa Righini Bonelli to the effect that the tables bear comparison with Galileo's "Bellosguardo" tables* At first sight the latter are given only to minutes of arc, whereas the *Jovilabe's* tables are to seconds. If we average over periods of 3000 days in the Bellosguardo tables, however, we find that the Galilean mean motions are of the same accuracy as those on the instrument. A comparison can be made from Table 1. Cassini's accuracy can obviously be easily exaggerated

According to Gassendi's biography of him,† Peiresc (d. 1637) had the idea of

---

*Bonelli[15] (p. 419).

†Paris (1641), translated by W. Rand as Gassendi.[59] See Book 2, p. 145: "Also he caused a mechanicall Theorie or Instrument to be made like theVulgar one of *Peurbachius*; that the Roots of the Motions being praesupposed, the Places of the *Medicean* Stars might be calculated for years, moneths, daies, and hours." Gassendi goes on to say that Peiresc gave satellites the names of members of the family of the Medici: Catharine (IV), Mary (III), this being the brightest, Cosimus Major (II), Cosimus Minor (I).

mechanical or graphical representation of the satellite positions. His journal, kept between 24 November 1610 and 21 June 1612, contains a record of numerous observations of the positions of the satellites. Gassendi tells us that he refrained from printing tables of the satellites' motions, lest he detract from Galileo's honor. They were made with the help of Jean Lombard and other assistants, who traveled to Marseilles, Malta, Cyprus, and Tripoli (Lebanon), where the positions of the satellites were recorded (in local time, of course), later to be compared with observations made by Peiresc in Aix-en-Provence.* The first accurate longitude differences by this method were thus obtained. Peiresc sustained his interest in this subject to the end of his life, and some idea of the importance to be attached to his work can be had from the value he assigned to the span of the Mediterranean in longitude, 41°30′ as against the very inaccurate 60° commonly assumed.

A third Jovilabe is described in the *Medicaeorum ephemerides* of Odierna (1656). The work, according to Delambre,† is very rare, but might well have transmitted the idea to later writers. Yet another Jovilabe, but this time one to reach a large audience through its publication in the *Philosophical Transactions* of the Royal Society, was described by Flamsteed[60] in 1685. This is easily understood from what has gone before, and so similar is it to the instrument now in Florence that one must suspect some sort of influence (Fig. 11).There is now no circle for the Earth's orbit, but a small parallactic arc is placed, as before, above the circles representing the orbits of the satellites. Flamsteed seems to take it that the user will have no difficulty in evaluating the annual parallax of Jupiter. I will not rehearse the procedure for using the instrument, which is meant only to supplement tables of eclipses of the satellites: their disposition is to be found with reference to the last eclipse. There are scales graduated in days and hours giving the angular movement of the satellites directly. There is again, however, no allowance for the varying angular velocity of the axis of Jupiter's shadow.

Another Jovilabe was described by William Whiston‡ in a work of 1738: *The Longitude Discovered by the Eclipses, Occultations and Conjunctions of Jupiter's Planets*, and no doubt there have been others.§ More to the point, however, is that Jovilabe which Cassini himself designed and used, made of sheets of card (Fig. 12). Weidler[64] gives a description of it, from details provided by Maraldi, whom he met on a visit to Paris. It comprises a graduated circle crossed by a transparent alidade (made of horn, and of a width such that it can represent Jupiter's shadow). There are four centrally pivoted movable disks in due proportion, representing the orbits of the satellites, and graduated around the edge at the appropriate mean motions with days

*Brown[35] (p. 490).

†Delambre[32] (2, p. 331). Note Odierna's use of a pendulum and meridian transits of stars for timing satellite phenomena—a very Galilean procedure. Note that there is a copy of this rare work in the British Library: 531.I.12(3).

‡Whiston[61] had more than 20 years previously published a proposal for finding longitude by the regular firing of explosive shells from vessels along the traffic lanes of the world's oceans. In 1721 he made a proposal to use the dip-circle. See Whiston.[62]

§Thus Jerome le Francais (Lalande)[63] mentions Cassini's instrument and also extends the basic principle in a graphical device of his own.

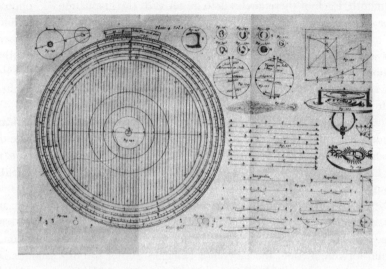

Fig. 11. The *Jovilabe* designed by John Flamsteed, 1685.

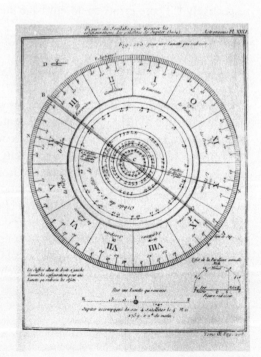

Fig. 12. Cassini's *Jovilabe*, as printed in Lalande's *Astronomie*, Vol. 3, Pl. xxxvi.

of the month. Each of these circles is to be set with the graduation "1" opposite the satellite mean motion on the first day of the month in question (this motion being taken from tables). The diameter through the origin of the outer scale represents the shadow axis. The positions of the satellites as seen from the Earth are their orthogonal projections on a line at right angles to the alidade, once this has been set to the geocentric longitude of Jupiter. In other words, the alidade is the line of sight from the Earth to Jupiter.

The advantages of Cassini's *Jovilabe* over its precursors lies in its use of the *geocentric* longitude of Jupiter. The angle between the shadow axis and the line of sight may therefore in principle be correctly set. On the Italian instrument this is not so, simply because Jupiter's position, whether heliocentric or geocentric, is left out of the reckoning.

Cassini's second inequality, which depends on the (heliocentric) position angle of Jupiter, is taken into account in none of these *Jovilabes*. In his 1693 tables he reckoned the inequality to be a maximum when Jupiter and the Sun are in opposition, and a minimum for conjunction, the difference being 14^m10^s for the first satellite, i.e., the time for it to travel $2°$; which is to say that eclipses occur 14^m10^s sooner in the former case than the latter. This, as already explained, is Rømer's equation of light, but Cassini (1) made it depend on the same parameter ($2°$) for all satellites;* and (2) made it independent of the eccentricity of Jupiter. Edmond Halley, in his London version of Cassini's tables, drew attention to the fact that they were not reliable, especially for the third and fourth satellite, for which he found errors of about 15 and 62 minutes, respectively.† These errors could not have been taken care of completely, however, in terms of the equation of light.

JAMES POUND AND JAMES BRADLEY, AND SATELLITE INTERACTIONS

It seems that James Pound and his nephew James Bradley were the first to introduce the equation of light into tables of the satellite motions. Bradley was born in the year of Cassini's tables, 1693, and as the discoverer of the aberration of light he was more than a little concerned with its velocity. Bradley had begun his astronomical work at his uncle's house in Wanstead, and from the very first he kept careful records of Jupiter's satellites, comparing his observations with predictions based on Cassini's tables. It was Pound who introduced him to Halley, and it is worth remembering that had Pound been prepared to give up his ecclesiastical living he would in all probability have been given the Savilian Chair of Astronomy at Oxford which went to Bradley in 1721. Halley had been Savilian Professor of Geometry until a year before, when he was created Astronomer Royal—a post which went to Bradley at Halley's death in 1742.

*Cassini.[65,66] The ideal equations in time units for the four satellites following Cassini's mean motions, on my calculation, are $14^m 02^s$ (IV). Cassini implicitly gives $14^m 10^s$ (I), $28^m 27^s$ (II), $57^m 22^s$ (III), $2^h 14^m 07^s$ (IV). (It is possible to deduce other parameters close to these.)
†Halley[18] (p. 255).

The satellite tables of the three men are almost as tangled as their careers. Pound's tables owed something of their style to Cassini's. They were sent to Halley, who printed them in the *Philosophical Transactions*,[67] and who also included them in his own collected tables, where they are added as an appendix to other tables of the satellites, sent to Halley by Bradley. (Halley's tables, printed in 1719, were not published until 1749, seven years after his death.)[68] Pound's tables (of the first satellite, like Halley's) are remarkably clear and well designed. He quotes his auxiliary parameters in millesimals, makes all his equations additive, and of course improves upon Cassini's mean motions. But his most memorable statement comes at the end of his explanations, where he tells the reader that many years of observation have convinced him "that the second inequality of this Satellite proceeds from the Progressive Propagation of Light, *and is common to all the rest of the Satellites*" (my italics).[58] He goes on to quote 7 minutes as the time for light to travel the mean radius of the Earth's orbit—a figure soon to be improved upon by his nephew's measurements of aberration. And he explains that he has added a *third* equation, to account for the varying distance of Jupiter from the Earth as a result of the eccentricity of the planet. (The second equation continues to take care of that part of the distance variation which is a function of the Jupiter–Sun–Earth angle. This might not seem to be the best of independent variables, but it was chosen for the historical reasons I have explained.)

In a sense, Pound's tables represent the end of the first great period in the history of Jupiter's satellites, a period of just over a century. With the complete incorporation of Rømer's light equations into the scheme for effecting a geometrical interpretation of the observations, the way was now open to the next phase, which might be said to have culminated in the celestial mechanics of Laplace. This new phase had begun with Newton—and I must at least mention in passing his attempted treatment of the motions of the nodes and apsides of the satellites by analogy with his theory of the motion of our own Moon.[69] Historical periods are not as neatly defined as those of the planets, however, and although I shall leave Newton aside, I would like to close by mentioning the work of Bradley, which in another sense marks the transition to the new style of research on the satellites of Jupiter.

Bradley's observations of the satellite eclipses, as listed in a Bodleian ms. (Bradley 35), run from 1712 to 1732. They are supplemented by lists of other astronomer's observations and many sheets of computations. There are also drafts (c. 1733?) of canons for the use of satellite tables. Bradley's relevant manuscripts (and I should mention here also Nos. 16 and 28) have the property of being assembled from documents of many different periods of his life. Although the greater part of the material is printed, Rigaud's edition will need to be supplemented by the manuscripts by anyone who wishes to form a clear picture of the development of Bradley's thought on the subject as a whole. This is more than I can do here, but I will indicate briefly the essence of his achievement.

Bradley's tables tend to follow the Cassini–Pound style. The published tables use the low figure of 7 minutes* (for the light time for the solar distance) rather than the

*Halley[68] (sig. Ffff 2v).

8^m13^s from aberration measurements, but only because those measurements were in the future when the tables were sent to Halley.* In Bradley's observations on his tables he shows that he is not a little perplexed by the satellite motions, which show inequalities possibly arising, he says, from orbital eccentricities of apsidal motions; "But by what we can collect from the motions of the second satellite, it is probable that they may be occasioned by the mutual action of the satellites on each other."† This passage is occasionally quoted in isolation as one of history's splendid conjectures, but the thinking behind it was far from casual. Bradley ruled out eccentricity of the orbit; it would have to be large to account for some of the rapid changes of velocity, and then it would be too large to be compatible with the remaining observations. He then asked about the *period of the errors*, and found that

> It nearly answers to the time the three inferior satellites take in returning to the same situation with respect to each other, and to the axis of the shadow of Jupiter, which is 437 days, or after 123 revolutions of the second satellite. After this period, the like errors return, nearly in the same order; but in the intermediate time, that is, after sixty revolutions, this satellite will deviate ten, twenty, thirty, and even sometimes forty minutes of time from its rate of motion during the seven preceding, or the seven following months. Now because the satellites are not found in the same place in the heavens after the aforesaid period is completed, it is possible these errors may vary somewhat on that account. And if the orbit of this satellite be likewise eccentric, as the late observations seem to make it, the inequalities arising from both causes must be very intricate and not easily to be separated by observation alone.‡

Bradley was mistaken to suppose an appreciable eccentricity, but he was right about the intricacy of the problem, and he was the first man to hit upon the *interrelations* between the satellites in their deviations from uniformity. He went on to say how the errors of satellites I and III are not so great as those of II, but seem to arise from the same causes. He expressed the hope that some geometer, "in imitation of the great Newton," would explain the irregularities on the basis of the theory of gravitation. The commensurability of the motions to which he had drawn attention was in fact to be one of the keys of the solution, arrived at by degrees in the later eighteenth century. Bradley had seen that in approximately $437^d.15$ the satellites make exactly 247 (I), 123 (II), and 61 (III) revolutions with respect to the shadow axis. The Swedish astronomer Pehr Wilhelm Wargentin, the last of the great purely empirical astronomers to work—in his case almost exclusively—with the problem of Jupiter's satellites, found the same result independently, and published it in 1741, before the Bradley–Halley publication.

It was left to Jean-Sylvain Bailly to show, in the 1760s, how the mutual attractions of the first three satellites gave rise to this periodicity in the inequalities.§ Bailly made much use of Clairaut's theory of the Moon. In 1766 the Académie Royale des Sciences offered a prize for a theoretical investigation of the satellite problem, and Lagrange's

---

*His ms. tables seem, even so, to preserve the maximum equation of $14^m\ 00^s$.
†Rigaud[28] (p. 81) there translated from Halley's Latin original.
‡Rigaud[28] (pp. 81–82).
§Bailly.[70] His initial findings were given to the Académie, and published by that body in 1762 and 1763.

successful memoir was so comprehensive in its analysis that even Delambre found it somewhat frightening.* Lagrange considered not only the attraction of Jupiter, but also the simultaneous effects of the Sun (which Newton had done),[69] of the three satellites (following Bailly, to some extent), and of the oblate mass of Jupiter (which Halley, in 1694, had conjectured might be of significance).† Laplace later added another factor: the plane of Jupiter's orbit does not coincide with its equator. The motion of each satellite was, as he showed, determined by three differential equations of the second order, involving six arbitrary constants—24 for all four satellites. There are seven other constants needed. He, Delambre, and Lalande spent much labor on the computation of the 31 elements, and when the tables which resulted were published in the third edition of Lalande's *Astronomie*, in 1792, they were far more accurate than anything produced by the dedicated Wargentin and Cassini, and all their empirical forebears.

The moral was clear: beyond a certain point of complexity, the astronomer can only observe what his theoretical scheme tells him is there. Knowledge of the inequalities in the satellite motions remained in a state of relative confusion until the theory of gravitation and mathematical analysis came to the rescue. Puzzles would even then have remained without Rømer. But without the help of Newtonian gravitation, the satellites had made a reputation for Galileo; they had provided a spectacle, for those fortunate enough to own a telescope; they had inspired thoughts which hovered between natural theology and space fiction; they had led to the invention of the pendulum clock, and to a limited solution to the problem of longitude; and with Ole Rømer they provided what was historically one of the two most important proofs of the finiteness of the velocity of light. And after Newton, and within two centuries of their discovery, they were to stimulate perhaps more intensive mathematical activity than any other object of astronomical study.

ACKNOWLEDGMENT

This chapter is based on a lecture given at the University of Aarhus, Denmark, to celebrate the tercentary of Ole Rømer's discovery of the finite velocity of light. Bearing in mind Wolfgang Yourgrau's interest in the mode of propagation of light, I hope it is an appropriate memorial to him. Its publication was much helped by Dr. J. J. Roche of Linacre College, Oxford.

REFERENCES AND NOTES

1. Galileo Galilei, *Le opere*, Antonio Favaro, editor, 20 Vols., (Barbera, Florence, 1890–1909; repr. with additions, Barbera, Florence, 1929–1939 and 1965).
2. S. Drake, *Discoveries and Opinions of Galileo* (Doubleday, New York, 1957).
3. J. Kepler, *Dioptrice* (Augustae Vindelicorum, 1611).

*Delambre[32] (2, p. 398).
†Halley[18] (p. 253).

4. P. Humberd, "Le baptême des satellites de Jupiter," *Rev. Questions Sci.* **117,** 171, 175 (1940).
5. J. Kepler, *Astronomi opera omnia,* Christian Frisch, editor, 8 Vols. (Frankfurt-Erlangen, 1858-1871; repr., Olms, Hildesheim, 1971-).
6. J. Kepler, *Conversation with Galileo's Sidereal Messenger,* Edward Rosen, editor (Johnson Reprint, New York & London, 1965).
7. A. M. Schyrlaeus de Rheita, *Oculus Enoch et Eliae* (Antwerp, 1645).
8. C. Huygens, *Cosmotheoros* (The Hague, 1698).
9. J. Kepler, *Narratio de observationibus . . .* (Frankfurt, 1611).
10. Galileo Galilei, *Istoria e dimonstrazioni intorno alle macchie solari e loro accidenti . . .* (Rome, 1613).
11. Stillman Drake, "Galileo and Satellite Predictions," *J. Hist. Astron.* **10,** 75-95 (1979).
12. J. W. Shirley, editor, *Thomas Harriot, Renaissance Scientist* (Clarendon Press, Oxford, 1974).
13. R. M. McKeon, "Les débuts de l'astronomie de précision," *Physis* **13,** 225-230 (1971).
14. A Borelli, *Theorica Mediceorum planetarum* (Florence, 1666), pp. 142-144.
15. M. L. Righini Bonelli, "Galileo, l'orologio, il giovilabio," *Physis* **13,** 412-420 (1971).
16. P. E. Ariotti, "Aspects of the Conception and Development of the Pendulum in the 17th Century," *Arch. Hist. Exact Sci.* **8,** 329-410 (1972).
17. R. T. Gould, *The Marine Chronometer, its History and Development* (J. D. Potter, London, 1923), pp. 11-17.
18. E. Halley, in *Phil. Trans. R. Soc.* **18** (214), 237-256 (1694).
19. R. Grant, *History of Physical Astronomy*(Society for the Diffusion of Useful Knowledge, London, 1852).
20. S. Mayr, *Mundus Jovialis* (Nuremberg, 1614). An English translation by A. O. Prickard is in *Observatory* **39,** 367-381, 403-412, 443-452, 498-503 (1916).
21. J. Klug, "Simon Marius aus Gunzenhausen und Galileo Galilei . . . ," *Abhandl. math.-phys. Kl. Königlich Bayerischen Akad. Wiss.* **22,** 385-526 (1906).
22. J. A. C. Oudemans and J. Bosscha, "Galilée et Marius," *Arch. Néerl. Sci. Exactes Nat.,* 2nd Ser., **8,** 115-189 (1903).
23. J. Bosscha, "Réhabilitation d'un astronome calumnié," *Arch. Néerl. Sci. Exactes Nat.,* 2nd Ser. **12,** 258-307, 490-528 (1907).
24. J. H. Johnson, "The Discovery of the First Four Satellites of Jupiter," *J. Brit. Astron. Assoc.* **41,** 164-171 (1930-1931).
25. P. Pagnini, "Galileo and Simon Mayr," *J. Brit. Astron. Assoc.* **41,** 415-422 (1930-1931).
26. Thomas Harriot, MS Petworth House HMC 241/IV.2.
27. J. O. Halliwell, ed., *Letters Illustrative of the Progress of Science* (Historical Society of Science, London, 1841), pp. 38-40.
28. S. P. Rigaud, *Miscellaneous Works and Correspondence of the Rev. James Bradley* (Clarendon Press, Oxford, 1832).
29. J. Roche, "Harriot, Galileo, and Jupiter's Satellites," *Archeion* **108,** (1982).
30. Christoph Scheiner, *De maculis solaribus* (Augsburg, 1612), pp. 27-31.
31. P. Gassendi, *Opera omnia,* Vol. 4 (Lyons, 1658).
32. J.-B. Delambre, *Historie de l'astronomie moderne,* 2 Vols. (V. Courcier, Paris, 1821).
33. G. Riccioli, *Almagestum novum . . .* (Bologna, 1651, 1653), p. 489.
34. P. Hérigone, *Cursus mathematicus,* 6 Vols. (Paris, 1634-1642), in particular Vol. 5.
35. H. Brown, "Nicolas Claude Fabri De Peiresc," *Dictionary of Scientific Biography,* Vol. 10, pp. 488-492 (Scribners, New York, 1974).
36. F. Fontana, *Observationes* (Naples, 1646), Tract 6, Cap. 2.
37. J. Hevelius, *Selenographia* (Dantzig, 1647), p. 526.
38. Galileo Galilei, *Il Saggiatore* (Rome, 1623), pp. 4-5.
39. A. Koyré, *La Révolution astronomique* (Vrin, Paris, 1961), Part 3.
40. J. D. Cassini, *Mémoires pour servir a l'histoire des sciences . . . suivi de la vie de J.-D. Cassini* (Bleuck, Paris, 1810.) The author is the great grandson of the subject Gian Domenico.
41. R. Taton, "Gian-Domenico Cassini," *Dictionary of Scientific Biography,* Vol. 3, pp. 100-104 (Scribners, New York, 1971).
42. G. D. Cassini, *Tabulae quotidianae revolutionis macularum Jovis . . .* (Rome, 1665), p. 328.
43. J. E. Montucla, *Histoire des mathématiques* (Paris, 1758), Vol. 2, p. 516.
44. J.-B. Du Hamel, *Regiae scientiarûm academia historia* (Paris, 1698), p. 145.

45. Anon., *Histoire de l'Académie Royale des Sciences* (Paris, 1707), p. 78.
46. I. B. Cohen, "Roemer and the First Determination of the Velocity of Light", *Isis* **32**, 327–79 (1940).
47. J. D. Cassini; "De l'origine et du progrès de l'astronomie," in *Recueil d'observations faites en plusieurs voyages. . .pour perfectionner l'astronomie et la géographie*, Messieurs de l'Académie Royale des Sciences, pp.38–39 (Paris, 1693).
48. R. Descartes, *Oeuvres*, C. Adam and P. Tannery, editors (Vrin, Paris, 1969), Vol. 1, letter 57, pp. 307–312.
49. C. Huygens, *Oeuvres completes* (La Société Hollandaise des Sciences, Amsterdam, 1967), pp. 463–469.
50. R. Hooke, *Micrographia* (London, 1665), p. 56.
51. J. D. Cassini, extract from a letter written to the *Journal des Sçavans* printed in *Phil. Trans. R. Soc.*, No. 128, 681–683 (September, 1676).
52. Isaac Newton, *Correspondence*, H. W. Turnbull, editor, p. 3 (Cambridge University Press, Cambridge, 1959).
53. N. L. De Lacaille, "Observations astronomiques faites à l'Isle de France pendant l'année 1753," *Mém. Acad. R. Sci.*, 44–54, especially 45 (1753).
54. N. L. De Lacaille, "Divers observations . . . etc.," *Mém. Acad. R. Sci.*, 94–130, especially 105, 120, 129, 130 (1754).
55. T. Eibe and K. Meyer, *Ole Rømers Adversaria* (Copenhagen, 1910).
56. O. Roemer, "Démonstration touchant le mouvement de la lumière.," *J. Sçavans* **4**, 233–236 (1676); translated in *Phil. Trans. R. Soc*, **12**, 803–804 (1677).
57. C. Huygens, *Oeuvres complètes* (La Société Hollandaise des Sciences, The Hague, 1899), Vol,8, pp. 30–35.
58. J. Pound, "New and Accurate Tables for the Ready Computing of the Eclipses of the First Satellite of Jupiter, by Addition Only," *Phil. Trans. R. S. 30*, No. 361, 1021-1034, especially 1034 (1719).
59. P. Gassendi (tr. W. Rand), *The Mirrour of True Nobility and Gentility* (London, 1657).
60. J. Flamsteed, in *Phil. Trans. R. Soc.* 15, No. 178, 1262–1265, and Table 2, Fig. 2, facing 1251 (1685).
61. W. Whiston, *A New Method for Discovering the Longitude* (London, 1715).
62. W. Whiston, *The Latitude and Longitude found by the. . . dipping needle*, (London, 1721).
63. J. Lalande, *Astronomie* (Paris, 1764), Vol. 3, pp. 197–202; second edition (Paris, 1771), Vol. 3, pp. 292–298.
64. J. F. Weidler, *Explicatio Jovilabii Cassiniani* (Wittenberg, 1727).
65. J. D. Cassini, "Tabulae motuum . . ," in *Recueil d'observations faites en plusieurs voyages*, Messieurs de l'Académie Royale des Sciences, pp. 9, 40, 103 (Paris, 1693).
66. J. D. Cassini, "Les Hypothèses . . ," in *Recueil d'observations faites en plusieurs voyages*, Messieurs de l'Académie Royale des Sciences, p. 52 (Paris, 1693).
67. E. Halley, in *Phil. Trans. R. Soc.* **30** (361), 1021–1034 (1717–1719).
68. E. Halley, *Tabulae astronomicae*, John Bevis, editor (London, 1749).
69. Isaac Newton, *Principia mathematica* (London, 1687), Book III, Prop. 23, Problem 5.
70. J.-S. Bailly, *Essai sur la theorie des satellites de Jupiter* (Paris, 1766).

9

GEORG MARKGRAF

AN ASTRONOMER IN THE NEW WORLD

Though he died at the age of 33, Georg Markgraf left behind two enviable reputations. The first of these rested on achievements clearly visible in the form of collections and publications on natural history and meteorology. The other reputation rested largely on promise: on the surviving title page of an astronomical book which was never written, on information abstracted from his papers by later French astronomers, and on the strength of a biographical sketch written by a younger brother. This brother, Christian, was only a year old when Georg began his university wandering, and was in any case anxious to exaggerate the value of an inheritance which had been largely dispersed before he even knew of its existence. This is not to say that Georg Markgraf's astronomical career is without interest, but its interest does not rest on extravagant claims as to a career which he might have had, but for an early death. He was neither a second Aristotle nor a second Ptolemy. I think there are signs, however, that it was his ambition to become a second Tycho Brahe,[1] with Johan Maurits' palace of Vrijburg his Uraniborg of the southern hemisphere. He was not quite the first European astronomer to make serious observations in the New World,[2] but he was perhaps the first truly systematic astronomer there, and he was almost certainly the first European to make systematic observations in South America, or indeed anywhere in the southern hemisphere.

Although Markgraf's life reflects the scientific ambitions of his time, the southern hemisphere was fortunate in him, and the tragedy is that he was not associated with a more settled and secure colonial enterprise. In its twenty-four year history, the Dutch colony in Brazil saw only three years of uninterrupted peace, namely from July 1642 to June 1645,[3] and during the five years or so he was there, Markgraf seems to have made observations for a total period of only about two years. His last brief spell of observing, of perhaps only six months in all, was done from the observatory in the Vrijburg—but by this time his patron, its proud builder, had already been asked to return to the Netherlands by the Heren XIX of the West India Company. Sent to Angola in 1644, under circumstances of which we know virtually nothing, Markgraf must have been sorely disheartened. Within a few weeks of his arrival he was dead,[4] and his astronomical efforts were to be without significant consequence for the advancement of the subject. The southern hemisphere was to wait more than a century for another first-class astronomer, this time N.-L. de Lacaille; and Brazil's only contribution to Lacaille's work was to repair the leaky French ship in which he travelled to the Cape of Good Hope.

THE BIOGRAPHICAL AND MANUSCRIPT SOURCES

If Markgraf's literary remains presented problems in the seventeenth century, they present even greater problems now. Journals covering the first part of the Brazilian period, once in the possession of his brother Christian, are apparently now lost. Christian was unable to find the journals covering the last three years of Georg's life.[5] The history of the

My thanks are due to two people in particular for help in obtaining Markgraf material, namely Dr. P. J. P. Whitehead and E. van den Boogaart. I must also thank the staff of the Gemeente Archief, Leiden, for temporarily transferring their Markgraf material to the Library of the Rijksuniversiteit Groningen.

1. See p. 223 below.
2. The southern constellations were being named by Dutch navigators from the end of the sixteenth century. Markgraf's observations antedate by several decades those of the Harvard observatory but Fr. Pierre Chastelain was recording isolated astronomical phenomena in Canada in a precise manner at least a few months before Markgraf's arrival in Brazil. See A.-G. Pingré, *Annales célestes du dix-septième siècle*, ed. M. G. Bigourdan, Paris, 1901, pp. 117, 133, 158, 180.
3. C. R. Boxer, *The Dutch in Brazil, 1624–1654*, Oxford, 1957, p. 127. Reprinted, with addenda, by Archon Books, Hamden, Connecticut, 1973.
4. At or near São Paolo de Loanda. See p. 222 below.

dispersal of Georg Markgraf's effects has been
most thoroughly told by Th. J. Meijer,[6] who has
made use of correspondence in the Leiden University Library,[7] and who has drawn attention to astronomical materials in the Gemeente Archief at
Leiden.[8] The main literary support for the present
chapter is this last collection of materials, namely a
bundle of 119 items which I shall here simply refer
to as the 'Leiden MSS'. With the kind permission
of the librarian of the Gemeente Archief I have
numbered these items, and shall incorporate bracketed references to them in my text, with the letter
L before the *item* number.[9] Short biographical notices of Markgraf are numerous, and I shall neither
list them nor show their interdependence. They all
rest very heavily on the (Latin) account by Christian Markgraf (published 1685), which was in turn
reproduced by J. J. Manget (1731), and which
exists in English in a Sloane MS (Add. MS. 333.9,
ff. 33–38) in the British Library.[10] A reasonably

thorough modern biography, that by E. W. Gudger,[11] rests more or less directly on this printed
source. Gudger made no pretentions to a complete
biography, his main interest being natural history.

If I may summarize the conclusions reached in
the works cited here, it seems that Markgraf's possessions reached the Netherlands in two ways: Johan Maurits left Brazil in May 1644 with the natural history collections—which from a monetary
point of view comprised by far his most valuable
possession, if indeed it was by this time still his—
while Markgraf had himself, earlier in the same
year, already taken his papers to Angola. From
there they were sent, at his death, to the Netherlands. Two astronomical manuscripts were then
given to Jacob Gool (1596–1667), arabist and astronomer at the University of Leiden [Pl. IX],
while the papers on natural history were given to
Johan de Laet—who published them.[12] This all appears from the testimony of Johan Maurits in 1655,
given at a time when the younger brother was
pressing for his inheritance. He had arrived in Leiden in June 1652, and was soon to insist that his
brother's chests had contained precious stones and
money. Despite a long wrangle, Christian seems to
have failed to get much out of the University. We
know that the curator of the collections there, the
mathematician and astronomer Samuel Kechel
[Pl. X], was close to Georg Markgraf,[13] and it is

5. J. J.Manget, *Bibliotheca scriptorum medicorum*, ii, Geneva,
1731, p. 263a. It is just possible that certain of the extant Leiden papers (L. 52 and 53—see notes 8 and 9 below for the
numbering), for the years 1637 and 1639, are what Christian
was referring to. They are journals of observations, in the
literal meaning of the word 'journal'.

It is perhaps worth adding here that although Christian
Markgraf's Leiden career does not seem to have run smooth-
ly, no doubt because of his quarrel with establishment figures,
he was no mean medical scientist himself. As J. W. van Spron-
sen notes in his contribution to *Leiden University in the Seventeenth Century* (Leiden, 1975; see pp. 338–341 and 343), although Christian had studied medicine at Leiden, he was refused permission to lecture there, and gave private chemistry
tuition. He was a follower of Sylvius and van Helmont. His
doctor's degree was taken at Franeker, the university in Fries-
land, and his medical writings were still in use in the eighteenth century.
6. 'De omstreden nalatenschap van een avontuurlijk geleerde',
Leids Jaarboekje, 1972, pp. 63–76.
7. UB Leiden, ASF 290, 15–27.
8. GA Leiden, Bibl. 12130.
9. My numbering (1–111, with 0 allocated to a letter dated
1891) is by items and not folios. Thus a sewn booklet is given a
single number. On four occasions items are given a single
number because they were associated in the bundle when I
first examined it (22 a/d; 24 a/b; 25 a/d; 69 a/b). I have arranged the items in a rough chronological sequence within nine
rather arbitrary subject headings, namely:

Eclipses (1–19); Sun (20–36); Moon (37–39); Planets
(40–51); Stars (52–62); Theoretical Astronomy (63–67); Instruments and Buildings (68–79); Astrology (80–101); Miscellaneous (102–111).

Very many items cover more than one subject, and no order of precedence has been imposed.

10. For an excellent bibliography with an explanation of the
relationships see the relevant items in that by P. J. P. Whitehead, 'The clupeoid fishes of the Guianas', *Bull. of the British
Museum (Natural History). Zoology*, Supplement 5, 1973. Note
the rarity of the 1685 biography. The Manget reference in
Meijer, op. cit., p. 75 should be to vol. ii (not xii): *Bibliotheca
scriptorum medicorum*, ii, Geneva, 1731, pp. 262a–264b. The
mistake seems to have originated with F. Markgraf (of the
Zurich botanical garden) who wrote on Georg in the *Dict. of
Scientific Biography*, vol. ix, New York, 1974. E. W. Gudger
(see n. 11 below) drew attention to Manget's reprinting of
Christian Markgraf's biography of his brother.
11. 'George Marcgrave, the first student of American natural
history', *Popular Science Monthly*, Sept, 1912, pp. 250–274;
'George Marcgrave, a postscript', *Science*, new series, xv,
1914, pp. 507–509. A. de E. Taunay's biography in the 1942
Portuguese translation of Markgraf's *Historia naturalis* bor-
rows much from Gudger's account. Gudger later made it clear
that his first article was written before he had seen more than a
transcript from a copy in Berlin.
12. See P.J.P. Whitehead in *Johann Maurits* (The Hague, 1979)
13. Christian calls him Georg's 'contubernalis', which might
mean simply 'comrade' or might be more specific ('resident
under the same roof').

likely that he felt that he had a better title to the collections than had Christian. He certainly passed on much of the information retailed by Christian, but this seems to have come from letters to Kechel, rather than from inherited papers.

Although Christian's biographical sketch forms a basis for all others, we must look elsewhere for whatever posthumous fame was achieved by Georg Markgraf as an astronomer. His observations, although not numerous, having been made in the southern hemisphere had a value which was largely unrelated to their accuracy. No doubt partly for this reason his papers were copied, and many of his observations were published, needed as they were for astronomical (and not historical) purposes. Copies based on many of the Leiden MSS, and on MSS which are now apparently lost, are in the library of the Observatoire de Paris, and have been in Paris since the seventeenth century. They were consulted by J. J. Le F. De Lalande (1732–1807), who says that copies were given by Guillaume Delisle (1675–1726) to the Dépôt des Plans, Cartes et Journaux de la Marine.[14] Lalande's sources of in-

formation included fairly certainly A.-G. Pingré (1711–1796) and J.-N. Delisle (1688–1768), who were both aware that the astronomer Ismael Boulliau (1605–1694) was one of the copyists. Lalande has it that the originals were in Cadiz with the MSS of J.-E. Louville (1671–1731), but this is hard to square with the fact that many of them are now in Leiden (note especially L. 52 and 53, of which there are Paris copies), that they are not known to have been out of the Netherlands since their arrival there in the seventeenth century, and that Boulliau had an excellent opportunity to copy them when he was librarian to the French ambassador in the Netherlands. He was probably making his copies, if this conjecture is correct, at about the time of the litigious activities of Christian Markgraf, perhaps by arrangement with him, or more probably with Samuel Kechel.[15] We know that Boulliau, who was a friend of Frans van Schooten and Christiaen Huygens, visited Kechel in Leiden some time before 1657.[16] The Paris copies add appreciably to what we can learn from the extant Leiden MSS, especially on the facilities and techniques of observation used by Markgraf, but there is no reason to suppose that they were made from copies other than were available in seventeenth-century Leiden. Boulliau is undoubtedly a key figure in the transmission, as evidence from the Flamsteed papers in London also strongly suggests.[17]

The question now arises: Have we lost irretrievably documents which might transform the picture of Markgraf the astronomer we get from a study of the Leiden and Paris MSS? I think the answer is a firm 'No'. Those who believe otherwise have in mind the promise of the title page mentioned by de Laet. Only at the end of a survey of

---

14. *Traité d'astronomie*, ii, 1771, p. 160. Of the MSS now in the Observatoire de Paris, the best is MS (fol.) A.B.4.5. This in the past carried the number 76, the main section (114 pp. long) being then 76A, and the remaining loose leaves, in a different hand, having been inscribed 76B to 76T. The loose pieces seem to be in the hand of J.-N. Delisle, judging by other Delisle material in the same archive (see A.A.7.1). The main section is a fine copy, perhaps done by an amanuensis for Delisle, and having the appearance of an edited work. It is partly based on missing material, and partly on material still extant at Leiden. The separate sheets include synopses and reductions of some of Markgraf's records, and are therefore of a rather different character. See n. 43 below. In these separate sheets page references are sometimes made to the main section. The main section does not seem to be the Boulliau copy, or to derive from it directly, if we are to judge by readings reported in Pingré's published work (compare op. cit., p. 141).

Other MSS at the Observatoire de Paris are numbered A.A.2.5 (with three sheets of Markgraf material, numbered 19, 2, A–C) and A.A.6.7 (see sheet 59, 15, F). The first of these is in several hands, and includes references to Newton and Halley. The three sheets mentioned have calculations of solar altitudes from 1639 on. The second includes much miscellaneous Mercury material, in several hands, seemingly pulled together by Delisle circa 1736. (I must thank Dr. E. Decker for help in tracing these manuscripts.) The Paris MSS have many independent misreadings, judged against L. 53, and probably derive from the originals by more than one route. More information as to their inter-relation might well be available in the vast Paris collections of Delisle material. See *Dictionary of Scientific Biography*, iv, pp. 24–25 for an outline of what survives.

15. There are forty unedited volumes of Boulliau's correspondence in Paris at the Bibliothèque Nationale (fonds français, 13019–13058), and the answer might well be found there. That the astronomical papers were not in Christian's hands is suggested below (p. 402).

16. NNBW ii, 1912, p. 652.

17. Flamsteed was—as Lalande somehow knew—one of the many later astronomers who took a passing interest in Markgraf's observations for their own sake, as is evident from a page of excerpted eclipse data in the Flamsteed papers (National Maritime Museum, Greenwich—P.R.O., vol. 41, f. 26r). At the top is a statement that the data were communicated by Edmond Halley, who got them from Thévenot at his houses at Issy in 1668 (see Pingré, p. 278). Furthermore, Flamsteed shares a small numerical copying error with the Paris copies.

the surviving material will it be possible to explain away this document. It seems clear that Markgraf simply did not live long enough to accumulate or assemble the materials for the promised book.

MARKGRAF AND HIS UNIVERSITIES

The main points of Christian's biography of Georg Markgraf are easily summarized. (I shall of course stress the astronomical connections, leaving the natural history to Dr Whitehead. I shall say little of the sustained polemic against Willem Pies, which, however, I am inclined to think was largely justified.) Georg was born at Liebstadt, near Pirna and Meissen in Saxony, in 1610 on September 20 (old style) at 7 p.m., as his horoscopes tell us.[18] The date and place are confirmed by the baptismal records of the local church.[19] His mother, Elizabeth Simon, was the pastor's own daughter, and his father, a member of an old Liebstadt family, was the head of the local school, well equipped to educate him in Latin, Greek, music, and drawing.[20] At sixteen, on 16 April 1627, he began his travels from university to university. He was to matriculate more than nine years later at the University of Leiden, as the records show, on 11 September 1636. By the time he did so, as Christian informs us, he had studied mathematics (under which we may include astronomy), botany and medicine,[21] at Strassburg, Basel, Ingolstadt, Altdorf, Erfurt, Wittenberg, Leipzig, Greifswald, Rostock and Stettin. The horoscopes among the Leiden MSS (see the Appendix below) contain items for Güstrow (near Rostock), and it is probable that he met Samuel Kechel at Stettin (for a place on roughly Stettin's latitude there are horoscopes for a 'family K').[22] At only two of those places[23] was there an astronomer

of any note during the period in question: Jacob Bartsch, who in 1630 became Kepler's son-in-law, and who collaborated with him in the production of tables of logarithms (published 1631), had earlier concentrated on the production of globes, planispheres, and star-charts generally. It seems that Markgraf possessed some of his work.[24] Eichstadt wrote some minor medical, astrological and astronomical works. At a much later date he was to help Hevelius in Dantzig. Perhaps his most useful contribution to astronomy was a set of planetary ephemerides, published in three parts at intervals (1634, 1639, 1644), and covering the years 1636 to 1665. These three published volumes of ephemerides were prefaced by very useful texts in spherical astronomy. Christian informs us that his brother helped in the preparation of these ephemerides, and he mentions an encomium by Eichstadt in the preface to them. In fact I can find no testimony to Georg in the preface to any of these three volumes.

It was almost certainly during the Stettin period that Georg developed his excellent computational skills and his ability to work with logarithms. Among the Leiden MSS (L. 63) there is a sewn booklet entitled 'De genesi ac ortu Canonis Log-[arithmorum]' excerpted from a treatise on logarithms, and headed 'Σὺν θεῷ d. 29 Sept./9 Oct. A.C. 1636 Lugduni Batavorum'. I take it that the Greek is to be construed 'Study carefully!'[25] On the quoted date, Markgraf had only recently arrived in Leiden. Early writers on logarithms tended to borrow shamelessly from their predecessors, especially, of course, from Napier and Kepler, but it is reasonably certain that Markgraf's immediate source for his booklet was a work by Eichstadt, included in his *second* book of ephemerides (cf. especially p. 2), which was not to be published until 1639. There is another document (L. 45) in which Markgraf lists some astronomically useful logarithms, but this is undated. A sewn booklet with some theorems on conic sections (L. 66) probably belongs to the Leiden period or earlier. The only firm manuscript evidence I have found (L. 22) referring to his pre-Leiden astronomical work dates

---

18. See p. 232 below. There are three relevant items in the Leiden MSS, (L. 88–9). Christian quotes the hour, which he might have got independently of the document(s) in question, and that this was so is suggested by his apparent ignorance of the story of the collapsing house (p. 221 below).
19. Meijer, op. cit., p. 75.
20. Manget, loc. cit. Cf. references to the later career of Georg's father, in P.J.P. Whitehead, op. cit., n. 12 above.
21. The words 'chemiae causa' added here presumably refer to a study of iatro-medicine. Cf. n. 5, on Christian's interests.
22. See Appendix, section 16. Kechel nevertheless entered the University of Leiden as early as 16 Oct. 1632. .
23. I ignore the Jesuits at Ingolstadt, whose work was third

rate, and who are not likely to have had much to do with Markgraf for religious reasons.
24. See p. 220 below.
25. The same phrase occurs on L. 28, above a list of errors in the Rudolphine Tables. The date was then 1637 July 8 (N.S.). See also n. 45 below.

Pl. IX. Jan de Vos, Portrait of Jacob Gool. Oil on Panel, n.d. Coll. Stichting
Familiearchief Van de Goes van Naters, Wassenaar. Gool was Markgraf's teacher in astronomy
and oriental languages at the University of Leiden.

Pl. X. D. Druyf, portrait of Samuel Kechel. Pencil drawing, n.d.. Coll. J.P. Crommelin, Amsterdam. Kechel, like Markgraf, was a central European refugee in the Netherlands, a mathematician and an astronomer.

Pl. XI. Z. Wagner, Johan Maurits' first house in Brazil. Drawing from his *Thierbuch*. Staatliche Kunstsammlungen, Dresden. This old Portuguese house on Antonio Vaz was probably Markgraf's observatory. If so, then it represents the first European purpose-built observatory in the New World and the first in the southern hemisphere.

Pl. XII. F. Post, a general view of the Vrijburg. Pen drawing, British Museum, London. This was the palace built by Johan Maurits, who arranged for an observatory to be built for Markgraf in one of the towers. Courtesy of the Trustees of the British Museum.

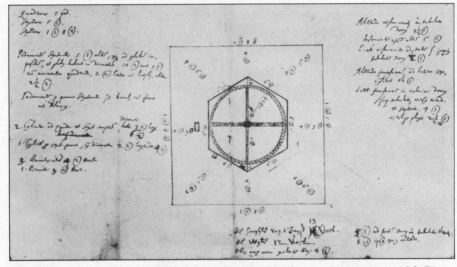

Pl. XIII. One of Markgraf's plans for the Vrijburg observatory (Cf. Pl. XII). Note the mixture of Dutch and Latin – the former presumably meant for the builder. Also on the plan are some details of Markgraf's instruments. Gemeente Archief, Leiden, MS 12130, f. 70.

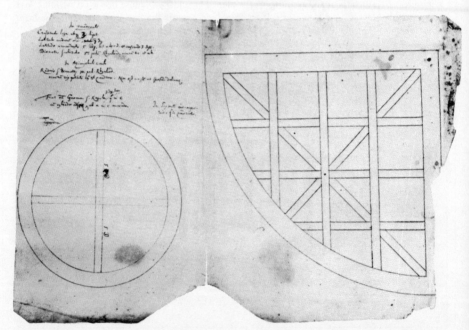

Pl. XIV. Markgraf's working drawings for the construction of a large quadrant. Gemeente Archief, Leiden, MS 12130, f. 71.

QUADRANS MAXIMUS CHALYBEUS
QUADRATO INCLUSUS, ET HORIZONTI AZIMUTHALI
CHALYBEO INSISTENS

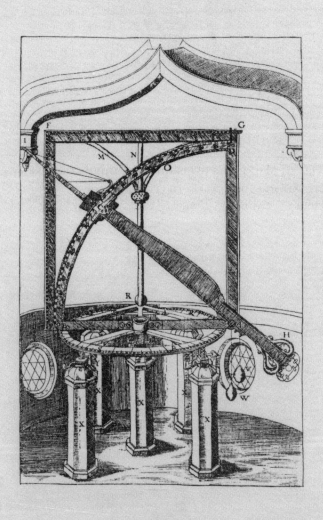

Pl. XV. An illustration from Tycho Brahe's *Astronomiae Instauratae Mechanica* (1598), giving an approximate idea of the appearance if not the ornamentation of Markgraf's quadrant, for which the sketch shown on Pl. XIV was drawn.

Pl. XVI. Markgraf's working drawings for a sextant (cf Pl. XIV).

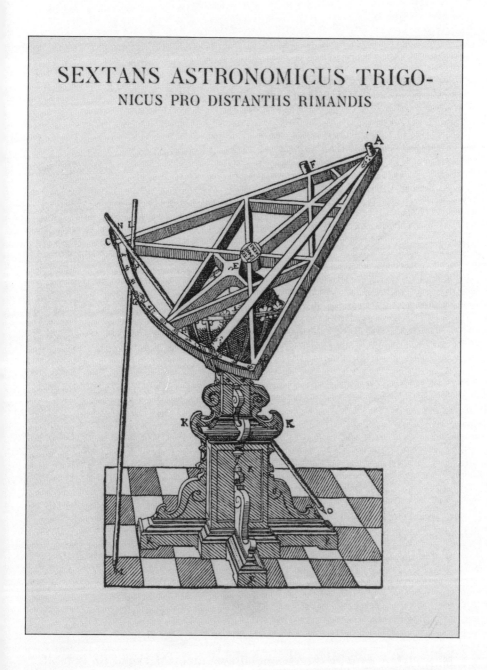

SEXTANS ASTRONOMICUS TRIGO-
NICUS PRO DISTANTIIS RIMANDIS

Pl. XVII. A sextant illustrated by Tycho Brahe (cf. Pl. XV) comparable
with one of Markgraf's instruments (Pl. XVI).

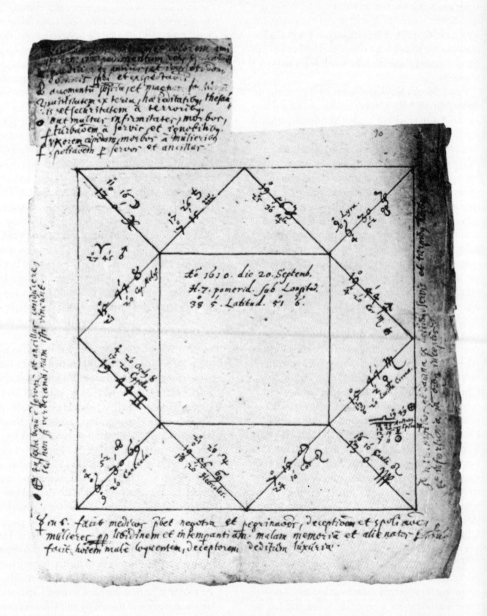

Pl. XVIII. A nativity (horoscope) drawn up by Markgraf for himself.
Note his date of birth in the middle (anno 1610, 20 September, at 7 p.m.), with the supposed
latitude and longitude of his place of birth. The positions of the planets, all carefully
calculated elsewhere in the Markgraf papers, are noted at the appropriate places in the twelve
'houses'. Some words of interpretation are written round the edge of the paper. Perhaps the
piece which has been torn away contained a prediction that was plainly false or embarrassingly
true. Gemeente Archief, Leiden, MS 12130, f. 90.

from 1637 or later, but contains reductions of sunspot observations made as early as 1635 Jan. 19 (new style). Neither the observations nor the draughtsmanship are in any way remarkable—but of course the document shows that he was using a telescope.[26]

From Stettin Markgraf went by ship across the Baltic, through the Danish straits, and down to Holland. On reaching Leiden he officially studied medicine, but the astronomy he also pursued there would have been reckoned by many an ancillary study—the link being astrology. Eichstadt's teaching would certainly have stressed this point of view, although the fortunes of astrological medicine were fast waning. At Leiden there was a good observatory, one of the first in a new European tradition whereby universities rather than private individuals took responsibility for such institutions. As I have already pointed out, the great Arabic and Hebrew scholar Jacob Gool (Golius) also taught astronomy and mathematics at Leiden. His interests were perhaps more textual than scientific, and his best work was an edition and Latin translation (published posthumously in 1669) of an astronomical work by the ninth-century astronomer Al-Farghānī. Gool did nevertheless leave some observational records to posterity. According to Christian, Georg Markgraf learned some Arabic from Gool, who after Markgraf's death took possession of an Arabic MS found among his effects. He might have been the man responsible for transferring the journals, now lost, to Christian.[27] Gool bequeathed money to Kechel, and again one is led to conjecture that Kechel took *personal* possession of Markgraf's astronomical papers—thus explaining how they now come to be in the Gemeente Archief rather than in the University Library. (If Christian had them, he certainly failed to make very good use of them.)

There are among the extant Leiden MSS seven documents which shed light on Markgraf's activity during his fourteen months at Leiden. Three of these, including one already discussed, suggest that he made frequent sunspot observations, but he does not seem to have drawn any significant conclusions from these (L. 22, 24a, 24b). A paper (L. 26) with a lunae-solar computation for 1637 Sept. 12 we find re-used for a Mercury observation and reduction on 1639 Sept. 18.[28] Astronomically more significant is a series of observations of the planets and stars (meridian altitudes in particular) running from 1637 Jan. 19 to 1637 Nov. 16, and showing him to have been an assiduous observer (L. 52). He remarks that he sees the Moon 'per tubum', that is, through a telescope, and he records sunspots and the satellites of Jupiter.[29] Biographically more interesting is a note at the end saying that clouds put an end to his observing, and that on November 18 he left for South America.[30] This is ostensibly at variance with Christian's statement, quoted often since, that he left on the first day of the new year. Johan Maurits' voyage to Brazil, via England, at roughly this time of year, took only 48 days from the time he left England. Had Markgraf's journey begun in mid-November and had it taken the same interval he would have arrived in early January. Christian— who seems to be referring to a journal—tells us that after two months at sea Georg arrived scarcely a month before fighting began. Johan Maurits left Recife to attack Salvador (Bahia) on April 8. He first tried to starve the garrison into submission, but finally attacked with his army on May 17–18.[31] As far as Georg's itinerary is concerned, we should presumably follow Christian's account, and assume that Brazil was reached in early or mid-April, the lost six weeks having been spent in Amsterdam before sailing. At all events, we know that Markgraf was making astronomical observations by June 1638 in Brazil. He tells us that he calculated a total lunar eclipse for June 25, but that rain, wind, and cloud prevented him from getting more than a glimpse of the partially eclipsed Moon (Paris MSS). He gives us no indication of his whereabouts at this time, but he was presumably in Mauritia, having survived the perils of the siege. More particularly, we are told that he first survived serious diarrhoea, and later was narrowly missed by an 18-pound cannon

---

26. On the first sunspot observations, made at about the time of Markgraf's birth, see my chapter (7) in *Thomas Harriot, Renaissance Scientist*, above, chapter 6, pp. 109ff.
27. Meijer, op.cit., p. 66.

28. Dates given without qualification are in new style. Markgraf used both styles, but rarely left any doubt as to which was intended.
29. He had at least one telescope in Brazil. See p. 226 below.
30. '... et ego die 8/18 Novembr. illic discessi in Americam Australem abiens.'
31. Boxer, op.cit., p. 87.

220 *The Universal Frame*

ball which spilled out the brains of a soldier at the door.

Before we turn to his work in the southern hemisphere we should consider a paper among the Leiden MSS (L. 27) which seems to belong with some solar parameters (for apogee and eccentricity) for 1637.[32] The paper contains, in a barely legible script, what looks like a list of Markgraf's books and charts. It is short, and contains two or three items to which Markgraf refers elsewhere in the Leiden papers. Perhaps it was a list of books he had accumulated as a wandering scholar, or perhaps it lists only what he took on his voyage. So far as I can make it out, it refers to the following items:[33]

1. Johannes Bayer, *Uranometria*, Augsburg, 1603. A star catalogue, with plates, that led to a reform of stellar nomenclature. In another place[34] Markgraf suggests that the stars on his globes were placed in accordance with Bayer's catalogue.

2. Johannes Kepler, *Tabulae Rudolphinae*, Ulm, 1627. The planetary tables begun by Tycho Brahe (1546–1601), whose assistant Kepler had been. Eichstadt's ephemerides had been calculated according to the Rudolphine tables and work by another pupil of Tycho's, Longomontanus (Christian Severin, author of *Astronomia Danica*, 1620).

3. Johannes Kepler, *Epitome astronomiae Copernicanae*, published in three parts, 1618, 1620, 1621. Perhaps the most widely read astronomical book of the time.

4. Willebrord Snel, *Tiphys Batavus*, Leiden, 1624. Lessons on navigation (Tiphys having been the pilot of the *Argo*).

5. A Freiberg edition of Archimedes (?).

6. 'De Castor', possibly Fracastoro, a sixteenth-century medical writer whose works were often reprinted.

7. Celestial and terrestrial planispheres, 1628 and 1629, apparently three of the four published by J. Bartsch in Strassburg with the title *Planiglobium*.

8. Guillaume de Nautonier, *Mecometrie de Leymant*, Venice, 1603. A work which it was vainly hoped would allow the determination of longitude at sea by magnetic observations.

To this list we can add other titles with which Markgraf was certainly familiar, whether or not he owned them. In particular, I have already mentioned Eichstadt's ephemerides, and my account of his instruments will show him to have known Tycho Brahe's *Astronomiae instauratae progymnasmata* and Kepler's *Ad Vitellionem paralipomena*.

We are ignorant of the circumstances under which Markgraf went to Brazil. The West India Company later denied that he had been engaged by them.[35] Christian said otherwise, and twice attributed to his brother the title 'Astronomus Societatis [Brasilianae]', under which title he says Georg's name appears in the Company's records in Amsterdam.[36] Boxer states very specifically that Johan Maurits gathered around him forty-six scholars.[37] Markgraf seems to have been in the Count's personal service at a fairly early date, but from the details given by Christian (using his brother's journals) it is clear that Markgraf did not go out to Brazil at the Prince's request. If he went out as one of the *vrijburghers* it might just conceivably have been—as Willem Pies wants us to believe —as a servant ('*domesticus*') to Pies, Johan Maurits' physician. (The Leiden papers (L. 106, 108) include recipes for medicines and an index to a dietary of some sort, and in any case we know that Markgraf had studied medicine, and that he set up a pharmacy in Mauritia.) We know that Markgraf had met both Pies and Jan de Laet in Leiden, either of whom might earlier have persuaded him of the advantages of an expedition to Brazil. According to Christian it was de Laet, Prefect of the West India Company, who engaged Georg Markgraf as astronomer to the company. It is hard to decide between the two accounts, each retailed with an ulterior motive. In Christian Markgraf's biographical essay there is an explanation of the move which can hardly be more than a partial one. In grandiloquent phrases he tells us that Georg burned with a strong desire to contemplate the southern stars, 'and above all Mercury', and that he knew of the great harvest of natural things to be had in America. (It would certainly have been easier to follow Mercury in Dutch Brazil than in high northern latitudes.) He therefore 'moved every stone', says

32. See p. 229 below.
33. The list is rudimentary (thus 'Tab.Rud.' is all that appears for item (2)), and I am responsible for its expansion and numbering.
34. See p. 226 below.

35. Meijer, op. cit., p. 65.
36. Manget, pp. 262b–263a and 264a.
37. Op.cit., p. 112.

Christian, 'and seized at every opportunity to go to America'.

LIFE IN BRAZIL

Of Markgraf's life in Brazil we know a little. As already explained, a month after his arrival there he went with the troops who were laying siege to Salvador (Bahía). He returned to the district of Pernambuco ('Farnambuca'), says Christian, where his fortunes improved. Johan Maurits had lost so many men and 'military architects' in the siege that he sought others skilled in the art, and of them Georg was so exceptional that he was accorded many honours. He was given the annual stipend of Captain, and told to establish himself in the new town of Mauritia. There he installed an astronomical observatory and a pharmacy, and practised medicine, meanwhile having the protection of soldiers. This all seems to come from the lost journals, and sounds authentic enough.[38] Did he ever make the fortune so many of the *vrijburghers* were seeking? We need not take too seriously Johan Maurits' statement of 1655 (made shamefacedly after Georg's possessions had been dispersed) that he was always broke ('bloot van Gelt') for reasons best known to those who were in Brazil with him. Pies suggests that he spent all his money on drink, which brought about his death.[39] Christian, however, was able to quote a letter from Pies' brother to Kechel, chiding him for not going to Brazil with Markgraf, who was said to be worth 60,000 guilders![40] Whatever the degree of exaggeration here, it is clear that Markgraf was well and truly in possession of the Prince's patronage before the time when an observatory was being planned for him in the new palace of Vrijburg, and the dedication he planned for his astronomical book suggests more than a promise of *munificentia* from Johan Maurits.[41]

Markgraf certainly could have been living in the house of Pies when, as all were sleeping, in the night of 1640 March 18, the building collapsed of its own accord (L. 53). This catastrophe put an end

to a long series of observations which had begun on 1639 Sept. 15, and which he was later to write up with extreme care. (Like the Leiden series, which in style and content they resemble, these could well have been meant for the printer.) He relates the story of the disaster in a space in the MS after the entry for 1640 Jan. 9. His brother Christian does not tell the story, which fact rather suggests that he was not in possession of the astronomical papers. Georg's manuscript refers to the house as 'domus nostra, ubi habitamus'—'our house, in which we are living'. He goes on, again using the present tense, to speak of the seven inmates whose lives were all spared, despite trapped limbs. His household possessions were lost or severely damaged, and it was three months before the house was rebuilt. In the meantime, as Markgraf tells us, he turned to mathematics, incapacitated as he was, and as he was to remain for more than two months, by a dislocated right shoulder.[42] It is pointless to speculate on which of the three or four undated mathematical pieces in the Leiden bundle belong to this period, but a possible candidate is a 14-page sewn booklet on the calculation of the parts of a spherical triangle (L. 64; cf. L. 65). As for the character of the work done in his first, and chief, period of observation, I shall discuss it later with that of the two shorter observing periods which followed.

Markgraf resumed observation on 1640 June 11. An analysis by Pingré[43] of 150 meridian altitudes recorded by Markgraf seemed to imply a change in the place from which observations were made, as judged by geographical latitude, in the first and second periods. The second period ended on February 7 of the following year. In both periods Markgraf was ostensibly based on the island of Antonio Váz. The earlier mean latitude deduced lay between (South) 8°15′ and 8°16′; the later between 8°5′ and 8°6′. A genuine reduction of 10′ in latitude corresponds to a move northwards of about 18.5 km, more than three times the distance between the Dutch fort on Antonio Váz and the Olinda settlement. I am sure that Pingré was right

38. Manget, p. 263a.
39. See the letter quoted by Meijer, op.cit., p. 68.
40. Manget, p. 263b.
41. For the dedication, see p. 224 below.

42. *Axilla*, lit. 'armpit'.
43. Op.cit., pp. 145–148. There are excerpted data on meridian altitudes in the Observatoire de Paris, with the other Markgraf material. See n. 14 above. They show that whoever was responsible for them was interested in them for a theory of refraction.

in supposing that instrumental error explains the discrepancy. (What Pingré could not know was that the lower value for the latitude was the nearer of the two to the truth.) But I think we can even find an explanation for the instrumental error. There is an illustration by Zacharias Wagner [Pl. XI] of the house in which Johan Maurits lived for most of his time in Brazil, that is, before his relatively short spell in the Vrijburg. The house, on Antonio Váz, was an old Portuguese house, but—as pointed out to me by Professor José Antonio Gonsalves de Mello (see also his *Tempo dos Flamengos*, 2nd edition, 1978, p. 84)—there is a superstructure above the roof which was not a part of the original house. This superstructure was remarkably like the observatory built for Markgraf at the top of one of the Vrijburg towers, both in style and proportions, as may be judged from my subsequent description of the later building. This being so, it seems very probable that the addition to the Portuguese house was intended as an observatory for Markgraf. It must have been extremely difficult to place such an edifice securely above an old building, and the poor values for the deduced latitudes are doubtless a reflection of this fact.

After the second period of astronomical observation, Markgraf resumed his expeditions. We find only an isolated astronomical record—that of a lunar eclipse seen from Fort Ceulen[44] on 1642 Apr. 14—before the beginning of the third period of continuous astronomical observation.[45] This third period, during which Markgraf observed from the newly completed Vrijburg palace [Pl. XII], is characterized by a value for the geographical latitude agreeing more or less with that derived from observations made during the first period. The third period (overlooking an eclipse observation of 1642 Oct. 7) began on 1642 Nov. 2. Markgraf's last recorded Brazilian observation seems to have been 1643 June 22, but there is no record of this among the Leiden papers.[46]

Before looking more closely into the astronomical records for their scientific content, we can piece together a few personal details relating to the last months of Markgraf's life. His brother tells us of letters to Kechel from Brazil in which Georg 'announced that he had packed up his possessions and was awaiting a favourable wind, so that by God's grace he might return with His Highness the Prince to his native land'. But, Christian goes on, Georg was sent to Angola unexpectedly, 'and others say the same', for reasons of which he was ignorant.[47] In fact the newly conquered African colony (1641), at the hub of which was São Paulo de Loanda, was of great strategic importance to the Brazilian enterprise. It was a slaving centre, the acquisition of which had at one blow strengthened the Dutch Brazilian economy and weakened that of Spanish America. Other forts were occupied in quick succession, including the last of the Portuguese Guinea settlements (Fort Axim, February 1642).[48] It is of some interest that one of the eclipse records in the Leiden MSS (for the lunar eclipse of 1642 Apr. 14, item L. 4) is in the form of a letter in Dutch from a ship's officer, neither signed nor addressed but headed, soon after this last conquest, 'Loanda 19 Apr. 1642'.[49]

No doubt Johan Maurits had some African scheme in view, which he thought Markgraf might help advance. But, says Christian, no sooner had Georg arrived in Angola than he died. A letter from Pies to Gool states more specifically that Markgraf lived only four or six weeks in Africa.[50] I know of no firm evidence for even the month of his death, although this probably occurred in 1644.[51]

---

47. Manget, p. 263b.
48. Boxer, p. 107.
49. A later hand adds the well-merited words 'Haec observatio nullius pretti est'.
50. Meijer, p. 68.
51. There is an entry amidst a jumble of rough notes on MS L. 7 which at first led me to think that we could set more precise limits to Markgraf's life, but I now think that the reference is to future eclipses, and not to observations by him or anyone else. Unfortunately the crucial verb endings are all badly written, but the heading seems to be 'Eclips. erunt in Brasilia observatas', followed by references to the solar eclipse of 1640 Nov. 13 (against which, presumably at a later date, he wrote 'observavi'), and the lunar eclipses of 1642 April 14 and October 7, and of 1643 April 3. Underneath this comes 'in Guinea et Angola', with mention of two eclipses, the first of 1643 September 27 (lunar, partial) and the second of 1644

---

44. At the mouth of the Rio Grande. Formerly Reis Magos, but renamed after Matthias van Ceulen, whose force captured it in December 1633.
45. See L. 39; Pingré, op.cit., p. 158; p. 99 in Paris MS A.B.4.5; and p. 230 below. (The page is headed 'Σ.Θ'. See p. 218 above.)
46. References to later eclipses (the last being that of 1644 March 8) seem to me to be in the nature of predictions. See n. 51 below.

A NEW WORLD TYCHO?
The Mauritian Tables

Prefacing the Markgraf section of the *Historia naturalis Brasiliae*[52] is the passage by Jan de Laet to which I have already referred. Here, writing as a sort of literary executor, de Laet tells of evidence, in the form of a draft outline of his plans, that Markgraf intended to publish a great astronomical work in three parts, which were meant to come *before* the *Historia naturalis*. The work, we are told, was to be called *Progymnastica mathematica Americana*. The first section was to have covered astronomy and optics, and was to have reviewed the southern stars between the tropic of Cancer [*sic*] and the south pole; planetary and eclipse observations referred to those star positions; new and true theories of Mercury and Venus based on his own observations; a theory of astronomical refraction and parallax; a new figure for the obliquity of the ecliptic; and something on sunspots and other 'Uranian rarities'. The second section of the astronomical work was to have been concerned with geography and geodesy, and to have included the doctrine of terrestrial longitude; the size of the Earth, from original observations; and an assessment of some previous geographical errors. The final section, based on the two preceding, would hav comprised astronomical tables, *Tabulae Mauritii astronomicae*.

It would be possible to go a considerable way towards a reconstruction of this unfinished work, al-

though the lacunae would relate to its most vital parts. We ought at least to note the similarities—beginning with the title—between it and Tycho Brahe's *Astronomiae instauratae progymnasmata*. (Did Jan de Laet transcribe Markgraf's '*Progymnastica*' correctly?) First published posthumously in Prague in 1602, this had been partly printed at Uraniborg on the island of Hven, where Tycho, under the patronage of Frederick II of Denmark, had built what was then the world's most ambitious astronomical observatory. There Tycho lived and worked for more than twenty years, until, having lost favour at the Danish court, he made his way by stages to Prague, where he arrived in June 1599. Tycho had enjoyed the patronage of the Emperor Rudolph II in Prague for less than two years when he died.

Tycho had connections with Rostock, where, more than thirty years later, Markgraf studied; Prague was not exactly far from Markgraf's own home; it was the birthplace of his friend Kechel and it was the place where an imperial clockmaker Christoph Markgraf was at work throughout Tycho's time at the court;[53] Johan Maurits might not have been the Holy Roman Emperor, but he did have the tower of a palace in the southern hemisphere to offer; we know that Markgraf had constant recourse to the *Tabulae Rudolphinae*, based on Tycho's observations, and published at length by Kepler;[54] and we know that Markgraf consciously designed his instruments along Tychonic lines.[55] Tycho's *Progymnasmata* was much concerned with the nova of 1572, which Markgraf could not of course match, although no doubt the heading 'raritates Uranicae' was waiting for the right phenomenon to turn up; but in many other respects (the systematic listing of observations as a basis for a new theory, the descriptions of the instruments used, the records of eclipse observations, and so on) the plan of the unwritten book resem-

---

March 8 (solar). Against the first is a note 'In Brasil non', and against the second 'In Brasil et videt (?)'. Also relating to the first is a note 'haec ad Cabo (*sic*) Verde non adparebit', and relating to the second what seems to read 'Cabo Verde (ad de Mina non)'. Cape Verde is of course the most westerly point of Africa, now in Senegal. That these are eclipse *predictions*, rather than observations, is shown by the fact that the 1644 eclipse began after sunset at Cape Verde, according to Th. von Oppolzer's *Canon der Finsternisse* (Vienna, 1887). São Jorge da Mina was a garrison captured from the Portuguese on Johan Maurits' initiative (by Colonel Coen—see Boxer, op.cit., p. 84) shortly after his return to Recife in March 1637. The town, now Elmina, is to the west of Cape Coast in what is now Ghana. On the general problem of settling the date of Markgraf's death, see Dr. Whitehead's analysis, op. cit., above n. 12, p. 453.

52. Leiden (apud Fr. Hackium) & Amsterdam (apud Lud. Elzevier), 1648. The first section is: Guilielmi Pisonis... *De medecina Brasiliensi.*

53. I have not been able to explore the precise relationship of Georg and the excellent craftsman of the previous generation, but it is conceivable that Georg heard of Tycho's doings at first hand from him. See H. von Bertele and E. Neumann, 'Der kaiserliche Kammeruhrmacher Christoph Margraf und die Erfindung der Kugellaufuhr', *Jahrbuch der kunsthistorischen Sammlungen in Wien*, Bd. 59 (1963), pp. 39–98). I am grateful to Mr. J. H. Leopold for this valuable reference.

54. See p. 220 above.

55. See pp. 225-8 below.

bles the plan of Tycho's work. Here, for good measure, is the dedication which was to have appeared under the title:

Opus desideratum a nemine adhuc tentatum, accedente munificentia Illustr. et Excellentiss. Herois J. MAURITII NASSOVIAE COMITIS, etc., Terrae et oceani Brasiliensis praefecti, Summi Inchoatum Feliciter et itidem cum bono Deo post multos labores absolutum, in nova civitate MAURICIA, sita in Brasilia Americae Australis regione, Auctore Georgio Marcgravio de Liebstad Germano.

Did Johan Maurits blush as he read all this?

As I have already pointed out,[56] one of the Paris manuscript copies (MS A.B.4.5) has been given the appearance of edited printer's copy. In fact there are corresponding sewn booklets among the Leiden papers, overlapping the copies substantially, about which the same might be said; and yet in both versions there is so little theoretical reduction of the observations as to make the material no more than a basis for further work. There are a few tables in the Leiden papers which might have found a place in the 'American astronomy',[57] but they are not, I feel sure, the sort of thing with which Markgraf hoped to surprise the world. All the evidence suggests that he planned to produce either a whole new theory, or at least new sets of parameters, for Mercury, Venus, and the Sun. As for Markgraf's remaining promises, we have the raw materials, but no indication that he had ever gone far with their reduction. And astronomical reputations are rarely made on the strength of observations alone.

The most puzzling of Markgraf's promises relates to the geodesy, for apart from some remarks on how he used his small sextant for geodetic work

(see below), and his solar eclipse observations—which when compared with others made in Europe allowed him to deduce a single relative terrestrial longitude—little beyond a garbled printed fragment survives to suggest that he did large-scale geodetic work. Of course there are the fine maps he is said to have drawn,[58] but where and when did he make the measurements which allowed him to state a figure for the size of the Earth? In the journal of observations, after the entry for 1641 Feb. 7 he tells us that on Feb. 9 he set out for nine months in the cause of 'geography', and that after his return, in the cause of 'chorography' and natural history he set out to the north of Brazil. Perhaps his geodetic observations were made during the first of these periods. As for the determined value for the Earth's size, this is to be found in an appendix, added by de Laet to the *Historia naturalis*.[59] Seldom can a more confused piece of writing have been committed to print. The printer (?) seems to have been under the impression that the order of arabic numerals making up a number is unimportant, and vulgar fractions defeated him totally. Without going into details, however, it is reasonably certain that embedded in the thinking behind the text there is a figure of 28,175 Rhineland rods per degree on the Earth's surface. This gives a radius of 6081.1 km for the Earth, to be compared with the modern figures of 6378.4 km (equatorial) and 6356.9 km (polar).[60] The figure was better than Willibrord Snel's (1617) and others earlier in the century, although Willem Blaeu had found better before Markgraf left Holland, and there is so much room for conjecture as to what part the many earlier Dutch measurements played in the growth of Markgraf's ideas on geodesy that it would be

56. See n. 14 above.
57. A table of the semidiurnal arc for Mauritia (L. 23), and another (not completed) of the altitude of the so-called nonagesimal point (L. 67). There are four booklets of ephemerides of the Sun and Moon for 5-day intervals during 1638—but these are calculated from the Rudolphine Tables (L. 29–31), and are the heritage of this collaboration with Eichstadt. A booklet of lunar rising times was for approximate and personal use (L. 38). A table of the times of high and low water for a period of a month, for an unspecified place, was probably not meant for this book (L. 103). A sewn MS book of tables of planetary equations (*Tabula anguli pro prosthaphaeresibus orbis in quinque planetis*) was derived from tables in the Rudolphine Tables (L. 104).

58. See Meijer, pp. 74 and 76 for further references.
59. At pp. 260–261.
160. The Rhineland rod is taken as 3.767 m. The unit was made a standard by the States General in 1604, on Simon Stevin's recommendation (*Principal Works*, iv, 1964, p. 24). The rod was divided into tenths and hundredths. Markgraf refers to the rod and its parts by encircled numerals, 0, 1, and 2 respectively. The rod is equivalent to 12 feet of 12 inches of a common Rhineland measure. (It is about 3 per cent greater than 12 feet in present-day Imperial measure.) Unfortunately for consistency, Markgraf's plans seem to imply that he used a rod of 12 'feet', each of 12 'inches' on some occasions and on others of 10. (We shall need to know his system when we come to consider his observatory plans.) See also n. 70 below.

pointless to try to take the matter any further here.[61]

A NEW WORLD URANIBORG?

Without doubt the most valuable section of the Paris material concerns the observatory in Mauritia and its equipment. Covering the first ten pages of the longest manuscript, this chapter can be related to a dozen different fragments among the Leiden papers, many of them very puzzling in isolation, but all with its help more or less intelligible.

The chapter opens in a way which shows that it was ready for publication:

Observatorii nostri Astronomici et Instrumentorum quae fabre fieri curavi munificentia Ill[ustrissimi] et Excellentissimi Herois J. Mauritii Comitis Nassovii Brevis Descriptio.

In nova civitate Mauritia in Insula Antonii Vaaz quae est in Brasilia Americae Australis Regione.

This all clearly belongs with the title page found by de Laet. 'Over the house of the most illustrious hero Maurice of Nassau, Governor of Brazil, etc.', the chapter now opens by telling us, 'we arranged (*curavimus*) an outside theatre of square section of side 20 Rhineland feet.' The fanfare of the titles is over, and the remainder of the chapter is as prosaic as any quantity-surveyor's notebook.

From the theatre—reached from the house by a flight of 43 steps—there was a wide prospect over land and sea. In the middle of the theatre stood a hexagonal building, containing two chambers, one above the other. Each of the six vertical walls was 6 Rhineland feet across and 13 Rhineland feet high. The top was surmounted by a pyramidal (hexagonal) roof, pierced by triangular trap-doors, which could be held open with stays. The trap-doors to north and south were larger than the other four (4 feet side as opposed to 2 feet 6 inches). The sides carried windows to the upper chamber, and these could be curtained. A door to the theatre platform was in the wall of the lower chamber, of course, and it concealed steps (10 in all) leading to the upper room, under which steps the sextant was stored (see below). Indoor observations could be made from horizon to zenith, in any direction, if need be. There was a walk at a height of 4 Rh. feet above the upper floor (presumably somewhat as Tycho had illustrated).[62] The lower (windowless) chamber was meant for optical work—for the projection of the solar image, for instance, in connection with sunspot observations. Surmounting the roof of the hexagonal tower was a pole carrying a standard of gilded copper bearing the arms of Johan Maurits.

There are plans of all this (but not elevations) among the Leiden papers (L. 69, 70) [Pl. XIII]. The observatory was ingeniously designed. Whether the building was as solid as a tower observatory requires to be is another matter. (The dark lower chamber had only a heavy wooden floor.) With a mural sundial on the north side of the tower, and a large iron clock within, 'whose bells can be heard all over the town',[63] the observatory, resplendent under the Prince's standard by day, and vibrant by day and night, must have been generally thought well worthy of John Maurits' short-lived Utopia.

The observatory incorporated features which Markgraf had seen elsewhere, in particular in the writings of Tycho and in the new buildings at Leiden. It is worth our while comparing the Vrijburg observatory with the central part of Tycho's Uraniborg. This was surmounted by an octagonal (rather than hexagonal) pavilion, a dome with clock dials east and west, and a further onion-spire topped by a gilt vane in the form of Pegasus. For observation, there were also octagonal towers, with

---

61. A small stylistic curiosity worth mentioning is that the passage in question uses Rhineland rods, Holland miles, and *hours* as linear measures. In fact a degree was at this period of history occasionally taken to be 18 hours walk at ordinary speed, following a commoner convention which made it 8000 paces, of 2.25 Rh. feet, to one hour. Navigators often quoted distances in units of time, a habit regularly deplored by some scientific writers. It seems to me that the de Laet appendix has not only been garbled but also 'navigationized' by some well-meaning person other than Markgraf.

62. *Astronomiae instauratae mechanica*, Nuremberg, 1602 (posthumous). Reprinted as vol. v of the *Opera Omnia* (ed. J. L. E. Dreyer), Copenhagen, 1913–1929. For an English translation, see *Tycho Brahe's Description of his Instruments and Scientific Work* by H. Ræder, Elis Strömgren and Bengt Strömgren, Copenhagen, 1946.

63. There are indications that the copyist of the Paris MS could not read the original near this point. He wrote hesitantly that 'S. Excitta' had a room beneath the theatre. It seems probable that this is a reference not to a saint (which would have suggested a chapel) but to 'Sijne Excellentie', Johan Maurits himself.

226 The Universal Frame

more straightforward pyramidal roofs. These had removable panels (rather than glazed windows). They were surrounded by observation galleries for use with smaller instruments. The earliest Leiden building seems to have been in the same tradition.[64] In 1632 Gool had bought for the university Snel's great quadrant, and by a resolution of 9 August 1632 an observation platform was built for its use. The platform was 19 by 15 feet, only marginally smaller than Markgraf's, and surrounded, like that, by a balustrade. A room below the platform was built for a pair of globes, and in 1633 it was decided that the platform should be surmounted by an 8-sided (or 6?) tower to house the quadrant. There is disagreement as to whether or not this had a retractable roof, but De Sitter was probably right when he suggested that the roof had rather fourteen hatches. Again this would be reminiscent of Markgraf's scheme. The earliest drawing of the structure at Leiden makes it look remarkably like Markgraf's observatory.

The instruments housed in and around the observatory were clearly modelled on Tycho's. They are described in the longest Paris MS, and sketched roughly on one or two leaves among the Leiden papers (L. 71–73). A summary of Markgraf's instruments is easily given (I follow his order):

1. Movable hardwood[65] quadrant, height 5 Rh. feet, with sights (see below) and mounted on an azimuthal circle of 10 Rh. feet diameter, with a 'gnomon' indicating azimuth. The azimuthal circle was supported on 12 pillars. In its general arrangement and size it corresponded to Tycho's large iron quadrant, which was supported, however, on only five pillars. The framework of the actual quadrant resembles other Tychonic instruments (e.g. the quadrans volubilis) [Pl. XIV, XV].

2. Sextant, height 5 Rh. feet, with sights as on the quadrant. Housed under the steps to the upper observatory. A 5 Rh. feet high mount for this, with ball joint, stood outside on the 'theatre'. Compare Tycho's sextant, which stood on Tycho's 'theatre'. [Pl. XVI, XVII].

3. Sextant, height 20 Rh. inches.[66] Said to be for geodetic work and astronomical fieldwork. There is a fixed mount for this on the theatre.[67]

4. Four globes, celestial and terrestrial, of different sizes, with stars placed according to Bayer's *Uranometria*. Perhaps the globes were paper covered, since they are said to be for use only *inside* the observatory.

5. Two clepsydrae.

6. A telescope ('tubus'), 7 Rh. feet long. There are drawings in the Leiden papers (L. 72, lower two sketches) suggesting smaller telescopes. There is no doubt about the uppermost sketch of the three, which corresponds to a description given in the Paris MS of the support for the telescope by which a solar image was to be observed (and also projected into the dark chamber?). A 9 foot beam with pierced brackets to hold the tube, which was fitted into the hole in the wall at its upper end, was adjustable in altitude by means of the cross piece at the lower end, the cross piece having a large number of holes for a stay.[68]

64. For the following information I am grateful to an unpublished thesis by Karen M. Heinemann of the University of Leiden: *Het Academiedak. De sterrenwacht, het uurwerk van 1589, de academietoren*. The De Sitter reference is to his *Short History of the Observatory of the University at Leiden*, 1633–1933, Haarlem, 1933. This, and further comprehensive references, will be found in the Leiden thesis, for my acquaintance with which I must thank Dr. E. Dekker.

I should here mention a drawing, purporting to be of Markgraf's Recife observatory, now in the Siegerland Museum (Oberen Schloss, Siegen). It is an unsigned and tinted pen and ink sketch, perhaps of the nineteenth century, showing a tower of square section surmounted by a 12-sided(?) dome and standing on a base conveniently hidden by a long hut. The sketch is without value, but it would be of interest to know by whom and under what circumstances it was made.
65. A 'very hard wood' called 'Pão Sancto Lusitani', as he explains here and elsewhere.

66. My designation, counting 144 such inches to the rod.
67. Markgraf explains that for the base of this sextant he has, when he travels, a little hammer in the Polish shape made of solid iron, approximately 9 inches long, with a head 1 inch square, to allow him to put the sextant horizontal, to measure distances, and vertical, for altitudes. This is fitted to ('applicatus est') a staff 4½ Rh. feet long of 'Pão Santo' hardwood, with a steel point so that it can be pushed firmly into the ground. (In the Leiden papers, (L. 70), he calls this a 'pedamentum', lit. 'stake'.) Not, surely, a very satisfactory mount. Lest the word 'hammer' (malleus) be thought a mistranslation, I note that there is a full-scale drawing of a claw-hammer (labelled 'marteau de fer') in the Leiden papers, and this is for use with the staff ('baston', i.e. modern Fr. 'bâton') carrying the *large* sextant. Since this had a permanent mount, the hammers are clearly not for knocking the staves into the ground. I assume that they are pivots on which are seated the sextants, pierced with square holes. Did Markgraf know much French? There is a sheet from a Latin-French phrase book in the Leiden papers (L. 111).
68. The evidence for the contents of the lower chamber is to

There was also a wooden mount, moved by a screw etc., on the theatre, expressly to support the 'tubus opticus'.

7. Pendulum, being a turned metal cylinder of 41¾ ounces, suspended by a chord 29 Rh. inches long, the whole to measure the mean solar day.[69] I calculate the length of the cylinder (using some of the astronomical observations made with its aid) as a little over 7 cm. Since he varied the length of chord used, a precise figure is out of the question.

8. Assorted objects: water level, plane table, two lanterns, three-legged stool with three steps.

Among the Leiden papers (L. 79) is a small piece of blackened parchment with a hole at its centre (diameter 10 parts of a scale on which 100 parts measure 3.1 cm).[70] Written on this parchment is also 'Dist. 7540 tabell.' This object can be related to places in the MSS in which Markgraf gives the results of his observations of a solar image. Thus on 1640 June 28 at 10 in the morning he set up his telescope stand (projecting into the dark chamber) and projected the solar image through the hole of 10 parts. The Sun's image was 74 or 73 parts in diameter, and was 7340 parts from the hole. The chamber was being used as a *camera obscura* in the now standard technical sense. Kepler had described the method of projection through a pin-hole (a single objective lens does not affect the calculation materially, but does improve image quality) in his *Ad Vitellionem paralipomena* (1604), and there is a reference to Kepler with some relevant sketches on a slip of paper in the Leiden bundle (L. 78).[71]

Among the Leiden MSS there are full-scale drawings of 'dioptrae' for the quadrants and sextants (L. 73). These are mentioned in the Paris chapter. Although the name is confusing, it refers to the sighting frames which run along the edges of the scales of the instruments. In his design of sights generally, Markgraf follows Tycho, and he actually refers to 'pinnacidia Thyconica'. The word 'cylindrum' occurs often in this connection, usually with measurements quoted. The word here has two different meanings. It may refer to a plummet, for levelling an instrument, or to a small cylinder on the sighting-rule, substituted for the objective pinnule in certain types of observation, exactly as Tycho had taught. Markgraf's notes on sighting methods can be best understood in the context of Tycho's 'supplementum de subdivisione et dioptris instrumentorum' in his *Astronomiae instauratae mechanica* of 1598.

In the Paris chapter, Markgraf gives us some idea of his ambitions as to accuracy in the use of his various instruments. The large quadrant was divided down to half minutes of arc, and Markgraf

---

be found in two separate passages in MS A.B.4.5. The first, at p. 2, speaks of the 'camera clausa... per (*sic*) speculationibus et praxi optica'. The second passage (pp. 6–7) gives the details of his instrument for observing sunspots and other solar phenomena:

In inferiori obscura camera, quae intra per rotunda foramina etiam luce illustrari potest, basis est facta pro tubo imponendo ad observationes deliquiorum solis, macularum solis, et aliarum rerum. Basis illa constat, primo ligno 4½ ① alto, 4 ② lato & 1 + ② crasso quod erectum statui potest, nam inferior transversale lignum habet, cui incumbit. Haec [?] transversim (per crucem) multa foramina habet, et in medio cavitatem, ubi inseritur regula 9 ① longa, quae hinc inde agi potest, et altior et humilior statui. Regula 3 ② lata, 1 ① plus crassa; et supra hanc orbis, cuius diameter 1 ① inseritur, qui et circum agi et demitti ac attolli potest. In regula interstitio 1½, 3, 3 & 1½ ① perpendicularia corpora erecta sunt, 8 ② alta, quibus tubus imponitur et superius clauduntur. Altera extremitas regulae foramini seu fenestrae incumbit.

'Tubus' is Markgraf's standard word for telescope, and there are several original records of his having used the telescope for sunspot observations—as was then customary, by cutting down the light intensity artificially or by natural means (cloud, haze). But did he use optical projection of the sun's image? The parchment (L. 79) with the hole of ten parts (approx. 3.1 mm) used at a distance of 10368 parts (as on the paper with the reference to Kepler's *Opera*, p. 341; see my next paragraph) suggests projection over a distance of about ten feet. This might have been with a single lens, or the pin-hole only; it is not likely that use was made of the advantages of projection using both the objective *and the eyepiece* of a telescope. Since the passage quoted above mentions the possibility of bolting the telescope into the 8-inch supports, it might be thought reasonable to suppose that the support was meant to be used in both modes, with and without the tube.
69. 'Pro νοχθημερω observando'. I assume that the Greek word intended was νυχθήμερον (nom.) The pendulum, incidentally, was obviously a free one.

70. There is no doubt that he is here using 1000 units to a Rhineland foot (of which there are 12 to the Rhineland rod). On one Leiden document (L. 73) he seems to refer to a 5 foot staff and a 1½ foot rule of ebony, both so graduated (1000 parts to the foot).
71. *Ad Vitellionem paralipomena, quibus astronomiae pars optica traditur*, Frankfurt, 1604. Chapter XI is entitled 'De artificiosa observatione diametrorum solis et lunae et deliquiorum utriusque', and Markgraf's work refers to the second 'problema', pp. 339–342. Kepler's drawing is at p. 338.

thought he was accurate to quarter minutes. The
one-minute divisions on the corresponding azimuth circle allowed him, he thought, to state azimuths to half minutes. The large sextant was divided as the quadrant, but the small one was divided
at intervals of only two minutes of arc. (This implies rather more than three divisions to the millimetre.) All that this tells us, however, is that the
instruments were ambitiously made. Of their real
quality we can only judge by an analysis of
Markgraf's work. This task can safely be left to
posterity, with the provisional remark that, judging by a casual selection of meridian altitudes for
medium altitude stars, standard deviations from
the later period seem to be of the order of a minute
of arc. Although not nearly as good as Tycho's,
Markgraf's results were comparable with those of
most of his contemporaries. By the middle of the
seventeenth century, astronomers had come to
realize that their mural quadrants gave much better
results than movable instruments. Of course, both
were susceptible to pivoting error, zero error, errors due to inaccurate engraving of the scale, colli-
mation errors, and so forth. Since it seems that the
errors introduced into Markgraf's results in these
ways were comparable with those in European data, it is natural to ask how many instruments, or
parts of instruments, Markgraf imported from Europe. How many artisans with the necessary skills
were to be found in the Dutch colony? Did Markgraf himself make instruments? We are never likely to have firm answers to these questions, but the
use of local wood, the fact that the buildings and
instruments seem to have been so well coordin-
ated, and the document with working sketches and
instructions in the French and Dutch languages,
suggest that Markgraf's—in manufacture if not in
design—was truly a New World observatory.

THE USE OF THE PENDULUM

Markgraf knew the advantages of a good meridian
reading, and the hazards of altazimuths. For the
positions of the fixed stars he took meridian altitudes, wherever possible, but he was usually obliged
to observe Mercury in altitude and azimuth. For
the local time of the central observation, he interpolated between the times deduced (T_1 and T_2)
from observations of a familiar star before and af-

ter the event (T). This interpolation he performed
by counting the number of oscillations of his free
pendulum in the two time intervals [n_1 in ($T - T_1$)
and n_2 in ($T_2 - T$)]. Then T is simply found, in
effect, as ($n_1 T_2 + n_2 T_1$) : ($n_1 + n_2$). It is not necessary to know the period of swing of the pendulum,
or to keep this the same on successive days. I assume
that an assistant counted the swings.

The method of finding the time corresponding
to the position of, say, Mercury, is a good one, but
it makes very great demands on the astronomer,
leaving him with a great deal of calculation to do.
One of the most conspicuous things about the aggregate of surviving material is that—despite extensive calculation—there is very little evidence
that theoretical use was made of it. As I have pointed out already, Pingré and others tried to make
something of the material, but without any important consequence. As for Markgraf himself, when
we look into the Leiden papers we find few signs of
his intentions. Let us consider, however, what
seems to have been his favourite planet, namely
Mercury.

THE PLANETS

Most of the documents among the Leiden papers
recording planetary phenomena (chiefly L. 40–51)
show a marked concern with Mercury. The Paris
MSS admittedly show that Markgraf spent some
time on the meridian altitudes of the superior planets; the distances from Venus to other stars or planets are often recorded; qualitatively interesting
phenomena (e.g. visibilities) are noted, and predicted; and yet in the end it becomes clear that no planet was observed with such care as that devoted to
Mercury. Observations of the planet fall into two
main categories:

1. Conjunctions (including occultations) with
other celestial bodies whose coordinates could be
calculated with greater or lesser accuracy. Thus
there are many observations of this sort in his continuous records (L. 52, 53), there is a note of a solar
conjunction for 1638 (L. 109), there are lunar occultations for 1639 (L. 41; cf. Pingré, p. 141), 1642
(L. 18), and lists of comparable phenomena for the
years 1638 (L. 40), 1640 (L. 46), 1642 (L. 48), and
1643 (L. 47).

2. Observations of altitude and azimuth, made

along the lines indicated in the last section, namely with the free pendulum. The best extant sources for this material are the Paris MSS, from which Pingré abstracted much.[72]

Such materials as these were meant to yield reliable planetary coordinates (ecliptic longitude and latitude), and to correlate them with the times of observation in order to improve the theory of planetary motion. Markgraf made many calculations of a sort that was necessary if he were to derive new parameters for the planetary orbits, but these calculations are disordered, incomplete and unexplained, and I can see no purpose in trying to edit them as he himself might have done. In order to integrate his material with that from classical and contemporary European astronomy he must needs use the same time system—hence the importance to him of an accurate figure for his terrestrial longitude.[73] In order to calculate longitudes by method (2), i.e. by reference to the positions of Sun, Moon, or stars, he had to use the best reference works available to him—the Rudolphine Tables and Bayer, supplemented by his own work—as well as a theory of atmospheric refraction—again Tycho's. With what results?

Sad to say, there is not the slightest evidence that Markgraf ever reached the stage of integrating his calculations of planetary positions with a comprehensive theory, new or old. Indeed, although he occasionally compares the latitudes and longitudes deduced from his observations with those calculated on the basis of the Rudolphine Tables (e.g. in L. 26 and L. 44), he takes the matter no further. The few ecliptic longitudes for Mercury that I have reworked from his observations are, as it happens, generally no more accurate than those he could calculate from the Rudolphine Tables, and only the most generous of optimists can see in Markgraf's work the seeds of a reformation of the Mercury theory. The situation with the remaining planets is no better. All that we can say is that Markgraf died too soon to prove himself a member of that élite on whom he so clearly modelled his activities.

SUN, MOON AND ECLIPSE

Much the same is to be said of the fate of the solar and lunar observations. Apart from those with a more or less immediate justification (in particular, those for terrestrial latitude mentioned earlier), surviving materials give us no indication of the theoretical use Markgraf intended to make of them. The Leiden sunspot drawings, for instance, are associated with no periodicity or rotation calculations (L. 22, 24), and no obvious use is made of some measurements of solar diameter (L. 32). Lunae-solar ephemerides and related calculations are drawn up on the basis of the Rudolphine Tables (thus L. 16, 23, 26, 28–31, 37, 39). There is a possibility that some solar parameters quoted for 1637 (L. 25: apogee 96°10′9″, eccentricity 2°1′10″) were based on observation; but the eccentricity in particular could have been improved had Markgraf consulted Kepler's *De Stella Martis* and been prepared to diverge more from Tycho. A shortage of books is not itself an explanation, since this document presumably dates from the Leiden period. The fact that Markgraf adopted a geocentric model in this document may or may not be significant, as an indication of his world-view.

There is no doubt, however, that Markgraf wished to improve on the theories he inherited. He might well use them to draw up lists of stars which the Moon might occult, but only as a step to observing the occultations themselves, and so obtaining precise lunar positions (see L. 40, 48, 55, 56, and 58). His interest in eclipses, of course, was based on this need for precise positional information.

The eclipse records have for some reason attracted more attention than any other part of Markgraf's legacy. Johan Maurits' order to the masters of his ships to record the details of eclipses was obviously given at Markgraf's request, and three returns survive in the Leiden papers (L. 1 for the solar eclipse of 1640 Nov. 13, the ship being 23 miles off the Brazilian coast at lat. 20°5′ S.; L. 2 for the lunar eclipse of 1642 April 14, the ship being near the entrance to the Panama river; and L. 4 for the same, from Loanda). Caspar Barlaeus[74] makes a great

---

72. Op. cit., p. 141 for 1639; pp. 145–148 for 1640; pp. 161–162 for 1642. Cf. the Leiden papers for 1637 (L. 52) and 1639 (L. 53).

73. Note that this difference is calculated on the same scrap of paper as the lunar occultations for 1639 (L. 41).

74. *Rerum per octennium in Brasilia et alibi gestarum, sub praefectura illustrissimi Comitis I. Mauritii Nassaviae etc. Comitis, Historia* (1st ed. J. Blaeu, Amst. 1647; 2nd ed. T. Silberling, Cleves,

deal of the 1640 eclipse, and sings the praises of Markgraf for his having so precisely recorded it ('deliniavit'—in fact Markgraf includes a couple of rough sketches in his observational record, L. 6, a much handled document). In this, Barlaeus was well justified: Markgraf's observations were detailed and well chosen. (In the unlikely event that a present-day astronomer wishes to use them, he should note that neither Barlaeus nor Pingré, nor the Paris MSS, are fully accurate.)

Many of Markgraf's papers show his concern with eclipses (L. 3, 5, 7, 8, 10, 11, 12, 14, 17–19, for example). Some of these I have already mentioned in connection with Markgraf's movements (L. 7 mentions Fort Ceulen) and his derivation of a terrestrial longitude (L. 10). There are calculations of anticipated eclipses, and closely written notes for the Rudolphine Tables, with page references, on appropriate methods (L. 14). It is perhaps worth mentioning again the document listing eclipses between 1640 and 1644, on which 'observavi' is written against the solar eclipse of 1640, and where two eclipses are related to Guinea and Angola, as well as Brazil.[75] In all this, it should not be overlooked that numerous European astronomers (including Gool, Holwarda, Hortensius, Langren, and Wendelin, here in the Netherlands) were keeping eclipse records; but that many eclipses are invisible in Europe. The solar eclipse of 1640 was one of these, and in Pingré's *Annales Célestes* Markgraf and the 'maître de navire' share the entry alone. As for the solar eclipse of 1644 March 8 ('in Brazil et vid.'—L. 7), there is no entry in Pingré.[76]

VIVIMUS INGENIO

The Leiden papers are leavened by only a few non-astronomical fragments. There is the torn and much altered draft title page for the *Natural History* (L. 110). There is—on a sheet with some reduc-

tions of stellar observations—an alphabet for 'secretary' script, with a short text (practice for a will?)[77] beginning 'Op heden den vyfden Meerte Anno Vyftienhundert twee-en-zeventich, stilo curio...' (L. 107). This script was probably not Markgraf's—but the sheet bears testimony to what is evident throughout the Leiden papers, namely the precious character of usable scraps of paper. We find a sheet with a classificatory scheme for musical instruments (L. 86) on one side re-used for a horoscope. Mixed in with astronomical jottings we find a note about an elephant's head from Angola and reindeer hide from Lappland (L. 109). I mentioned earlier two medical items (L. 106, 108). Highly interesting are some brief notes (L. 102) on the already legendary Nova Zembla expedition (notes which mention years between 1594 and 1597), which had attracted astronomers' attention because of the excessive atmospheric refraction. This—for example—caused the Sun to stay above the horizon for three days more in November than was expected.[78] It is clear that Markgraf's curiosity was selective and that it ran chiefly along such scientific lines.

In one of the Paris MS copies we find an interesting statement concerning the Magellanic clouds, which are said to be without stars, and of the same matter ('eadem materia') as the Milky Way.[79] (Forty years later, Edmond Halley at St Helena was to resolve stars in those clouds, with the help of a much better telescope than was available to Markgraf.) Scientific as his approach to the world might have been, there is the same sort of rift between his work and contemporary natural philosophy as we find with so many other astronomers and 'eadem materia' is the closest approach to natural philosophy I have found in his writings. Markgraf was cast intellectually in a more or less traditional mould. His numerous astrological pieces support this contention; and the number of surviving horoscopes, not to say the care with

1660, pp. 329–332, pp. 330–331 being set aside for the circumstances of the eclipse).
75. L. 7. See n. 46 above.
76. This eclipse was visible only in parts of the Pacific, South America, and the Atlantic. For further details of eclipses, and charts showing visibility, the standard work is Th. von Oppolzer, *Canon der Finsternisse*, Vienna, 1887 (repr. with tr. and editorial material by O. Gingerich and D. H. Menzel as *Canon of Eclipses*, New York, 1962).

77. The name of Selleninck occurs.
78. G. de Veer, *Reizen van Willem Barents, Jacob van Heemskerck, Jan Cornelisz. Rijp en anderen naar het Noorden (1594–1597)*, ed. S. P. L'Honoré Naber (Linschoten Vereniging, xiv, 1917). (In English by the Hakluyt Society, 1876.) For Simon Stevin's interest in the matter, and further references, see *The Principal Works*, ed. E. Crone et al., iii, Amst., 1961, pp. 314–317.
79. A.B.4.5, p. 12.

which they were prepared, vouch for his sincerity in that connection.[80]

Markgraf's life reflects many of the scholarly ambitions of his age. He was fortunate in his patron, less so in the West India Company and its short-sighted policies. There has, of course, always been more to patronage than merely giving a disinterested scholar an opportunity to develop his talents. The prince must pay for his artistic and scholarly tastes, whether indulged for their own sake or for the sake of dazzling others. Johan Maurits was able to build and to collect in South America in ways which would have been impossible to a minor princeling in Europe, where the *objets d'art* of the great collections were more costly by far than the exotic flora and fauna of Brazil—at their Brazilian price of course. But why an astronomer in his retinue? Did he consider that his court needed its astrologer? There is little or no evidence that Markgraf served him in this way.[81] Did Johan Maurits put a high value on grandiloquent title pages in his honour? I have suggested that Markgraf saw himself as a southern reflection of Tycho. Did Johan Maurits consciously liken himself to Tycho's noble patrons? Whatever the answers, George Markgraf surely knew the exchange value of his talents. He, like his patron, lived a life in Brazil in many ways richer than any available to him in Europe. As he wrote, when trying out his pen in the midst of a sheet of calculations, including some of the width of a river and the height of a palm tree (L. 105), 'Vivimus ingenio'—'We live by our wits'!

---

80. I discuss this material in an appendix below. Note also a booklet with principles for interpreting horoscopes (L. 80, possibly taken from a preface to Eichstadt's ephemerides) and two aspect calculations (L. 81–82).

81. See the following appendix.

APPENDIX

THE HOROSCOPIC MATERIAL IN THE LEIDEN MANUSCRIPTS

Gemeente Archief Leiden, Bibl. 12130

The horoscopic material, like the Leiden MSS as a whole, is almost completely lacking in organisation. Whatever plans Markgraf might have had for the bulk of the astronomical material, it is unlikely that he ever contemplated publishing the horoscopes, and since most of them are uninterpreted, we may treat them only as the astronomer's 'file copies', the bare mathematical bones of the predictions he presumably supplied to his friends— and perhaps also to his patrons. (There is no sign here of a horoscope either for Johan Maurits or for Willem Pies.) The horoscopes are of interest for the small fragments of information they reveal, information which might be thought incommensurate with the difficulty of extracting it and with the length of the following summary. I give this summary in the hope that it might serve some biographical purpose. It is necessary to add a warning, however. Most of the horoscopes are for people identified either through their initials or not at all. We can make conjectures as to their identities, especially through their birth places, which are occasionally recorded and which may occasionally be inferred by mathematical means. Speculation is a great temptation, but we should remember that Markgraf's potential acquaintance was very large, and the fact that three people (at least) have surnames beginning with the letter K does not imply a family connection, any more than does the fact of our finding two sets of initials 'D.K.' imply that they belong to one and the same person.

For convenience of reference I shall itemize my findings. The documents are now numbered, besides which any horoscope is identifiable by its date, and this information I give in the following form: [L. 88: 1610.9.30, n.s.]. This refers to Markgraf's own birthdate, 1610 September 30, new style. He used the new and old styles of the calendar quite indiscriminately. To convert, add 10 days to the old style date to obtain the date in the new style.

1. There are 45 sets of calculations relating to horoscopes among MSS L. 80–101. (Often the

calculations are very full, but occasionally they are incomplete. Markgraf's accuracy in calculation is exemplary, and the lacunae do not reflect on his skill.) Since many of the calculations are reworked or copied out more than once, the number of persons involved is much smaller than 45. There are the results of 28 distinct sets of calculations, possibly relating to 27 different people. In short, there seem to be materials for 26 birth horoscopes (nativities), and two specimens of a 'revolutio thematis', cast for a later period in a person's life.

2. One of these themata is cast for I.B.B. [L. 85, 100, 101: 1645.6.27, n.s.], who is undoubtedly the same man as was born—according to a nativity discussed below—on [L. 84, 85: 1607.6.28, n.s.]. Both are for the latitude of Loosdrecht, named in some of the five relevant documents. See below. The thema was therefore for a birthday. The second thema was for D.K. [L. 94, 95, 96: 1632.9.12, o.s.]. I deduce from the thema a different latitude from that in the birth horoscope for a D.K. [L. 7, 87: 1617.7.20, n.s.], and noting that the thema would not have been for a birthday anniversary, feel reasonably sure that the two D.K.s stand for different people.

3. Only three of the nativities relate with any certainty to people born in Brazil, and the subject of one of those (Fitiola Helde—see 18 below) was born before Markgraf arrived there. The most interesting group is of subjects born before 1617 (eleven in all), and we do not know where Markgraf made the acquaintance of most of these. The other conspicuous group is seemingly of children of a single family, possibly the children of a branch of the 'K' family, and born perhaps in Stettin or Holstein (16 below). Note the very wide spread of birth places, even discounting the southern hemisphere. Note that the two Mauritia births were apparently to French families.

4. The latitudes of towns are occasionally given, but the longitudes on only five occasions. The longitudes are based on the meridian of a place approximately 24° W of Greenwich. It is highly probable that Markgraf was using Willem Blaeu's atlas (in one of its many editions), or longitudes to be found with its aid. (The place of reference is a meridian further west than the Fortunate Islands (Canaries), through which Ptolemy's prime meridian passed. The more westerly meridian, used in Arab geography, is made standard in Europe by

Mercator's world chart of 1569.) As is to be expected, the latitudes quoted are accurate to 6' arc, or better, while the longitudes may be in error by more than 2 degrees. The internal consistency of latitudes deduced by me (from the ascendent point and *imum medium coelum* in each case) for what I suppose to be a family group (see 3 above) is such that I am reasonably sure that my deduced latitudes are all within 2' of what Markgraf took the latitudes to be.

An outline of the horoscopes, more or less in chronological order, is as follows:

5. [L. 83: 1594.5.26, o.s.] for J.F., written also J. Faber (?). Place: 'Husii (Hol.)' From deduced latitude this is not in Holland, as it might have been thought, but is Husum (Holstein). See 16 below.

6. [L. 84, 85: 1607.6.28, n.s.] for J.B.B., born Loosdrecht. Two documents, from the very end of Markgraf's life. It is tempting to suppose that J.B.B. was a member of the Bronckhorst-en-Batenburg family, but Loosdrecht is not near the family seats (Rimburg and Gronsveld).

7. [L. 86: 1607.9.21, o.s.] for M. Christof. Notvayne (?), born Hilperhusum. Quoted latitude and longitude make it clear that the place is the modern Hildburghausen, about 60 km north of Bamberg. 'Aliquis Coburgi'.

8. [L. 87: 1609.2.11, o.s.] for G.H. à B., born Altenburg, modern Altenberge, about 12 km northwest of Münster.

9. [L. 87: 1610.2.22, o.s.] for A.F. von H., born Gustrovius, modern Güstrow, about 35 km south of Rostock. Could this be 'von Holstein' (see 5 above & 16 below)?

10. [L. 88, 89, 90: 1610.9.30, n.s.] for Markgraf himself, born Liebstadt, south of Dresden [Pl. XVII]
    Neither his name, nor that of the place, is entered on the horoscope. Three documents, one being a trial calculation only; one being neatly written out, but having the top corner deliberately torn away; and the third containing very careful calculations, both according to the Rudolphine Tables and 'according to the rule of Kepler' (for a thema). Since a few details are worked out for the 28th and 29th years of his life, we may suppose a date of around 1638 for the third document. Perhaps it was worked out during his idle moments on board ship. There is a surprising discrepancy between the latitudes used in the three documents, as though he has confused the lat. of Liebstadt (taken

y him once as + 50°54', present town centre
+ 50°52') with the latitude of Gera (see 15 below).

11. [L. 91: 1611.6.15, o.s.], no name or initials,
birthplace Rewalia, modern Revallia or Tallinn
Estonia, opposite Helsinki in the Gulf of Finland).

12. [L. 87: 1613.1.20, o.s.], for G.K., born Stet-
in, modern Szczecin.

13. [L. 92: 1614.5.27, o.s.], for Th. Pursen (?),
born Hafnia, modern Copenhagen. The name,
which is difficult to read, might in any case be a
phonetic rendering of Pedersen.

14. [L. 91: 1615.2.7, n.s.], for S.K., born Prague.
s this Samuel Kechel, Markgraf's fellow lodger at
Leiden, and later mathematical teacher there? In
NNBW II (1912), p. 651, Samuel Carolus Keche-
ius van Hollesteyn (to adopt the spelling used in
university documents) is said to have been born in
Prague in 1611. See remarks under 16 below.

15. [L. 7, 87: 1617.7.20, n.s.], for D.K., so named
in one of the two documents with material for
this horoscope. The name of the place is difficult to
decipher, but I deduce the modern Gera, while
what is written certainly reads G[...]â. On one document there are eclipse notes which I presume to
have been written in Angola, at the very end of
Markgraf's life.

16. A series of horoscopes, and associated calcu-
ations, all repeated in two documents (L. 93, 94),
with neither names nor places. The subjects are
given simply as 'filius' or 'filia' (four of each, spread
over 14 years, with no two closer than 13 months).
The latitudes I deduce all strongly suggest Markgraf's figure for the town of Stettin. The same is
true of the latitude deduced for the thema for
D.K., 1636, for which there are three documents
one admittedly quoting a latitude which seems to
read 48°). This thema is to be found in a bound
booklet containing only this 'family', and I assume
that the family was therefore 'family K'. Can these
be siblings of Samuel Kechel? They cannot be his
children. One might conjecture that all the K's are
Kechels, the family having originated in Holstein,
possibly with J. Faber (b. 1594) of Husum and
A. F. von H. as members by marriage. The Kechels
might have travelled much (Stettin, Prague, Gera,
Stettin). On the other hand, where the town of
Stettin is not named, a birthplace might be in Hol-
tein, or at any reasonably close latitude, for we
know that Markgraf would have had access to
Eichstadt's Stettin tables for astrological work, and

that he might have accepted them as sufficient for
other places. Husum is rather far (a degree) to the
north of the town.

Dr. Nicolette van Santen-Mout has very kindly
drawn my attention to a number of different references to Samuel Kechel and his father (Jan or
Hans). The father, who took the appellation 'von
Hollenstein', matriculated at the university of Leiden on 23 November 1598 under the name of
Joannes Kecchel Pragensis. He became a rich Prague burger, but after 1620 went into voluntary
exile, presumably as a result of Calvinist leanings.
(Note that Markgraf's own father went northwards to Dobra, about 50 km from Stettin.) See
N. Mout, *Bohemen en de Nederlanden in de zestiende
eeuw*, Leiden, 1975; C. Merhout, *O Malé Straně*,
Prague, 1956, p. 135; A. Ernstberger, *Hans de Wit-
te, Finanzmann Wallensteins*, Wiesbaden, 1954,
p. 73.

Details of the 'family' of nativities are as follows
(old style throughout; s = son, d = daughter):
[1622.9.16, d], [1623.11.6, s], [1624.12.10, d],
[1626.9.30, s], [1627.12.7, d], [1629.1.18, s],
[1630.6.15, s], [1635.3.12, d]; the thema for D.K. is
for [L. 94, 95, 96: 1632.9.12].

17. Chronologically mixing with these, but not
otherwise connected, are calculations for nativities
(no calendar style given): [1624.5.24],
[1625.11.25], [1631.3.15]. These are all on one document (L. 89), which also includes a calculation
for his own nativity. See 19 below.

18. [L. 93, 98: 1636.7.1, n.s.], for Fitiola Helde,
born (before Markgraf reached Brazil) on the
island of Antonio Váz. Two documents, one
(L. 98) stating that the child died in 1639.

19. [L. 89: 1639.11.22, n.s.], no name, but for
Brazil. Associated with nativities listed under 17,
above.

20. [L. 99: 1643.3.12, n.s.] for Freder. Valloy,
born Mauritia.

21. Record of birth of Henry Bonjour, Mauritia
[L. 88: 1643.5.3, n.s.].

22. 'Revolutio thematis' for J.B.B. (see 6 above)
at [L. 85, 100, 101: 1645.6.27, n.s.], and at a very
precise moment of time (18h 25m 19s)! Three documents, of which one shows that Markgraf calculated 'directiones', or significant astrological events
for J.B.B. at ages 5, 9, 10, ... 37; it is thus of no significance that 1645 is after Markgraf's own death.

23. An important conclusion to be drawn from

this horoscopic material is that it was as meticulously calculated for subjects in a none too elevated social position as it could have been for any prince. It
was as carefully calculated—or almost so—for his
friends as for Markgraf himself. And there is never
any shadow of doubt about the sincerity of the calculator, who undoubtedly thought his efforts meaningful.

Addenda

For Christian Markgraf's biography of his brother, see now P.J.P. Whitehead. 'The biography of Georg
Marcgraf (1610-1643/4) by his brother Christian, translated by James Petiver', *Jrl. Soc. Bibliog. Nat. Hist.* 9
(1979), 301-14.

The horoscope I assigned to 'DK' is, as Mr W. Downer points out to me, for 'D. à K.', and therefore the
chance that it refers to a Kechel is much reduced.

Mr Jorge Polman has raised in correspondence with me the interesting possibility that the papers now in
the Paris Observatory describe Markgraf's observatory as depicted in the Wagner painting, that is, that they
might relate to the Portuguese house, rather than the Vrijburg palace. His arguments are well set out in a
pamphlet *Markgraf e o Recife de Nassau*, Clube Estudantil de Astronomia, Recife, 1984. It is very difficult to
decide between the alternatives, for the uppermost sections of the towers on both buildings were roughly
similar, as far as we can judge from the surviving drawings. The height implied by the 43 steps mentioned in
the text is an important clue, except that we cannot be sure of the point from which they commenced. I see no
problem in the curvature of the 'pyramids' on the Vrijburg towers, for observing hatches were often made in
curved roofs, as they were at Uraniborg and still are today. The problem with resting a hypothesis on the
unsuitable character of the Vrijburg towers as an observatory is that it rules out too much, for where, then
was Markgraf's second observatory? It would be surprising, too, if the second observatory were radically
different in plan from the first. The architectural historian J.J. Terwen found my explanation well suited to
his reconstruction of the plan of the Vrijburg. There are also the problems of the large clock in the tower, and
the Prince's chamber (see p. 255 above). Despite these difficulties, Mr Polman's suggestion offers distinct
possibilities.

Since much interest attaches to the question of the introduction of western astronomy to the New World,
a reference should be added to Nicholas Tyacke, 'Science and religion at Oxford before the Civil War', in D.
Pennington and K. Thomas (eds), *Puritans and Revolutionaries*, Oxford, 1980, pp. 73-93. In 1631 John
Bainbridge, Savilian professor of astronomy at Oxford, laid plans for an astronomical expedition to South
America, and employed Roger Fry, a sea captain. The expedition was attacked by the Portuguese, and Fry
was taken as a prisoner to Brazil. There, undaunted, he made a series of astronomical observations, the first
of which were duly sent back to Oxford, to Bainbridge's delight. This was all a few years before Markgraf's
arrival. Fry's observations included one of a solar eclipse that eventually allowed a longitude determination
of São Luis do Maranhão, relative to Oxford.

10

FINITE AND OTHERWISE:

ARISTOTLE AND SOME SEVENTEENTH-CENTURY VIEWS

It is unwise to discuss so pervasive a cluster of concepts from 17th century natural philosophy as space, matter, extension, and infinity, without referring to Aristotelian and scholastic doctrine. One may argue far into the night over the question of whether Newton's words reveal a dependance on Burthogge, rather than Herveus Natalis, and whether Descartes at a particular place was using Anselm or John Damascene; but one thing we must surely accept is that the *language* used was that of the Aristotelian inheritance. Of course it was changed in many ways; but just as when trees are turned into telegraph poles, the grain of the wood remaining visible, so with the language of Aristotle. Since we are particularly concerned with doctrines of the infinite, I will give the merest outline of those parts of Aristotle which seem to me to have a bearing on the 17th century discussion. I will then place rather more emphasis than Ted McGuire is inclined to do on the common element in the thought of Descartes, Locke, and Newton. In the time available I shall not be able to make more than passing reference to scholastic discussions of the same themes, although I am sure that to do so would show that, whether he liked it or not, 17th century man was a lineal descendent of the scholastics. As it happened, Georg Cantor gave the palm to the seventeenth century for its arguments against the actual infinite — arguments which he judged to be more cogent than Aristotle's, but arguments which he believed he could refute. He commended Locke, Descartes, Spinoza, and Leibniz, while suggesting Hobbes and Berkeley as additional reading.

ARISTOTLE ON THE INFINITE

Aristotle opens his discussion of the infinite with a sentence that could hardly have been bettered as a source of questions for debate:

The science of nature is concerned with spatial magnitudes and motion and time, each of these being necessarily either infinite or finite — even though some things dealt with by the science (e.g. a quality or a point) are not, and should perhaps not be, put necessarily under either heading.[1]

After a brief survey of the views of the Pythagoreans, Anaxagoras, the atomists, and Plato (of whom more anon), Aristotle sets out his own position. He distinguishes (i) the infinite in respect of addition, and (ii) the infinite in respect of division – number being infinite in the first sense, space in the second, and time in both.[2] Even more fundamental, however, is Aristotle's (perfectly natural but also perfectly dangerous) insistence that he takes the word "infinite" in the sense "*that which cannot be gone through*". It might be of the nature of a thing that it is "incapable of being gone through" (and "this is the sense in which the voice is invisible" – an example repeated at *Met*. 1066a. 35–); the process of traversing might be possible, but have no termination; the traversal might present difficulties; or it might not be *actually* begun or ended. There are many echoes of these distinctions in the middle ages. Aquinas and Ockham accepted them, more or less – to name only two influential writers.[3]

Here, then, is one statement amounting to a definition of what Aristotle meant by "infinite". Even though the *Physica* is a work chiefly devoted to the study of the objects of sense, Aristotle clearly wanted the word to take its primary meaning from mathematics – where the integers provide a paradigm of a sequence which "cannot be gone through". In *Metaphysica* (K. 10), he shows that he has seen, and studiously left alone, two concepts beloved of later theologians. His infinite is not a "separate, independent thing", nor is it "the indivisible". The first of these would reduce to the second – or so he appears to think – and he is simply "not examining this sort of infinite, but the infinite as untraversable". For Aristotle, the infinite must be "of a certain quantity", and it cannot, in consequence, be indivisible.[4] It is emphatically not The One.

Another concept that Aristotle seems to be carefully avoiding is *space*, at least in the sense of χώρα. He has much to say about *place*, and *spatial magnitude*, and the *void*, however. Place is the limit of the container, the first (i.e. working outwards) unmoved body that bounds the thing with whose place we are concerned.[5] Everything in the universe is in place, but not the universe itself.[6] Clearly what he says about *body* has implications for place – and thus (since for Aristotle there can be no infinite body) it is misleading to say, as is occasionally said, that "Aristotle supposed space to be infinite in extent". Again, on the question of the void, it is too simplistic by far to say only that Aristotle "rejected the notion of void". Anticipating a little, in order to put the problem of the void to one side, I should like to draw attention to an extremely interesting analogy between the void and the infinite, as drawn by Aristotle in the *Metaphysica*. There is, he explains, no actual void (whether separate from bodies, occupied by bodies, or in the interstices in bodies), just as there is no actual infinite; but just as a line is infinitely divisible, so we can always imagine a body of lesser density than any specified density. Matter, for Aristotle, is a continuum, but there is no lower limit to the density of matter.[7] I imagine that the typical 19th century exponent of the physics of the continuum would have said much the same sort of thing.

Aristotle's discussion of the infinite is not one of his most systematic, but running through it all are two important conceptual distinctions. The first, between "infinite in respect of addition" and "infinite in respect of division" – between infinite and infinitesimal, as some would later have said – has already been mentioned. The second is the notorious distinction between *potential* and *actual* infinite. Since he is especially concerned with the concepts of number, space, time, and body, in this connection, we can see that such a question as "Is there an actually infinite body with respect to addition?" is one of sixteen questions that Aristotle deemed important (four topics with two twofold distinctions). Needless to say, there are other topics of interest than the four named; and all the questions of existence can be turned into questions of consistency – "Is the notion of an infinite body inconsistent?", and so on. In fact Aristotle attaches great importance to this last sort of question, as we shall see.

Consider first the *Physica*, where, although making passing reference to such general mathematical questions as "whether the infinite can be present in mathematical objects and things which are intelligible and do not have extension", he is mainly concerned with objects of sense. Is there an infinitely great body? Aristotle argues at length that there is not. His arguments take many forms, and are often unsatisfactory to the extent that they depend on his theory of the natural places of the four elements and the "balance" of contraries. He makes several strong philosophical points, even so. If, for instance, we mean by "body" something that is "bounded by a surface", then the notion of "unlimited body" is inconsistent.[8] On the other hand, to suppose that the infinite does not exist in any sense leads to difficulties: (1) there will be a beginning and an end of time; (2) a magnitude will not be divisible into magnitudes; (3) number will not be infinite. "Clearly there is a sense in which the infinite exists and another in which it does not".[9] It would be easy, but misleading, to say that for Aristotle the existent infinite is potential and the illegitimate infinite the actual (or better "completed"). He will certainly allow potential infinity to number, space, and time (with respect to addition) and to space and time (with respect to division). They are potentially infinite in the sense that they may grow beyond any given bound.

There were tensions here, of course, in the Christian commentaries, for obvious reasons: Christians and many Muslims supposed the world to have been created at a specific time, and to be destined to end at another definite time. But there were tensions too in Aristotle's own thought, for his universe was, after all, limited in spatial extent. Could there not be *something* potentially infinite beyond the bounds of the finite universe? The scholastic, no less than the Christian philosopher of later centuries, found problems here — problems of the heavens beyond the stars, problems of the implied limitation to God's power in saying that his creation is only finite, and so forth — but they were problems of a different sort from those of Aristotle, who does not seem to me to have come completely to terms with his potential/actual distinction. I will indicate briefly some of the difficulties.

"In respect of addition", he writes," there cannot be an infinite which even potentially exceeds every assignable magnitude unless it has the attribute of being actually infinite, as the physicists hold to be true of the body which is outside the world, whose essential nature [they hold to be] air or something of the kind".[10] We have seen why he rejects infinite body; but surely the quotation just given concerns more than the case of infinite body. Suppose we accept his various arguments for the finiteness of the universe, and his claim that no sensible magnitude is infinite.[11] In view of the quotation, why, however, will he not grant actual infinity ("with respect to addition") to *time*, to take an example of something to which he is prepared to grant potential infinity? It has potential infinity, we are told, in that it may grow beyond any given bound; but it is said to be no real or actual infinity in the sense of an infinite given whole, since *its parts do not coexist*. I think it is clear from this remark that Aristotle is putting a great deal of weight on an analogy — in the words of the scholastic, his "actually infinite" is "the infinite of the heap". We might be inclined to associate a number with each point of time, but the number is not a "magnitude" in the sense used in the quotation. This is not at all strange; but it does seem strange to us to refuse to call an *interval* of time a "whole", and therefore presumably to deny that it is a magnitude. Aristotle, like every other physicist of significance, was prepared to think of time intervals as something that could halved, doubled, and so on, whether or not the parts of time could be said to coexist.

One of the most tempting interpretations of Aristotle's "potentially infinite" is as simply "conceivably infinite". But what does this mean? It would be plainly trivial to say that it means "conceivably potentially infinite". The danger in the most obvious alternative is that it be treated as a visualisable thing, a conceptual object in a crude sense. There are remarks by Aristotle with a bearing on this. The geometers are said to neither need nor use the infinite; and yet — adds Aristotle — they postulate "that the finite straight line may be produced as far as they wish".[12] Aristotle's concession to the geometers has suggested to some that he accepted a potential infinite in the sense of a special sort of separately existing thing. The only "thing" it can be is the possibility of a certain sort of mental activity which is in principle unending. (Note how he speaks of mathematicians' thoughts as an actuality.)[13] In an unsophisticated sense of the word "actual", it is Aristotle's main purpose to put an end to talk of an actual infinite, and at its very heart his analysis rests on an analysis of meanings. "A quantity is infinite if it is such that we can always take a part outside what has already been taken", and yet "whole" and "complete" are such that there is nothing more to be taken (at least in the respect considered as making the whole a whole). An actual infinite is a contradiction in terms.

Without modifying such definitions, no "completed infinite" is possible, and the most interesting parts of Ted McGuire's paper seem to me to be where he is concerned with 17th century efforts at revising definitions. In the end, the revisions seem to have been more apparent than real. Before turning to them, however, let me mention a close analysis of many Aristotelian texts made by Jaako Hintikka, who asked why Aristotle, who certainly seems to have firmly believed that no genuine possibility can remain unactualised through an infinity of time ("the principle of plenitude", as Lovejoy called it), gave infinity potential but not actual existence.[14] His conclusion, if I may summarize it in an excessively brief way, is that the principle of plenitude *is* satisfied, in the sense that "the endless coming-to-be of further and further divisions is 'ensured' (Aristotle's verb is *apodidômi*) by the potential existence of the activity of diving".[15]

There are other ways of sidestepping Aristotle's conclusions, as I have indicated, by the expedient of modifying his definitions, but it is worth drawing attention to the stipulation that we must be able to "take a part *outside what has already been taken*". To those who, with Riemannian hindsight, would throw at him the modern notion of finite but unbounded, his answer is waiting in the middle of our *Physica* III. 6, where he considers the case of an endless ring. With such a thing "it is always possible to take a part which is outside a given part"; but — despite the similarity with the unacceptable notion of a "completed infinite" — the cases are not fully alike, for "it is necessary that the next part which is taken should never be the same". He looks for endless enumeration in the supposed counterexample, and does not find it. His ring is indeed finite and unbounded, and had he been more concerned with the property of boundedness, he would no doubt have said as much.

As Aristotle's doctrine of the infinite was handed on to the 17th century, it was modified, disjointed, and mixed with an extraordinary variety of extraneous ideas. There was the Platonic identification of the infinite with the indeterminate and formless — which could hardly therefore be identified with a Christian's God. We find a line of thought coming from Plotinus, who takes the Platonic doctrine (in connection with matter, which could be supposed evil with impunity, since it could not answer back), and adds a theory of infinite Mind, complete, self-sufficient, and all-powerful. Whenever God (or some mind-like entity somehow involved in the world) entered the discussion, it became doubly complicated, for not only were there extra classifications to be made (one might even say "hierarchical" ones, since God must always come out better than his creation), and problems of *relating* the different sorts of entity (God, the world, space, time, and so on), but there was also need for a revision of the very *concepts* of the infinite. *Number* tended to take a secondary place, elbowed out by the ineffable — whether it be essence, the unquantifiable act, reason, power, goodness, or whatever. All the mystery of the numerical infinite, forever eluding the human grasp, became transferred to the divine infinite — but in logical structure the two subjects were very discrepant, however hard the scholastics tried to run them together. The ever present danger was of course that of pantheism — with the world sharing in the divine infinity. This is found in forms as different as Giordano Bruno's and Spinoza's. (And neither of these seems to me to have handled the notion of the infinite in a conspicuously consistent way.)

DESCARTES AND THE INDEFINITE

One of the mysteries of the history of concepts of the infinite is why Descartes's distinction between the positively infinite and the indefinite has been adjudged — by his contemporaries no less than by latter-day historians of philosophy — so momentous. Some have gone the other way, and have seen it — but without good reason — as a reflection of Aristotle's distinction between the actual and potential infinites. Of course Descartes places God in a central position in his philosophy, and restricts the positively infinite to God; and Aristotle could never have been persuaded to do anything comparable with his "actual infinite", a contradiction in terms as far as the corresponding infinite by addition" is concerned. But an even more important difference is that where Aristotle has what one might call a *constructive* theory of the potentially infinite, Descartes inclines to the phrase "I do not know". To quote yet again his much quoted remark to Henry More:

> God alone do I understand to be positively infinite: of other things, such as the extension of the world, the number of parts into which matter is divisible, and so forth, I confess I do not know whether or not they are straightforwardly infinite; I know only that I can recognize no end [*finis*] to them, and so — in respect of myself — I call them indefinite.[16]

It is easy to see in this passage from a letter of 1649 (which can be matched by many similar passages in his writings) a protestation of orthodoxy. Was the infinity of the physical world theologically acceptable to his contemporaries? Christina of Sweden, through Chanut, expressed her fears, but Descartes was able to cite Nicolas of Cusa, a Cardinal, who had got away with the idea.[17] Whether he understood Cusanus aright is immaterial, for — as we see — he confesses to being *ignorant* of whether the extension of the world is absolutely infinite. In fact he is here following the Cardinal rather closely, for just as Descartes distinguished infinite from indefinite, so Cusanus distinguished *infinitum* from *interminatum*. For him the world was simply *interminatum*, without termination, lacking bounds, undetermined — and therefore, according to him, subject only to partial and conjectural knowledge.[18] Resisting the temptation here to go back from the fifteenth century to the fourteenth in search of the genealogy of the concept *interminatum* (we might have looked at Ockham's *illimitatum*, but we might continue and come perilously close to Aristotle himself)[19] we ought at least to note that Nicolas of Cusa seems to have a tripartite classification in mind: "though the world is not infinite, yet it cannot be conceived as finite ... ".[20] It must, he imagined, be the third thing, interminate. His proto-Hegelian synthesis of opposites coinciding in a transcendent absolute, all grasped through a "learned ignorance" which rises above discursive rational thought, is not Cartesian philosophy, but it does seem possible that Cusanus' conceptual distinction might have had some sort of influence on Descartes. (In a letter to Descartes, Burman credited him, Descartes, with having been the first to apply the infinite/indefinite distinction. Descartes does not appear to have offered any comment.) As I have already said, what seems to me puzzling is why the distinction should have seemed so world-shattering. I think some of its appeal must be put down to its appropriateness to a philosophy of doubt — which I will touch on shortly. Descartes' readers were certainly often in doubt as to its meaning.[21]

FEARS OF THE CHARGE OF IMPIETY

If Descartes was indeed worried by the thought that his utterances on to subject of infinity might lead to an accusation of impiety, he was not alone in his fear. The religious temper might vary as between different countries and different generations, but the eternity problem was felt keenly in most of them. Sir Walter Raleigh, for example, was not a dangerous intellectual heretic, and yet at his trial, in 1603, when his enemies made capital out of his atheism, Justice Popham addressed him in these terms: "Let no devil persuade you there is no eternity in heaven; if you think thus, you shall find eternity in Hell fire".[22] In one manuscript version of the speech, the word "devil" is replaced by "Hariot nor any such doctor". As for the more learned, and by inference more dangerous, Thomas Harriot, the story got around — to be repeated in Aubrey's *Lives*, that he rejected the creation of the world on the grounds that *ex nihilo nihil fit*. No doubt Harriot was influenced, directly or otherwise, by Bruno's lecturing in London (1583–5). At all events, the suspect Latin formula is to be found at several places in Harriot's manuscripts, not the least interesting being where it is said not to contradict *omnia fiunt ex nihilo*. What would Justice Popham have made of *that*? In his Preface to the *History of the World* (written and published during his many years in prison after the trial) Raleigh protested orthodoxy, and censured those who argued from "out of nothing nothing is made" to the world's eternity.[23] Their problem, in brief, is well expressed by Jean Jacquot: "Either you followed Lucretius and denied God and Creation, or you denied Creation in time and deprived God of his freedom".[24] This, at least, is how the problem was seen. Atomism went hand in glove with these two sorts of impiety, more by historical association than by logic; and when, in a correspondence in 1606–7, Harriot tried to persuade Kepler of the virtues of the concepts of atom and void, Kepler fought shy. Even Harriot's friend Nathaniel Torporley tried his hand at a refutation of atomism, fearing that the Christian creation dogma was at stake. No wonder that Harriot complained to Kepler that he could not philosophize freely.[25]

Descartes was closer to Bruno than were Locke and Newton, and if I may be excused for labouring what is obvious to some but not all historians of philosophy, the social pressures on the "philosopher of the infinite" were very different at the end of the century from what they had been in Harriot's day. Even the bold Thomas Hobbes was content to call the question of the duration of the world, and its twin, the question of its extent in space, finite or infinite, questions "unlawful to dispute of". He was, he said, content to leave them in the hands, not of philosophers, but of "those lawfully authorized to order the worship of God".[26] (Seth Ward's *In Thomae Hobbii exercitatio epistolica*[27] reveals a marked dislike of Hobbes' stance. This was all after mid-century.) Three or four generations of philosophers were still, however, destined to debate fervently the *ex nihilo* principle, and to probe the concept of causation in the process, without serious fear of civil redress – although there is a quavering in Leibniz's voice when in or around 1676 he shows concern for the prejudice to piety caused by abuse of "la philosophie nouvelle".[28] Bentley, in his third Boyle lecture, argued that something must have existed from all eternity because nothing can emerge from nothing; but then he went on to say that the "something" was of course God, and thus *ex nihilo nihil fit* was made to look innocuous. The same sort of thing must have been said many times over. It is, for instance, virtually the same as Locke's statement in the *Essay*, IV. 10.4, published a couple of years earlier. Others turned the debate in a more interesting direction. Could a creature be consistently said to be "coeternal with its creator"? You could sling mud at those who thought so, call them Arians and so forth, but they didn't write their books in the Tower of London. Locke, incidentally, prepared his answer to those who thought matter eternal because it could not be made from nothing, and it is in this connection that he hints at a mechanism for the creation of matter by God, a mechanism that seems to have been suggested by Newton, and one to which I shall refer later.

To return to Descartes: his account of the infinite might well have served to placate a few theologians, but it would be quite wrong to present that as its main philosophical purpose. It is in fact close to the very heart of Cartesian epistemology. To put the matter very simply, God and infinity come almost at the top of Descartes' list of certainties. Once again, how very different from Aristotle! We are tempted to suppose that we perceive the infinite as a negation of the finite. No, says Descartes, on the contrary, "I see that there is manifestly more reality in infinite substance than in finite, and therefore that in some way I have in me the notion of the infinite earlier than the finite – namely, the notion of God before that of myself".[29] It is almost as though, not content with putting God before things, in developing his philosophy from the *cogito*, he has begun to feel that somehow he should have put God higher on the list than the thinking self! We are still in a period of history when philosophers could play casuistical games with arguments as to which of two opposed concepts was the positive and which the negative. "Infinite" is grammatically negative, a fact which disturbed Descartes as it had disturbed other Christian writers before him. The remedy was to stress the limitation of the finite, by which it differs from the infinite, and is therefore "a non-entity or negation of being".[30] In the end, as well all know, Descartes made what was seen as an audacious claim, namely that we *can* have a clear and distinct idea of God, but that "our mind, being finite, cannot *comprehend* the infinite" – or all that is in it.[31]

Before going further consider what McGuire calls "Newton's comment on this kind of reasoning", namely Descartes' argument (against Hyperaspistes) that to negate presupposes positive knowledge of what is negated.[32] In fact when we consider the quotation from *De gravitatione*,[33] there is not only "much here that Descartes can agree with"; the drift of Newton's remarks is straightforwardly Cartesian. In playing with words, making *limit* a negative concept and therefore *infinite* a positive one, he is following in Descartes' footsteps. Locke, in his 1671 draft of the *Essay*[34] criticizes this line of argument, but rather oddly suggests that those "who contend for a positivie Idea of the Infinite" (and surely Descartes was one of those intended) were in this matter concerned with *quantity* (extension, duration, power etc.). In examining the question, Locke becomes involved in a discussion that had greatly occupied the thoughts of Aristotle and many scholastic philosophers, namely whether the "end" of a "duration" is the last moment of existence or the first of non-existence.

Even on the question of the relation of God to space, Descartes and Newton differ less than is apparent at first sight. We can place too much emphasis on the passage (from *De gravitatione*) in which Newton charges Descartes with having feared that infinite space "would perhaps constitute God because of the perfection of infinity".[35] Newton counters that infinity of extension is only as perfect as that which is extended – an estimable proposition, but one that misrepresents Descartes' views, not to say his fears. "I dare not call the world infinite", wrote Descartes to More, "for I perceive that God is greater than the world", but "not by virtue of his extension . . . which does not properly exist in God at all, but by virtue of his perfection".[36] Descartes does not set up extension as a rival to God. He avoided doing so by having two separate concepts of infinity. He is quite prepared to *admit* that the extension of the world, the number of parts into which matter may be divided, and so on, might in principle be infinite, and *then* "infinite" seems to mean only "without limit". To emphasize again the letter to More: Descartes confesses that he does not *know* whether these things are infinite or not. To Chanut he admitted that *God* might know limits to the world.[37] The indefiniteness is emphatically a reflection of the failure of our own faculty of knowing. The important thing is that we should not try to match up the infinity of God with that of extension — which "does not properly exist in God at all". Newton must have agreed. For all Leibniz's misunderstandings, due in large measure to Samuel Clarke's having mixed Newton's views on an omnipresent Deity with his own on space as an attribute of God, Newton nowhere (so far as I am aware) speaks of God as extended — infinitely or otherwise.[38]

EXTENSION, PERFECTION, AND DIVINITY

So much for Descartes' fears — which are too easily exaggerated. As for his way of speaking of perfections, here again he and Newton are closer than appears at first sight — despite the (fundamentally scholastic) way he has of occasionally using the word "perfection" as a synonym for "reality".[39] Like Newton, he thought himself able to pronounce on what are to be classed as perfections. They say similar things about the perfection of the intellect. Leaving aside a comparison of their pronouncements, however, consider what Descartes says about the limitation of perfection — which he often speaks of as having degrees. In a letter to Clerselier (23 April 1649) he quotes from the Latin edition of the *Meditationes*, treating of defects in substances, that is, limits to their perfection.[40] These limits are said to be accidents; but in the infinite substance (God's is clearly the only one envisaged) the infinitude of perfections is treated as something other than an accident of the infinite substance which "has" it. What else can it be? It must be the very essence of the substance![41] The convolutions of Cartesian metaphysics are seductive, and there is a temptation to play the neo-scholastic game. Is it not a tenet of the *Meditationes*, for instance, that all lesser perfections (such as I have, and including my free will) are a part of the supreme perfection of God? If a limit to perfection is an accident, is not absence of limit also an accident? In the case under discussion, however (the nature of God) infinitude must consequently be something other than absence of limit, since according to Descartes infinitude is not an accident. In fact his "positive" view of infinity-as-essence represents infinity as "a real thing . . . *incomparably* greater than all those that have some end".[42] Here he lays down a route which Newton will not be able to persuade himself to follow. Descartes may claim that he *never* uses the word "infinite" to signify "having no end-point", but note such remarks as are made in the letter to Chanut, where he spoke of God's knowing limits to the world. There are several comparable statements in his works. For much of the time he writes as though he is obliged to draw a sharp line between two categorically distinct concepts of the infinite — the positive, "essential", non-mathematical, and the negative, absence-of-limit concept. And yet there is a tension imposed not only by his mathematical vocabulary but by his mathematical way of looking at problems.

I want to suggest that there is no real conflict between the views of New-ton and Descartes on the relationship between the concept of extension and that of divinity, for the simple reason that neither man was anxious to allow the two concepts to interrelate significantly. In McGuire's words, for Newton "the infinity of extension in no way competes with the reality of the Divine being".[43] As explained, Descartes would surely have agreed (although he would not have put it like this without first being vouchsafed the knowledge that extension is indeed infinite). As I have said, the two infinites were for Descartes supposedly categorically distinct. Faced with the statement that the world is indefinite, and that we do not know enough to say anything more positive, we naturally assume that Descartes would have allowed that it is *conceivable* (in some sense of that word) that the world be infinite — that is, infinite in extension. Is this so? Is there, in other words, what we might call a "positive concept of infinite extension", in *some* sense resembling the infinity of God? McGuire argues at length that "extension contradicts the nature of infinity, because extension is an imperfection, and infinity cannot be admitted where there is not perfection".[44] Why then should Descartes have introduced the concept of *indefinite*? Did he really wish to say that when he described the world as indefinite, meaning that he did not know whether or not it was infinite, in one sense of the word "know", he could adduce *a priori* arguments to show the impossibility of an infinite world, namely arguments proving the self-contradictory nature of the concept *infinite extension*? Would one ever say "I simply do not know whether or not the window is both broken and unbroken", or the like?

The argument as spelled out by MicGuire is essentially this: imperfection is incompatible with infinitude, and extension, a created and dependent being, is imperfect. The text quoted (reply to the Second Objections) in fact talks only of *body*, in which *all* perfections are found, rather than of infinite extension. In the paragraph preceding, Descartes has actually been talking of ideas we may have of "infinite attributes"; and one of the infinite attributes he will admit to be (eminently as opposed to formally) contained in the idea of God is length. Not, admittedly, extension; but if infinite length is possible, so is infinite extension, at least in thought. There is much mystery spun out of Descartes' conception of extension, but following its development in his writings it is clearly possible to say that the *basic* part of it is something like "the volume (a function of lengths, breadths and heights) [of a body] in combination with the distribution of that volume".[45] Thoughts and extensions, he explains, are modes found in substance — and he explains what he means in regard to thoughts by reference to extension, which he obviously considers clear enough. In *Le Monde*, extension is plainly meant to be understood as measurable: it is the property matter has of occupying space, not an accident of matter but its "true form or essence".[46] Extension is a mathematical concept. In the tenth of his Principles of Philosophy he talks of conceiving the same extension as space, emptied of body, and as constituting body. (His arguments elsewhere against a void need to be reconciled with this, but that is not our job for the moment.) I know of no text which would support the claim that the "characteristics of unity and simplicity cannot be found in the nature of extension"[47] or that "material extension in itself is an imperfection". In Principle *X* Descartes speaks of the "generic unity" of extension considered as space. Considered on the other hand as the nature of body, extension was admittedly considered "as particular" and conceived to "change just as body changes". Presumably he would have said that extension is divisible in this sense; but does that mean that the corresponding concept of infinite extension is, for him, self-contradictory? I think not; but had Descartes given more serious attention to non-theological concepts of infinity, as Aristotle had done, we might have been more confident.

ON UNDERSTANDING AND NOT UNDERSTANDING THE INFINITE

The difficulties in the Cartesian position are not made any smaller by Descartes' distinction between two sorts of knowing — between *intelligo* and *comprehendo*. This is paralleled in many scholastic writers, not to say in Augustine. By "comprehension" Descartes meant to stress the *completeness* of our grasp of something. This idea of totally enclosing something (knowledge) within ourselves, an idea which sprang from a metaphorical usage, was codified by medieval philosophers (thus Ockham gives five meanings) without any overt appeal to metaphor.[48] When Descartes writes that "it is a manifest contradiction that, when I comprehend anything, that thing should be infinite",[49] it is natural to think in terms of a container metaphor of mind. You can't put an infinite number of pints in a finite pot, so to speak. In fact Descartes' words at this particular point seem to me to suggest that his argument was a different one and that he was saying rather that it is by reason of the fact that the definition of "infinite" contains the concept "incomprehensibility" that the infinite cannot be comprehended. (He certainly uses the other argument elsewhere — in the third meditation, for instance —

but in the same place he *will* allow that we have the *idea* of infinite substance, not from our finite selves, but proceeding from some truly infinite substance with "manifestly more reality" than is in finite substance.) If this reading is correct, we can ignore the *non sequitur* only because it does not occupy a central place in his argument. (If, as he suggests, it is impossible to comprehend "infinite", that is, a concept "incomprehensible and F, G, H, .. ", where F, G, H, .. offer no difficulty, then the concept "incomprehensible" — and hence also "comprehensible" — is incomprehensible, and the discussion cannot even begin.) The upshot of it all is only that our ideas — even of God — are inherently imperfect. Descartes is far from clear on these questions. He calls our idea "complete", for instance, but the tenor of the paragraph in question (the 7th in his Reply to Objections V) is that our idea of the infinite *represents* the whole of infinity. Consider his final sentence. " . . . it is enough to understand a thing bounded by no limits in order to have a true and complete idea of the whole of infinity". He goes on to make a similar point in regard to our idea of God. We shall very soon get into deep water if we try to impose total consistency on Descartes' several utterances on these themes. The "true" in the last quotation, for instance, is confusing in view of his repeated emphasis on the imperfection of our idea; the suggestion that our idea is one of the *whole* of infinity does not immediately seem to square with his emphasis on the incomprehensible character of the infinite; and finally there is something rather puzzling about the statement as a whole, bearing in mind his remarks (quoted earlier) to More, to the effect that he never uses the word "infinite" to signify only "having no end-point". Descartes was replying to objections raised by Gassendi (V, 7), and I cannot help feeling that he would have benfitted by the infusion of some of Gassendi's thoughts in his own. Gassendi, briefly, can be said to have argued that our idea of infinite substance can come from the elaboration of limited principles concerning the visible world — in the language of the schools, by "composition and amplification".[50] This looks at once more medieval and more modern than Descartes' talk of the idea of infinite substance proceeding from the infinite. Whatever the merits of the two sides of this debate within a debate, it is enough for my present purposes to point out that Descartes did in a reasonably strong sense *agree* with Newton (if we may be allowed to turn the god of chronology on his head) that infinity can be understood. We can have a complete idea of it, but not comprehend it, in the peculiar sense of that word.

In saying that infinity can be understood, does this apply only to the concept in its solitary proven application, namely its application to God? We must come back to the supposition that infinite extension is a self-contradictory notion. In arguing for this view, McGuire reminds us of the utter dependence and contingency of created things (because of their re-creation by God at successive times), and passes straight on to extension, as though the same thing holds for that. "It is in this strong sense, then," he writes," that Descartes conceives of extension as a contingent and dependent being".[51] In sum, we are told that extension is for Descartes (1) divisible and hence imperfect; (2) contingent and dependent; and hence that on both scores it is impossible that it be positively infinite. My own view is that the style of argument is not entirely true to Descartes'; but since he clearly does *not* want to say that extension is positively infinite, I will be as brief as possible in saying why the style is wrong.

Certainly extension is divisible — and in fact Descartes makes its divisibility even basic to the divisibility of matter.[52] Certainly infinite extension cannot be found respectability along theological lines, for it is not a property of God, as we have already seen. Certainly it is divisible, but to say that it is — on this account — imperfect seems to me to misrepresent Descartes' way of speaking about imperfection. Imperfection may be revealed through limitation — as when we say a piece of matter is bounded, or our knowledge is limited — but the imperfection of the bounded thing derives from its reduced value, its loss of power to do what it is its function to do. It is *inappropriate* to talk of extension's limitation in this way. I realise that this is a rather impressionistic reconstruction of the case, and that it is always possible to make the rejoinder "whatever you say about the matter has to be said of extension, in the sense of occupied space". This will be the rejoinder of one who will not allow my argument to continue along the lines " in limiting extension I am no more rendering it imperfect than is a geometer who marks off a segment on a line" — an argument that will be denied because (although as quoted earlier, Descartes does talk of empty space) extension without matter does not exist in the Cartesian world. Descartes does talk freely enough of imagining limits to extension. Why then should he not speak of imagining the *absence* of such limits? Consider that very clearly-and-distinctly explained imagining of *Principles* II.21: if, he says, we imagine a limit to extension, we shall find that we are still "able to imagine beyond that limit spaces indefinitely extended", and these we perceive to be filled with corporeal substance. Indefinitely extended. We *are*, it seems, able to imagine a situation where there are no limits to extension. Indefinitely, not infinitely, extended: but — to repeat an earlier quotation — "it is enough to understand a thing bounded by no limits in order to have a true and complete idea of the whole of infinity",[53] so although we cannot be expected to imagine the whole of infinity, it does not seem that we are meant to question the consistency of the concept "infinite extension". We are driven back, not by an inconsistency, but by the sheer limitations of our knowedge.

THE CREATION OF EXTENSION, AND THE INFINITY OF IT AND THE WORLD.

On the question of the creation of extension, Descartes is not as explicit as he might have been, and indeed in all such matters we have to remember that we are on the verge of indulging in a species of philosophical necromancy in order to promote a dialogue between Descartes and Newton. Extension is not in God, and therefore must either be (a) outside God, and have been so from before the Creation — we ought really to say from before the *first* creation, on Descartes' view of continual creation; or (b) created by God *ex nihilo*.[54] The first would detract from God's omnipotence. The second must be accepted, thought Descartes; but why did he not consider a third possibility, namely that (c) extension is created necessarily *with* matter, as its essence?

Would this have been to presume to limit God's freedom? One might circumvent such an objection along the same lines as Descartes tried to circumvent his trouble with the eternal verities. (God did not necessarily will them, but he willed them to be necessary.)[55] Of course Descartes' religious style will not allow him to have extension competing with God, but if for a moment we allow infinity in different *respects* (e.g. in length as well as perfection) there is no reason why infinity in a limited number of respects should imply and challenge to God, who is infinite in all perfections.

We are now back to Descartes' distinction between indefinite and infinite. Inevitably he must admit that infinitude is only specifiable in certain respects, although, as we have seen, he finally settles on a unique infinite, God himself, the sum of all perfections. But it does not seem to me that his philosophy forces this position on him. I suggest that in the last analysis his infinite/ indefinite distinction is rather trivial and that it was an obstacle to an easy dialogue involving later philosophers on this topic. Consider the somewhat abject tone of *Principles* I.26: we must not try to dispute about the infinite, for it would be absurd to comprehend it in our finite minds. Is the half of an infinite line infinite, he asks, or an infinite number odd or even? Two perfectly good mathematical questions, you might think; but he will not allow those without infinite minds to consider them. How disappointing, in view of the start already made by scholastic philosophers and Galileo to think constructively about a potential mathematics of the infinite. Not, of course, that they had made much progress, but they seem to have sensed the riches in store, riches which do not seem to have interested Descartes very seriously. "For our part, while we regard things in which, in a certain sense, we observe no limits, we shall not for all that state that they are infinite, but merely hold them to be indefinite". We observe no limits; but might we not speculate on the way "limits" might behave? Descartes mentions indefinite extensions, subdivisions of bodies, and stars, and he goes on to explain his distinction between an apparent absence of limits due to our lack of understanding, and one due to the nature of what is considered.

Now of course there is nothing wrong with such a cognitive distinction, as long as it is not used to prevent us from asking awkward questions. What would it have meant for Descartes to have said, quite categorically, that the world *is* limited (in regard to addition, in Aristotle's sense)? Since empty extension is not conceivable,[56] he would have been obliged to work out a geometry with a concept of limit very different from any previously developed. (His would not have been a finite and unbounded world, but a world finite and bounded by non-extension, whatever that may be made to mean.) We do not need to imagine the horror he would have had of a *bounded* world, because on more than one occasion he expressed it himself, in response to More's Lucretian-style challenge.[57] It involves a contradiction, thought Descartes, to suppose that the world is finite or that it is bounded. In the discussion with Burman he tentatively acknowledged that we may perhaps have to say that the world is infinite.[58] In the face of all this evidence I fail to see how Descartes can be said to have thought infinite extension a self-contradictory notion, and I think we must assume that he would have chosen from the following alternatives:

(1) The world, like the totality of extension, is infinite. There is no inconsistency involved in saying as much. We are, as it happens, unable to say whether the assertion about the world is true, and therefore, being cautious, we call it "indefinite", but it *is* infinite in the (authentic Cartesian) sense of "bounded by no limits".

(2) The world, and extension, are finite. As it happens, we do not know whether this is so, and so call the world "indefinite".

(3) The world and extension are finite. As it happens, we do not know whether this is so, and what is worse, we cannot even conceive of the possibility. We call the world "indefinite".

He would presumably have wished to say *either* "either (1) or (2)" *or* "either (1) or (3)". The problem with (3) is that it requires the word "in-definite" to have a different meaning from that of (1) and (2). (Briefly, "inconceivable" rather than "unknown".) As it happens, everything Descartes wrote, where relevant, points to his having preferred (1). Had this not been so, the potential ambiguity of the word "indefinite" might have attracted more attention. As things stand, the word is subject to another sort of ambi-guity, the roots of which are hidden in the origins of the word "definition" (or rather its Latin ancestor). If P is a proposition which conveys information about the world and the information happens to concern the spatial limits to the world, and if I say "I know P", then I might say that the world is defined; and if I say "I do not know P", then I might say that the world is indefinite in the sense of "not defined" – here, as before, "defined" meaning that a definition is available in regard to the limiting of the world, I might, however, equally well translate "not-P" with the phrase "the world is not defined", meaning by this that *limits do not exist*. A case might be made out for saying that by his word "défini" (and its equivalents) Descartes was subconsciously trying to catch both meanings of my "defined" simultaneously; but of course he could not do so without making a category mistake.

NEWTON AFTER DESCARTES

There is a measure of irritation in Newton's charge that Descartes slides, in effect, from an epistemological thesis of the indeterminacy of (the limits of) extension to the thesis that it is an indefinite nature.[59] If this is a correct interpretation of Newton, then perhaps he was making the point I have just made about the category mistake. Perhaps Newton was not so much saying that the "indefiniteness thesis" amounts to an ontological claim as that he is impatient with Descartes because he will *not* make a genuine ontological claim, and insists on sitting on the fence.

Despite the fact that I have said so much about them, I find Descartes' writings on infinity (and related questions) disappointing. He might not have been afraid of the theologians, but something certainly petrified his thoughts. Newton's comments show that he did not always represent Descartes' views accurately; but there is a fresh breeze blowing through his comments – and these comments so often seem to be what Descartes himself should have said, on his own premises. For both, the notion of limit involves there being something to both sides of the limit. They would have agreed that they could not conceive of a point in space with space to one side and not the other.[60] Both assume that "infinity" has a basic meaning of "not limited". Newton can simply conclude that space is infinite, because he is not racked by Descartes' doubts:

... God understands that there are no limits, not merely indefinitely but certainly and positively, and because although we negatively imagine [space] to transcend all limits, yet we positively and most certainly understand that it does so.[61]

There is something comforting about a plain application of the principle of the excluded middle: "either [real being] has limits or not, and so is either finite or infinite".[62] The imagination may be unequal to the task of coping with the infinite, but the understanding – and *a fortiori* God's understanding – can do so.

I would agree with McGuire that Newton's distinction between the imagination and the understanding is more or less Descartes' – although I think there are important differences. Descartes' position is, after all, much more elaborately worked out, philosophically speaking. The two ways of speaking of thoughts, as representations and as contents of the mind, were common property, and are incidental to the imagination/understanding dichotomy. The danger here is in making Descartes (and, for that matter, Newton) too Kantian. Descartes refers to the faculties of imagination and pure intellection (in the sixth meditation), the first a "power and inward vision of my mind", requiring a particular *effort* of mind for its use, the second (*intellectio* in Latin but *conception* in the French translations) something the mind must always be doing. The important thing is not to make too sharp a distinction betwen faculties, in the traditional sense of the term. Thinking has many modes: doubting, understanding, willing, imagining, feeling, conceiving (intellection), and so forth. He makes intellect and volition the two essential modes, and intellect the more fundamental. (A mind must always be thinking – not so always judging or imagining or willing.)[63] In judging, the understanding conceives ideas of things, which may be affirmed or denied, and if they were really clear false affirmation would not arise. Free will is also involved in judgement, a faculty of choice, and error arises when we fail to control our volitions.[64] Whatever its antecedents, this cannot be called a "traditional" doctrine, and so far as I can see, was not taken over by Newton in any but a superficial sense. (It has had followers, of course, between Descartes' day and our own, but that is another matter.) And in passing, allow me to note that Descartes talks of the *infinity* of the human will: "As a matter of fact I am conscious of a will so extended (*sic*) as to be subject to no limits ... [a] faculty ... even infinite". Seeing, he goes on to say, that he can form the idea of it, he recognises "from this very fact that it pertains to the nature of God".[65] Again, I am not aware of any Newtonian parallel, although, as McGuire has pointed out, for Newton the understanding can apprehend that *extension* is "an external, infinite, [and] uncreated" entity.[66]

How does it do so, and is Newton's epistemology really "closer to Descartes" than it is to Locke, especially on the question of 'how the mind grasps the notion of the infinite"?[67] To make Newton sound more like Descartes, McGuire[68] sets side by side Newton's remarks about "the understanding grasping the infinite" and Descartes' about "having the notion of the infinite in me before the finite". The two contexts are, however, utterly different. Newton is talking of *extension*, while Descartes (writing to More) is talking about *being*, in a theological setting. One has to remember that (in the third meditation, which is what he is discussing with More) Descartes is extending his philosophy from the *cogito*, yielding the thinking self, to prove God's existence (i.e. the existence of a complete and independent being).[69] There is a sense in which both Descartes and Newton are prepared to talk of "the finite as the negation of the infinite", as opposed to Locke, who can be said to put it the other way around; but that by itself does not seem to me to be a particularly deep point of agreement or disagreement, philosophically speaking.

SOCIAL LINKS IN THE INTELLECTUAL CHAIN

Before making a closer comparison between the three writers it is worth considering some of the social aspects of the network of scholars to which they belonged. Kenelm Digby was in the Paris group surrounding Mersenne in 1637, the year of publication of the *Discours*, and some of his admiration was conveyed to Hobbes, with whom he corresponded. Most of the Digby correspondence has disappeared. Sancroft copied letters relating to Hobbes, Herbert of Cherbury, and Descartes, and these survive.[70] Hobbes was actually in Paris when the manuscript of the *Meditationes* arrived, and Mersenne gave it to him. A set of objections Hobbes prepared was sent by Mersenne to Descartes, without telling him who the author was, and Descartes thought the objections implausible.

The objections were received, and the book published, in the same year, 1641.[71] The two philosophers were so very far apart in their basic tenets that no dialogue of importance was likely to ensue, but with Henry More it was different. His *Infinitie of Worlds* (1647) shows the influence of Descartes' thinking; but from the position of disciple he grew progressively more critical. Like those other Cambridge Platonists Culverwel and Cudworth, he made much use of Cartesian philosophy, and especially its physico-mechanical aspects — which influenced Newton in ways too well known to require more than mention. On the other hand, they saw in Cartesianism a number of tendencies which they judged to be hostile to religion. By a piece of extraordinarily bad exegesis the Cambridge group managed to run together Cartesian philosophy and Democritean atomism — the latter, as I have mentioned in connection with Harriot, generally feared as atheistic. The Cambridge Platonists feared an alliance of Hobbesian principles with Cartesianism; and in what they saw as the answer to materialism, they coined the philosophy of an *anima mundi*, the soul of nature, the immanent cause of organic and inorganic nature. From the affable tone of the correspondence with Descartes from which I have been quoting (1648–9), More has so changed his tone by 1668 that in the Preface to his *Divine Dialogues* of that year he actually charges Descartes with atheism. Spirit, said Cudworth and More, is not "substance without extension". Extension does not have to be material; and if spirit were without it, God would be nowhere. Space is extension, not of matter, but (therefore) of spirit, and in infinite space we have the *extension of infinite spirit*.

The link with Newton has long been recognized, although McGuire has added several qualifications to the accepted account, and has argued that "Newton is largely a Gassendist on the problems of space, time, and existence".[72] Whatever the line of influence (McGuire suggests that Walter Charleton might have been the intermediary between Gassendi and Newton),[73] the broad outline of Newton's arguments resembles More's: (1) God's existence is eternal and uncreated; (2) everything — including God — must exist with respect to space and time if it is to count as an existent; (3) infinite space and time are consequences of God's external existence (Newton) as intrinsic attributes of God's nature (More).

The rivalry between Hobbes and the Cambridge philosophers could well be at the root of his expressed determination to leave the problem of infinity in the hands of "those that are lawfully authorized to order the worship of God".[74] The Platonists opposition to Hobbes extended to moral and political philosophy. They preached a personal religion, opposed to religious and civil centralization, and opposed therefore to the puritans and various arms of the official English church (which came and went with small predictability at about this time). Although in fact Hobbes does have a number of interesting things to say, despite his protestation, on the subject he said he would avoid, it is worth remembering how easily a philosopher's views may be distorted by sectarian antipathies.

AND LOCKE

Influence did not always run along social lines. John Locke's general agreement with the Cambridge Platonists on questions of political liberty and
religious toleration (and note his long friendship with the family of Cudworth's
daughter, Lady Masham) did not do much to bring their epistemological
positions together. But to return to a comparison of Locke's "philosophy
of the infinite" with Newton's and Descartes': on the face of things, their
views, too, could hardly have been more different. Leaving God out of the
picture, we have seen that Descartes was agnostic while Newton claimed that
the understanding can get a clear idea of extension (eternal, infinite, and
uncreated though it is), based as it is on a nature "real and intelligible". This
was easy to say, but it all seems rather vague when we look for an explanation
of the *workings* of the understanding – along the lines of Descartes' or
Locke's accounts, for example. Newton's positive idea of actually infinite
space looks at times almost like an innate idea, a datum so fundamental that
there is no point in showing how it is built up. To Locke this is anathema,
and he hopes to show how we synthesise the idea. Portions of extension
are deemed to affect our senses (*sic*), carrying with them the idea of the
finite. (The same goes for durations.) We have the idea of repeated addition
(or doubling) of extensions, and the idea of doing this without stopping,
and this, according to Locke, is how we are led to an idea of infinite space.[75]

In the same spirit Locke writes of utmost *divisibility*, that our thoughts
cannot reach an end in the boundless progression. In all this there is no very
clear sign of any influence from Newton's quarter. Locke was of course ten
years older, and had more or less completed the *Essay* by 1686, three or four
years before he first met Newton and several months before he first read
Newton's *Principia*.[76] [See "The empiricism of Locke and Newton", by
G. A. J. Rogers, in G. C. Brown (ed.), *Philosophers of the Enlightenment*,
1980, pp. 1–30] The *Essay* finally appeared in 1690. Before considering the
Newton-Locke relationship further, I shall look further into the *Essay*,
explain very briefly why Locke comes out against an actual idea of a completed infinite, and examine his theology.

I have already noted Locke's "idea of infinity" considered as "an endless
growing idea".[77] This he contrasts with "the idea of any quantity", a "terminated" idea (at the time the mind has the idea). It is not, he thinks, "an
insignificant subtilty" to say that we must carefully distinguish between "the
idea of the infinity of space and the idea of a space infinite". The first is a
supposed *endless progression* in the mind, the second an actual idea in the

mind of a *completed endless repetition*; in short, a contradiction in terms. Berkeley commended the distinction in the closing words of his youthful essay on the infinite.[78] It is occasionally said to have been adopted from Locke by Kant, who does indeed seem to make use of it for the proof of the thesis of the First Antinomy; but there is ample evidence that Kant did not *accept* it, that he thought that those who rejected the "actual mathematical infinite" did so "in a very casual manner", and that he had a trick up his sleeve rather reminiscent of certain passages in the writings of Descartes, namely in introducing an intellect which may exist — "though not indeed a human intellect" — capable of perceiving a multiplicity in a single intuition without the successive application of a measure.[79]

Whether or not Locke's account should be called "Aristotelian" is a question of emphasis. For Aristotle the infinite was something that could not be gone through; for Locke in his published version of the *Essay* it was something which required you to keep adding. There is more to Locke's account, but at this fundamental level he is less precise, less careful, than Aristotle, to the extent that he does not consider the type of endlessness exemplified by the ring. Nevertheless, both rely heavily on metaphor when they discuss what we might call "going on without stopping", for both speak of it as though it was a process requiring time, as walking, or calculating, requires time. As for Aristotle, number was basic for Locke — it "furnishes us with the clearest and most distinct idea of infinity we are capable of" — , but even when we number intervals of space and time and make use "of the ideas and repetitions of numbers", the "clearest idea any mind can get of infinity, is the confused incomprehensible remainder of endless addible numbers, whch affords no prospect of stop or boundary".[80] The infinity of number "lies only in a power still of adding any combination of units to any former number, and that as long and as much as one will".[81] This phenomenological approach to the subject is appreciably different from Descartes'. When Descartes spoke of an absence of limits he treated the case as one that might be judged instantly: we simply "perceive no limitation in the thing".[82] His other kind of infinite, unlimited perfection — in a word, God — , was likewise something that could be the object of "clear and intuitive knowledge". "In some way I have the notion of the infinite prior to the notion of the finite".[83] Those who are looking for a mathematics of the infinite will consider neither approach very promising. A completely different approach can be detected in some scholastic writers, and in Leibniz it is clearly expressed: it requires a complete separation of the conceptual apparatus of the infinite from that to which it may be applied — or not,

as the case may be. (The real may be governed by the ideal, according to
Leibniz, and "infinites and infinitesimals are grounded in such a way that
everything in geometry, and even in nature, takes place *as if* they were perfect
realities")[84]

Accepting that Locke's programme was of a different sort, we shall still
understand it properly only if we first understand his notion of *positive
idea*. We saw how Descartes and Newton in their different ways made the
infinite "positive", and the finite its negation. Locke takes the common-sense
reversal of labels. In an example of the mariner who fails to find the sea-bed
with his sounding line (cf. Ockham), he says the mariner *knows* that the
depth is so much as the line is long (a positive idea) and more (another positive idea).[85] On the other hand, the idea of "so much greater as cannot be
comprehended" is not positive but negative. The act of numbering is fundamental to an understanding of his "primary" meaning of finite and infinite —
referred to, following tradition, as "modes of quantity". His argument so
far leads him to say that we can have no positive idea of infinite number.[86]
He later extends the doctrine to numerable things. Besides his primary meaning for "infinite" Locke gives a secondary meaning relating to God's power,
wisdom, and so on.[87] As he explains, this is *figurative*, and is arrived at by
analogy with number or extent of the acts or objects of God's power, and
the like. In this matter the *Essay* is poles apart from Descartes, who is
made to look all the more a scholastic by contrast.

Starting from the basic principle that there is no acceptable positive idea
of actual infinite number, Locke extends his argument to numerable things.
Some ideas are repeatable and some are not. Two idea of white, for instance,
when combined "run into one". Space, duration, and number alone have
parts. (So much for the intension and remission of forms!).[88] Locke therefore draws the same conclusion as to the unavailability of positive ideas of
infinite *duration* (eternity being the "number of duration" looking forwards
or backwards from the present),[89] and of infinite *space*.[90]

Like Aristotle, then, Locke rejected the idea of the completed infinite,
and also turned his back on the idea of the infinite as a "separate, independent, thing" and an "indivisible thing". These were phrases beloved
of theologians and metaphysicians — and yet was Locke's empiricism not
influenced by Christian theology, by his Christian (latitudinarian) belief? We
know that Descartes reserved positive infinity to God, and that he made much
of God's *unity* and *independence*. We have seen the Cambridge Platonists'
view of God as an infinitely extended entity, "One, Simple, and Immovable",
and also "not only Eternal but Independent ... of anything whatever".[91]

We might have expected Locke to have adopted some similar attitude to such widely diffused 17th century ideas. What sort of God was Locke's?

In the chapter of the published *Essay* dealing with the idea of infinity, using almost identical phrases to those of the 1671 draft, he says that he thinks it "unavoidable" that every "considering rational creature" should have a notion of an eternal wise Being who had no beginning: "and such an idea of infinite duration", he adds, "I am sure I have".[92] This is at first sight an astonishing admission, for is not the duration assigned to God of the same kind as is experienced on earth? There were other theological opinions than this; did Locke have them in mind in writing this common-sensical passage? Turning back to the 1671 draft we find him referring to "the unconceivable *punctum stans* of the schools" which is the idea some might claim to have of duration of eternal beings having "noe succession".[93] Note how almost all leading philosophers inveigh against the "*nunc stans*". In the published *Essay*, however, he faces up to the problem of God's power to see "all things past and to come". "They all lie under the same view ... To conclude, Expansion and Duration do mutually imbrace, and comprehend each other; every part of Space, being in every part of Duration; and every part of Duration, in every part of Expansion".[94] Such an omnitemporal and omnipresent God sounds rather like the God of Newton's *Principia*.

Putting aside for a moment Locke's difficulties with the causal argument for God's existence, what about the concept of God's "having no beginning"? Surely this means that it is *innappropriate* to speak of God's beginning in time. This is not suggested by the following *volte face*, however:

But this negation of a beginning, being but the negation of a positive thing, scarce gives me a positive idea of infinity; which, whenever I endeavour to extend my thoughts to, I confess myself at a loss, and I find I cannot attain any clear comprehension of it.[95]

Elsewhere we find him slipping back into forbidden ways. "We easily conceive in God infinite duration";[96] and God is immovable "not because he is immaterial, but because he is an infinite spirit".[97] The second remark is reminiscent of a proof offered by Henry More that "no Infinite extended entity which is not co-augmented from parts, or in any way condensed or compressed, can be moved ... ".[98] Locke's favoured argument for the existence of God is the causal argument, though he occasionally has the argument from design. The cause of my existence must be other than a material being — it must be a "cogitative being"; and since it is inconceivable that there was a time when nothing existed, it being impossible that the real being of God came from a non-entity, therefore "from eternity there has been something".[99] He does not describe this Being as infinitely wise and powerful, but as "most powerful and most knowing", and indeed "most perfect", but on the score of extension, "it is hard to find a reason why any one should doubt that [God] ... fills immensity". There is again a touch of More in the next sentence: "His infinite being is certainly as boundless one way as another; and methinks it ascribes a little too much to matter to say, where there is no body, there is nothing".[100] Where there is no body there is God; but does he have Newton's omnipresent God in mind, or More's extended deity? Space was taken earlier to be a possibility for extended beings to exist, but that is not enough to allow us to place Locke alongside More on this question and to do so would very probably be mistaken, if we are to judge by Locke's 1671 draft of the *Essay*. Let us ask someone who thinks he has a positive idea of infinity whether it is of infinite local extension or infinite duration. "If it be local extension that cannot belong to God, who is a spirit &is therefor uncapable of local or corporeal extension, & is therefor proper only to body".[101]

In connection with the extension/omnipresence question there is a curious theory for the creation of matter at which Locke hints at one point in his *Essay*. The theory is that God by his power prevents anything from entering a particular region of space, and thereby (endowing it with impenetrability) creates matter there. To move the matter he simply creates, at a later time, the same impenetrability in another region. Again this tells us very little about God's relation to space; but since it seems that the theory was Newton's brain-child, perhaps we may assume that Locke and Newton both remained inclined to the view that God is omnipresent in space, without being extended.[102]

It would lead too far afield to consider Locke's application of the causal argument for God's existence, but its involvement in the problem of infinite regress is worth mentioning. The argument, briefly, is that the cause of my existence cannot be a mere nothing, since *ex nihilo nihil fit*, and by the same principle its own cause cannot be a mere nothing — wherefore the cause of my existence is eternal. On the principle (much used by the Cambridge Platonists) that there must be more reality in the cause than in its effects, and indeed more of all powers, the cause, which we call God, is the most real, the most powerful, the most intelligent, and so forth. Locke spends the later part of the relevant chapter [103] demonstrating the immateriality of God. His causal proof was not likely to attract much notice, since it was already part of an established Christian (neo-Platonic) tradition. The problem, for Locke's philosophy of the non-availability of a positive idea of the infinite, is that by the same token he should never claim to have an idea of the first term in his causal series; and of course the first term is none other than God. But he does so claim. In the words of the 1671 draft, "I acknowledge I have the notion of a first being or a being without a beginning & I thinke it unavoidable for every considering rational creature that will trace his owne thoughts to have such an one". But, he goes on to say, this is "note a positive Idea of Infinitie" rather "the negation of a positive thing (as the beginning of any thing is)".[104] He is clearly not above playing Descartes' game of positives and negatives, but I cannot see how it can have been thought to help his case.

The three vertices of the McGuire triangle are Descartes, Locke, and Newton — all of them compared on the basis of their writings, but not in any "strongly historical" style of the sort that might reveal the paths by which the ideas of one developed from (or in dialectical opposition to) those of another. Of course he has often written in such a style, and one paper in particular — "Existence, actuality and necessity: Newton on space and time" — provides much of the background to his present essay, so to some extent my strictures may seem unfair. Consider, however, the last sentence of the essay: "If I am right about how Newton claims to know infinitely extended space, any kinship with Lockean empiricism must be denied". This may be read in different ways. A philosopher might have a very clear idea of that timeless entity "Lockean empiricism", and care little for chronological relationships. It might be argued that kinship does not imply interdependence, but merely a common inheritance. A historian might point out that the very early works of Newton could hardly be expected to show strong Lockean influence, for they were written at a time when Locke's epistemological interests were barely formed, let alone committed to print. Regarded as a comment on early (i.e., before the late 1680s) influence from Locke to Newton, the statement does not look very bold; but as a denial of influence in the opposite direction, it adds a little weight, certainly, to current criticisms of the old view that Locke was simply the philosophical interpreter of the Newtonian physics. No-one is likely to regard Newton and Locke as identical twins, but it does seem to me so very obviously right to think of them as kindred philosophical spirits that I think we should resist Ted McGuire's over-subtle separation of their natures. By the same token we shall have to say that the early Newton and the late Newton were not kindred spirits. Both may be called Cartesian dualists. The qualification is entirely justifiable in the 17th century context, for in both writers we find passages lifted virtually straight out of the work of Descartes. Mind and matter were for them separate substances, related in a causal theory of perception. Both of them accepted a broadly Cartesian language of ideas, and both drew the line at innate ideas. Their views of space and time were in their elementary aspects rather similar — although Locke changed his mind on that subject, for he had flirted with a relational doctrine during his travels on the continent. And of course there are the differences (dare I call them small?) on the question of "positive ideas of infinity". The number of passages from Newton's writings with a bearing on what might be called the "metaphysical infinite" *is* rather small. Both were avowedly anti-scholastic over the various doctrines of substance, and both combined a continuum view of space and time with a corpuscular view (in fact with a variety of corpuscular views) of matter. There are differences in plenty, but are these points of fundamental agreement evidence enough for kinship?

NEWTON AND LOCKE

Should we be looking for something deeper? I think Newton's geometrical fragments on the "infinite in fact and not only imagination" are extremely interesting, but they do not seem to me to put any more distance between him and Locke than we have already found. Before commenting on them, let me quote Locke again, from the *Essay*:

Some mathematicians perhaps, of advanced speculations, may have other ways to introduce into their minds ideas of infinity. But this hinders not but that they themselves, as well as all other men, got the first ideas which they had of infinity from sensation and reflection, in the method we have here set down.[105]

Such generosity, on Locke's part, puts the status of his earlier argument in doubt. In developing his account of what he calls the "primary meaning" of infinity, was he providing nothing more than a naive "commonsense" account, consistent, perhaps, but weak, and in principle capable of "mathematical" refinement? There is a sense in which he will have fulfilled his purpose if he can get his opponents, believers in our innate idea of the infinite, to acknowledge that indeed the mind works as he says, and receives its (possibly naive) ideas of infinity from sensation and reflection; but is this only a contingent truth about the ordering of experience in the seventeenth century mind? Might we not catch our philosophers young, and instil more refined ideas in them before they have had a chance to think along naive lines? After all, the typical child does not hear of the infinite at its mother's knee. We may grant Locke the contradictory character of the notion of an actual idea of an incomplete completed repetition; but what guarantee have we — and he himself can see none — that there are not other, more powerful, and consistent theories of the infinite? None at all, of course. The most that Locke can hope for is that his opponents will grant him his opening definition of "infinite".

How then do the "geometrical" quotations from Newton square with Locke's meanings?[106] One side of a triangle is swung round a vertex until eventually it is parallel to the side opposite that vertex. In the first quotation, overlooking a phrase where he seems to be suggesting that there exists a last point of intersection on the line opposite the vertex, he is in fact arguing that *there is no last point* on the line. It is a curious example, since it introduces elements (other lines, angles, the notion of monotonic increase, intersection, existence, parallelism) any of which might be the first point of attack by a modern geometer, aware of geometries alternative to Euclid. Could he not have spoken simply of proceeding to greater and greater distances along a line? Why did he not do so? I think the answer is that he wanted to make the obscure concept "actually more than finite" understandable, by making it a function of more readily intuited concepts. (In particular, the angle increases to a finite and intuitable limit.) In his other triangle example (from the earlier *De gravitatione*), where he talks about the point towards which the two sides of a triangle are directed and where they meet (although this meeting point might have to be pretended to lie "beyond the limits of the corporeal universe"), Newton is again ultimately referring simply to the points of a line. Why not talk about the line and nothing else? It seems to me that he is offering yet again what a recent generation of philosophers would have called an "operational definition" of the line in terms of things supposedly easily understood. Rather as before, the "actual" line "extends beyond all distance". Unlike Locke's infinite process, a process without an end, Newton's is a process in which the infinite *is* specified, at least *"fingatur"*. It would be anachronistic to expect him to offer an existence proof. By the lights of the time, the point existed. Can it be that when he wrote of "many things concerning numbers and magnitudes which to men not learned in mathematics will appear paradoxical" he was thinking of the words I quoted from the end of Locke's chapter on the infinite, and was consciously offering "ways of introducing into the mind ideas of infinity" of a sort that Locke suspected might be open to mathematicians of "advanced speculations"?

WHAT THE SEVENTEENTH CENTURY CONTRIBUTION WAS NOT

It was a commonplace in philosophy and mathematics before the nineteenth century that what was infinite was *quantity*. It is to Locke's great credit that he saw the confusion bred by a failure to distinguish between infinite quantity regarded (1) as a unique, completed, thing; and (2) as something of which we have an "endless growing idea".[107] Bolzano would later make the same point more effectively, and — like Locke — emphasize the logical priority of *number* over quantity. From number, Bolzano and Dedekind would shift the focus of attention to *classes*, equivalent classes sharing the property of having the same power or cardinality. The seventeenth century can offer at least a handful of hesitant attempts to come to terms with the problem of comparing infinite classes — setting up a one-to-one correspondence between their members in the now familiar way. There is nothing new about this problem, which has in any case been aside from the main themes I have been discussing. It was quite commonly discussed in fourteeth century Oxford, for example, and was subsequently never totally forgotten by European scholarship. Neither Descartes nor Locke, even so, seems to have made any attempt to grapple with the problem of the "inequality of infinites", and the apparent paradox which results, for instance, when each term in the series of natural numbers is made to correspond with its double. Galileo had given the problem publicity and Harriot had struggled with it at much the same time. Descartes shows from a letter to Mersenne of 28 January 1641 that he *knew* perfectly well of the problem, for in his letter he commented on the work *Quod Deus sit mundusque ab ipso creatus fuerit in tempore* (1635) by the mathematics professor at the Collège Royal in Paris, J. -B. Morin. Morin had recently attacked Descartes, despite which fact Descartes' reply was restrained:

I have read M. Morin's book. Its main fault is that he always discusses the infinite as if he had completely mastered it and could comprehend all its properties. This is an almost universal fault, which I have tried carefully to avoid. I have never written about the infinite except to submit myself to it, and have not written to determine what it is not . . . He supposes also that there cannot be an infinite number — which he could never prove.[108]

Fragmentary enough; but if it is to be believed, this suggests that Christian humility overcame the mathematician's curiosity. Hobbes was probably referring to Morin — or at least to the sorts of argument Morin introduced — when he too agreed to leave such problems in the hands of the established

Church; but Hobbes' language is so loose as to make it seem unlikely that he would have made much progress had he tried.[109] Leibniz makes an amusing aside about Hobbes' attempts to apply the exact methods of geometry to morals and physics: "Il y a un mélange chez Hobbes d'un esprit merveilleusement pénétrant et estrangement foible incontinent à pres. c'est qu'il n'avoit pas assez profité des Mathematiques pour se garantir des paralogismes".[110] As it happens, one of Leibniz's earliest disputations (Leipzig, 7 March 1666), known only from a list of theses to be defended, was on the theme "One infinite is greater than another",[111] so there at least we have a mathematician who did not brush aside the problem.

One could mention others who touched on the problem of the equivalence of an infinite set with a proper sub-set, but the story would be always the same, namely that there was never any suggestion that the infinite be *defined* in terms of the paradoxical property. Where the infinite does enter seventeenth century mathematics significantly is in connection with infinite series (seen clearly by mid-century to define a limit), often as the basis for quadratures (some introducing their own "paradoxes of the infinite", as when an infinite line contains a finite area). There are mathematical uses of the word in connection with almost every tract on the differential and integral calculus, and to abstract the many nuances of its use would induce something approaching infinite tedium, but one rough generalization is in order here. The mathematician almost invariably thought of his infinite collections (lines, spaces, numbers, times, etc.) as continuous magnitudes rather than as sets; and insofar as he tried to set up a "mathematics of the infinite" it is in the form of a calculus of 0 and ∞. Number was seen as a ratio of quantities (an obscure idea, but thought meaningful enough by those who used it). Leibniz was one of many who regarded the "infinite quantity" as a quotient of 1 and 0; and taking 0 as analogous (or even equal to) an infinitesimal quantity, it became perfectly natural in the fulness of time for Euler to speak of orders of infinity corresponding to orders of differentials, namely their reciprocals. The notion has respectable modern descendents, as have so many historical characters who behaved in their day with wild abandon.

I mention these tendencies only to underline the fact that the Descartes-Locke-Newton stream of thought we have been considering is a separate stream from what might be crudely called the "inverse infinitesimal" stream (with its vast network of tributaries), and different again from the Bolzano-Dedekind-Cantor network. This is all put very crudely, I know, but is needed by anyone who is tempted to regard the concept of infinity as monolithic. (No-one is likely to treat it as *simple*, I imagine.) Whether or not "number is prior to quantity", is it possible to frame a meaningful doctrine of the infinite without reference to the concept of number, and if so, does it resemble anything from 17th century metaphysics? And if not, is it true to say, with E. T. Bell, that "The 'infinites' of religion and philosophy are irrelevant for mathematics"? The biography of Cantor alone is enough to controvert Bell as far as historical influence is concerned.[112] What about their relevance to physics? Space, matter, void, time, motion, . . . Does it matter to physics that we include an "infinite" God in the list, as all our characters from the 17th century wished to do? I think such counterfactual hypothetical questions are an important part of philosophical history; and in this case I think my answer would be that metaphysics did *not* help physics very much; but I am equally sure that I am not qualified to give an answer to this particular question.

Finally, what did the 17th century metaphysics of the infinite do for the rest of philosophy? We have seen how it prompted Descartes to think carefully about the difference between comprehending, having an idea, and knowing, in different senses. Descartes' words in this vein reverberated through the world of philosophy. Take Hobbes: we have in our mind no idea of the infinite and eternal except that of our insufficiency to comprehend them. Take Malebranche's "intelligible extension", what we are supposed to think about when we think of the infinite. And Newton — understanding rather than imagining infinite space. And Locke — for whom the infinite was a testing ground for his way of ideas. And Leibniz — whose thoughts on infinity led to some influential ideas on necessity. A few examples only, but not a bad score sheet.

NOTES

[1] *Phys.*, II.4; 202b.30—.
[2] 204a.6.
[3] L. Baudry, *Lexique philosophique de Guillaume d'Occam*, Paris, 1938, p. 122.
[4] *Met.*, 1066b.17—18.
[5] *Phys.*, 212a.20.
[6] *Ibid.*, 212b.20—21.
[7] *Met.*, 1048b.9—17.
[8] *Phys.*, 204b.5—6.
[9] *Phys.* III.6, 7 is the source of the remainder of this paragraph and the next.

[10] *Phys.*, 206b.20–24.

[11] *Ibid.*, 207b.18–20. These propositions are accepted at several places in Aristotle's writings.

[12] Some remarks were added here which are open to several interpretations, but they are beside my present point. See *Phys.*, 207b.27–34. One interpretation is that if a geometer is embarrassed by a figure which bids to stretch beyond the universe (in his thoughts) then he may use in his proofs a smaller, similar, figure.

[13] Cf. *Met.*, 1051a.21–33.

[14] "Aristotelian infinity", *Phil. Rev.* 75 (1966), 197–218.

[15] *Ibid.*, pp. 217–8.

[16] [C. Adam and P. Tannery (eds.), *Œuvres de Descartes*, Cerf, Paris, 13 v., 1897–1913 (reprinted Vrin, Paris, 1957–8) will be cited in the usual way as "AT".] AT. V, p. 274.

[17] AT.V, pp. 50–1.

[18] For references, and more detail, see A. Koyré, *From the Closed World to the Infinite Universe*, Baltimore, 1957, Chapter 1 and the notes to it.

[19] Notes, however, the case of the endless ring, previously discussed. The ring is not limited, but it can be "gone through" in the sense that all its parts can be surveyed in a finite operation.

[20] *Op. cit.*, 1970 edition, p. 11.

[21] AT.V, p. 167, for the Burman letter. Henry More, to take an example of confusion, *approves* the indefiniteness concept of Descartes but speaks of God (and thus, for him, space) as infinite and the world as indefinite, meaning by this *finite*. See Koyré, *op. cit.*, p. 140. Joseph Addison, in *The Evidences of the Christian Religion*, 6th ed., 1776, p. 107 (reprinted from the *Spectator* (c. 1714)), writes of the extent of the throne of God as follows: "Though it is not infinite, it may be indefinite; and though not immeasurable in itself, it may be so with regard to any created eye or imagination". Comment would be superfluous.

[22] Jean Jacquot, "Thomas Harriot's reputation for impiety", *Notes and Records of the Royal Society* 9 (1952), 164–87 at p. 167.

[23] *Ibid.*, p. 179.

[24] *Ibid.*, p. 180.

[25] *Ibid.*, pp. 181–2.

[26] *Elements of Philosophy*, IV.26.1; W. Molesworth (ed.), *The English Works of Thomas Hobbes*, London, Vol. 1, 1839, pp. 410–14. The date of the epistle dedicatory is 1655.

[27] Oxford, 1656, pp. 116 ff.

[28] [*Die philosophischen Schriften von G. W. Leibniz*, Gerhardt (ed.), 7 Vols, Berlin, 1875–90 = "G".] G. VIII, pp. 64–5; G. IV, pp. 343–9.

[29] *Meditatio* III; AT.VII, pp. 45–6.

[30] To "Hyperaspistes", August 1641; AT. III, pp. 426–7.

[31] This is often reiterated. See, for instance, to Mersenne, AT.III, p. 233, and Gassendi's objection to the third meditation, AT.VII, pp. 296–7, for a contrary view.

[32] Cf. E. McGuire, in *Nature Mathematized*, ed. W. Shea, Dordrecht, 1983, pp. 92-3.

[33] *Ed. cit.*, pp. 101–2.

[34] P. H. Nidditch (ed.), *Draft A of Locke's Essay Concerning Human Understanding. The Earliest Extant Autograph Version.* University of Sheffield, 1980, pp. 163–5.

[35] *De grav.*, pp. 102–3; quoted by McGuire, *op. cit.*, p. 94.

[36] AT.V, p. 344.

37 AT. V, pp. 51–2.

.38 A passage from Des Maiseaux's preface to the 1720 edition of the *Clarke-Leibniz Correspondence* (see H. G. Alexander's edition, Manchester U.P., 1956, pp. xxviii–xxix) shows that Clarke was not altogether happy with his own theory, which also appears in his Boyle lectures. For a typical occurrence of it see *ed. cit.*, p. 23.

39 For example at AT. VII, p. 40.

40 AT. V, pp. 355–6.

41 *Loc. cit.*

42 *Ibid.*, p. 366. My italics. Note the language of comparison, and compare the mathematical language of ratios.

43 This is one of several references which I can no longer give, since the printed version of the paper to which I am replying differs somewhat from that version to which I actually replied.

44 See n. 43 above.

45 *Princ.* I, pr. 64; AT. VIII, p. 31.

46 AT. XI, p. 36.

47 See n. 43 above.

48 Baudry, *op. cit.*, pp. 49–50, lists Ockham's meanings.

49 AT. VII, p. 368; see n. 43 above.

50 AT. VII, p. 295.

51 See n. 43 above.

52 Cf. AT. VII, p. 163.

53 For the context, see above. Note that Koyré, *op. cit.*, pp. 104–6, turns *Princ.* II.21, 22 to wrong use, saying that the infinity of the word "seems thus to be established beyond doubt and beyond dispute", but that "as a matter of fact Descartes never assers it". *Princ.* II.22 is in fact aimed only at showing the unity of the world – and that it admits of no gaps between one "world" and another.

54 See the interview with Burman, AT. V, p. 155.

55 AT. IV, p. 118 (to Mesland).

56 Cf., for instance, *Princ.* II.16.

57 AT. V, pp. 311–12 (More), p. 345 (Descartes).

58 J. Cottingham (tr.), *Descartes Conversations with Burman*, Oxford U.P., 1976, p. 33. For the Latin originals see AT. V, especially p. 167, and cf. the letter to Chanut, where he admits that although the putative limits to the world are *incomprehensible* to him, they may be known to God. AT. V, pp. 51–2.

59 McGuire, but see n. 43.

60 *De grav.*, pp. 100–01.

61 *Ibid.*, p. 102.

62 *Loc. cit.*

63 AT. VIII, p. 25.

64 AT. VIII, p. 17.

65 AT. VII, p. 56.

66 See n. 43 above.

67 See n. 43 above.

68 See n. 43 above.

69 AT. VIII, pp. 16, 53.

70 See Marjorie Nicolson, 'The early stage of Cartesianism in England', *Studies in Philology* 26 (1929), 356–74.

71 See L. J. Beck, *The Metaphysics of Descartes*, Oxford U.P., 1965, especially pp. 17, 41.

72 "Existence, actuality and necessity: Newton on space and time", *Annals of Science* 35 (1978), 463–508. The Gassendi connection was explained and defended in R. S. Westfall, "The foundation of Newton's philosophy of nature", *British Journal for the History of Science* 1 (1962), 171–82.

73 *Op. cit.*, p. 469, n. 25.

74 See pp. 242-3 above, where I quoted other words to the same effect, from the same place.

75 *Essay*, Bk. II, Chapter 17.

76 See G. A. J. Rogers, "The empiricism of Locke and Newton", in G. C. Brown (ed.), *Philosophers of the Enlightenment*, London, 1980, pp. 1–30.

77 *Essay*, II.17.7.

78 A. C. Fraser (ed.) *The Works of George Berkeley*, Vol. 1, 1901.

79 For the most apt text see N. Kemp Smith, *A Commentary to Kant's "Critique of Pure Reason"*, 2nd ed., Macmillan, London, 1923, p. 486.

80 *Essay*, II.17.9. In the 1671 draft Locke expressed himself only marginally differently. On number, there also taken as basic, he says the same thing: there is no other infinity than that of numbers, "which is never actual but is always capable of addition". As for infinite power and knowledge he expressed the hope that on a future occasion he may show that thereby we mean no more than "such a power or knowledge which cannot by any thing that doth or can possibly exist be limitted or resisted. Which is not any notion of a positive actual Infinity, but only potential as of numbers, to the bounds whereof even in thinking we cannot arrive; & soe that is Infinitum to us ad cuius finem pervenire non possumus". *Ed. cit.* p. 165, my punctuation. Apparatus omitted. Compare this with my next paragraph, p. 136.

81 *Essay*, II.17.13.

82 AT.VII, p. 113.

83 AT.VII, p. 45.

84 Letter to Varignon, 2.02.1702.G (Math.) IV, pp. 91–5. Loemker (tr.), pp. 342–6. My italics.

85 *Essay*, II.17.15.

86 *Essay*, II.17.8.

87 *Essay*, II.17.1.

88 *Essay*, II.17.7.

89 *Essay*, II.17.10–13.

90 *Essay*, II.17.18.

91 Quoted from Koyré, *op. cit.*, pp. 148–50. (More, *Enchiridium metaphysicum*, c. VIII, 9.) Note More's proud boast that Aristotle made immovability the highest attribute of the First Being. Such ideas continued to be influential. Joseph Raphson, for instance, although he criticized Spinoza generally, followed the essentially Spinozist position in taking God to be extended; and Berkeley (*Of Infinites*) took him to task for this. Raphson quotes More in his own support. (See his *De Spatio Reali seu Ente Infinito*, 1697.)

92 *Essay*, II.17.17. *Ed. cit.*, p. 168, for "Draft A".

93 *Ed. cit.*, p. 167.

94 *Essay*, II.15.12.

95 Compare his use of the word "comprehension" with Descartes".

96 *Essay*, II.15.4.

97 *Essay*, II.23.21.

98 Koyré, *op. cit.*, p. 149.

99 *Essay*, II.10.4.

100 *Essay*, II.15.3.

101 *Ed. cit.*, p. 166, my punctuation.

102 I am not aware of any recent comment on the passage in question, or on its elucidation by P. Coste, the French translator of the *Essay*, a man who had lived with Locke during the last seven years of his life. I owe the reference to W. Hamilton, *Discussions on Philosophy and Literature, Education and University Reform*, 3rd. ed., Edinburgh and London, 1866, pp. 199–200.

103 *Essay*, IV.10.

104 *Ed. cit.*, p. 168. Cf. my earlier quotations from the same.

105 The closing words of II.17.

106 Leibniz, Harriot, and a number of medieval writers use the same figure to a rather similar, if not identical, purpose.

107 *Essay*, II.17.7; see above, p. 258.

108 AT.III, p. 292.

109 *Op. cit. (English Works)*, Vol. 1, p. 413.

110 L. Couturat (ed.), *Opuscules et Fragments inédits de Leibniz*, Paris, 1903, p. 178. The date was c.1686.

111 He refers to Cardan's *Pract. Arith.*, c.66, nn. 165 & 260. Leibniz was confusing Ward with John Wallis when he said that Seth Ward dissented, in his *Arithmetic of Infinites*. G.IV, p. 42; Loemker, p. 75.

112 *Development of Mathematics*, 2nd ed., New York, 1945, 545.

11

THE *PRINCIPIA* IN THE MAKING

In 1858 the streets of Grantham – where Newton was at school – were crowded by "persons to whom the word 'respectable' in all its peculiarly English signification could be so properly applied", all out to celebrate the unveiling of a statue to the great man. Cheered by a "sympathising audience seated on a platform which had been most appropriately divided into compartments, each ornamented with a different colour of the prism", Lord Brougham gave the first of several lengthy orations. The shortest was by the master of Grantham School. "We are," he said, "like those golden argosies, of which poets sing, labouring beneath the weight of their precious burden, and tottering to the port." One wonders whether he saw the ambiguity. At all events, he was determined not to "prattle long on so stupendous a theme", adding, "The ingenious poet, who sang the praises of Newton, contracted his thoughts within two nervous lines. Let us imitate his brevity."

Grantham, it seems, was for that brief day a rather accurate mirror of the way the English have tended to treat Newton. There have been the monuments and the empty rhetoric, both more often symptoms of corporate pride than of any historical urge. There have been numerous occasions on which "two nervous lines" were as much as the historian had the courage or the good sense to utter. But, in spite of it all, there have always been the few capable of giving informed exegesis, and not always within an artless historical frame. From small beginnings, mostly having to do with Cambridge mathematics, this tradition of scholarship has come to full maturity in two great Newton tributes of recent years, *The Mathematical Papers*, edited by D. T. Whiteside with some assistance from M. A. Hoskin and A. Prag, and *The Correspondence*, now edited by A. R. Hall and Laura Tilling. The first volume of the mathematical papers appeared in 1967, while the first volume of the correspondence appeared as long ago as 1959. (The first three volumes were edited by H. W. Turnbull and the fourth by J. F. Scott.) The two editions are

HALL, A. R. and L. TILLING (editors) [1975], *The Correspondence of Isaac Newton*, volume 5, Cambridge: Cambridge University Press, pp. 439.
WHITESIDE, D. T. (editor) [1975], *The Mathematical Papers of Isaac Newton*, volume 6 (1684–1691), Cambridge: Cambridge University Press, pp. 614.

produced by the Cambridge University Press in a similar and quite magnificent format. Since there are only two more volumes of the Whiteside edition to come, and since the correspondence has now reached beyond seventy of Newton's eighty-five years, these two great monuments are at last beginning to look secure.

Volume 5 of the *Mathematical Papers* left Newton in the summer of 1684, working on a variety of mathematical (and chemical) topics. Halley tells us, in a now famous passage, how in that year he and Wren and Hooke had met and discussed how to determine the law of force from which the known laws of celestial motion (essentially Kepler's) could be derived. Halley had rightly decided on an inverse square law of attraction, which Hooke indeed said he could prove to be what was needed. Wren doubted Hooke's word, and wagered a forty-shilling book. In August, as Halley reminds Newton, he found on a visit to Cambridge "the good news that you had brought this demonstration to perfection". The paper was not to be found, the problem was reworked (if, that is, Newton was not using Hooke's method of playing for time), and Newton was consumed in the process with an almost obsessive interest in the dynamics of material bodies. This was a subject on which he and Hooke had corresponded some years previously, and a subject in which he at first owed more to Descartes than (as was supposed before Dr Whiteside drastically revised the accepted picture) to Kepler and Galileo. For long he had believed in the Cartesian vortices. There is little doubt, however, that the correspondence with Hooke was the catalyst Newton needed, even though Hooke's pretentious claims to the inverse square law later caused Newton to reject virtually any association with Hooke's ideas.

It was Newton's renewal of interest in dynamics which led, within three years, to the publication of his greatest work, *Philosophiae Naturalis Principia Mathematica*. Since this work is the centre of attraction around which all Newtonian scholarship tends to revolve, whether knowingly or otherwise, any study of its genesis is doubly interesting. Volume 6 of the *Mathematical Papers* contains what is essentially the product of Newton's mathematical pen between Halley's visit and the appearance of *Principia*. It begins with the first tract "De motu corporum" (expertly translated, like most of the Latin works, *en face*), a work which was first published by J. W. Herivel in 1966. Here Newton gets down to business quickly, and with a minimum of axiomatic support proves Kepler's law that "all orbiting bodies describe, by radii drawn to their centre, areas proportional to their times". (The force is central, but otherwise of more or less arbitrary law.) After a number of similarly economical moves he arrives at a model for a planetary movement under the standard Newtonian inverse square law, and from it proves Kepler's third law. He ends the tract with a brilliant study of the motion of a projectile resisted in proportion to its speed.

Edward Paget took a copy of the short tract to London in November 1684, where Halley transcribed it before taking another journey to Cambridge. He arrived to find Newton hard at work on an enlarged treatise, and with his encouragement the treatise grew and divided and grew, until it had become the tripartite *Principia*, published at Halley's expense in 1687. The first enlargement of the initial tract is clearly recognizable as the direct ancestor of the great work. (It was first published in 1962 by A. R. and M. B. Hall, and again by J. W. Herivel, with whose treatment of the text Dr Whiteside takes issue, however.) It opens in much the same way as the *Principia*, with recognizable definitions and laws of motion. The interdependence of the works has been explored in I. B. Cohen's recent *Introduction to Newton's* PRINCIPIA. The revision of the first tract extends the potential planetary movements in hyperbolic and parabolic cases, and includes among many other deep mathematical insights a proof that uniform spherical mass may be treated for his purposes as a point mass.

The third item in the sixth volume of the *Mathematical Papers* is the ensuing revision, "Liber primus", which can be superficially equated with about a hundred pages of *Principia*, and which takes up well over twice as many pages in Dr Whiteside's edition, where it is of course fully annotated and supplemented.

No review can possibly convey the richness of those modern annotations; and lest any mathematical reader wish to test Newton's mettle, and try his hand at annotation, I recommend Lemma XXVIII: "There exists no oval figure whose area cut off at will by straight lines might generally be found by means of equations finite in the number of their terms and dimensions." We are reminded all along of the *concessions* Newton made to his reader, who was expected to have little beyond the elements of Euclid and Apollonius, apart from a willingness to be led by the hand of Newton himself through his own analytical methods, proved *ab initio*.

The myth that Newton was perversely obscure is one which Dr Whiteside is anxious to explode; and from the scholar perhaps most familiar with Newton's manuscript drafts and worksheets we can accept this judgement. Inevitably, Newton ventured into problems for which his mathematical techniques were not adequate. Virtually all the documentation necessary for us to follow the valiant struggle with the inequalities of the Moon's motion (as deriving from gravitational perturbations) is now judiciously assembled in one place, with three further revisions, so that Newton's strength can be seen as plainly as his sleight of hand. Of the theory of the Moon, Newton is purported to have said to Halley, "it has broke my rest so often I will think of it no more"; and not for twenty years after Newton's death were better solutions to be found.

Newton's neat (but often much corrected) manuscripts, finely illustrated at intervals throughout Dr Whiteside's edition, present a different problem from that inherited by Professor Hall and Dr Tilling from J. F. Scott. At Dr Scott's death the roughly assembled materials for Volue 5 of *The Correspondence* proved to be as problematical in their way as the originals on which they were based. Many of these, in turn, survive in multiple drafts, and some of them Newton might never have dispatched. Several of the letters relate to the routine business of the Mint, and provide us with an insight into the somewhat humdrum life of Newton the civil servant, where tin farthings gave him as many headaches as the Moon had done years before.

The main intellectual interest of the period 1709–13 lies, however, in the fact that these were the years during which Newton, with Roger Cotes's help, was devising the *Principia* and seeing it through the press. Considered as a single group of letters, the Newton-Cotes correspondence (as its editors explain in their introduction) was

> the largest and most important section of Newton's scientific correspondence that we have; nowhere else can one witness Newton in a detailed debate about scientific argument and scientific conclusions – a debate from which he did not always emerge victorious. Nowhere else does Newton write in detail about the text of the *Principia*. And all scholars would agree that this text which was hammered out between Cotes and Newton was the most important of all the versions, printed and unprinted; this was (to all intents and purposes) the *Principia* of subsequent history.

Newton was fortunate in having Cotes as his editor, for – as the correspondence shows – Cotes worked systematically through the proofs, drew attention to what others had done or tried to do in the same connection, sometimes overrode Newton's wishes, and handled relations with the printer. (A full account of Bentley's patronage of Cotes, and the frustrations of others who wanted to play an editorial role, can be had from Cohen's *Introduction to Newton's* PRINCIPIA.) It is hard to imagine where Newton could have found a more able and diplomatic editor. Cotes was not only an excellent mathematician but had a feel for "mere" numerical data that seems in much of the correspondence to have been surer than Newton's own.

Occasionally Cotes wandered off into academic politics, as when he tried in vain to secure the Lucasian chair (Newton's, before it was Whiston's) for Christopher Hussey. Like other friends he kept Newton informed of any who purported to criticize his principles. There was the idiotic Robert Green, for example, a Cambridge circle-squarer whose work was said in a letter from Nicholas Saunderson (a blind tutor at Christ's) to contain nothing "but ill manners and elaborate nonsense from one end to the other". A greater threat to Newton's ideas came from the Continent, above all (in the period 1709–13) from Leibniz, and it is very strange that such a small part of the correspondence relates to this threat.

Newton is known to have given much thought during 1711 and 1712 to the compilation of the *Commercium Epistolicum*, which embodied the Royal Society's case for Newton's priority in the matter of the calculus, but the many drafts he composed against Leibniz were not, it seems, linked with any Maachiavellian corespondence, unless such has been lost. Volume 5 does include many leters which illustrate the gathering storm of abuse and debate. There is here, for example, the letter in which Leibniz, writing to Hans Sloane, sought a remedy from the Royal Society for the wrong he had thought he had suffered at the hands of John Keill, who had charged him with plagiarism. (This letter is printed, like all the extended passages of Latin, in the original as well as in English translations, expertly done.)

It is, no doubt, with Leibniz on his mind that Newton's reactions to Cotes's minute criticisms become progressively colder and more terse. There is the problem as to why Newton deleted a tribute which he planned to make to Cotes. Was it (as Professor Hall and Dr Tilling suggest) that it would have necessitated a similar tribute to Johann Bernoulli, who had discovered a notorious error in the first edition of *Principia* (Book II, Proposition 10)? Or was it (as Dr Whiteside suggests) that Newton secretly blamed Cotes for failing to detect the error? Whatever the answer, Cotes's part was evident to the world only by his preface to the new edition, a preface he wrote half against his will and with precious little help from Newton, who asks specifically that he be not shown it, lest he "be examined about it".

The resulting essay was until recently looked upon as a focus for Newton's views on natural philosophy. It remains essential reading for anyone who wishes to understand the philosophical ferment of the time. Was gravity a primary quality? Had Newton rendered it an occult quality? Was the universe a complex of vortices? However we judge Cotes's preface, it is now clear that it is not a piece of disguised Newton. For a more authentic Newtonian riposte to Leibniz from this period we now have a draft of a letter to the editor of the *Memoirs of Literature*.

This draft does not cast any light on the problem of Samuel Clarke's collaboration with Newton in the composition of the "theological" General Scholium added to the second edition of *Principia*, a problem recently raised in the *TLS* (letter from I. B. Cohen, October 17, 1975). The draft does, nevertheless, cover much of the same ground as the famous Leibniz-Clarke correspondence of 1715 and 1716, terminated by Leibniz's death and printed by Clarke in 1717; Clarke is shown by one letter (presumably printed by Joseph Edleston) to have "corrected" Cotes's preface on the question of the essential nature of gravity.

Other letters in Volume 5 of *The Correspondence* touch on Royal Society gossip and scandal, and on relations between it, Newton, and Flamsteed, the Astronomer Royal. Few signs of Newton's personal involvement in the forcible official publication (1712) of Flamsteed's observations remain in the correpondence. As Professor Hall and Dr Tilling remark, "the humiliation of Flamsteed was not a misfortune that would deter Newton from pursuing a course that he believed to be incumbent upon him as a loyal servant of the Crown and President of the Royal Society".

Newton had been brought to the Mint in 1696 by Halifax – whose relationship with Catherine Barton, Newton's niece, is inevitably touched upon; although none of the letters throws light where it is most wanted. As to Newton's duties at the Mint, Sir John Craig's standard work on the subject is now supplemented by more printed letters which prove yet again what their editors call "Newton's excruciatingly conscientious attention to every detail of government business". They are – not surprisingly – unable to find any evidence among the Mint papers for the reputed £6,000 bribe from a copper-coinage contractor, or for the £2,000 annual pension which Bolingbroke is reputed to have offered Newton if he would retire. The evidence, if it ever was, has gone, and Newton continues to "maintain the dignity of history".

When Flamsteed saw the corrected sheets of his catalogue he "found more faults in it, and greater", than he "imagined the impudent editor either could, or durst have committed". His editor was Halley, in whom, had he only known it, he was even more fortunate than Newton in Cotes. But neither Halley nor Cotes was faced with the doubly taxing duty of the historian–editor who must make both sense and history out of his materials. This task Professor Hall and Dr Tilling perform magnificently. Very rarely do they falter. (There is some slight confusion over the background to lunar theory; but as Newton said in a work of 1702, "The Irregularity of the Moon's Motion hath been all along the just Complaint of Astronomers".) They have not only resuscitated a dormant publication, but have breathed more life into it than it had before. Misquoting Cotes's preface, one might say of their work, as of that by Dr Whiteside, that "si nunc revivisceret, Isaacus Newtonus vel simplicitatem vel harmoniae gratiam in ea desideraret". Could one say as much for the Grantham statue?

12

A TIMELY CATALYST

The dust that falls to earth after those short-lived meteor showers which mark our passage through the trails of comets is as nothing to what will soon begin to descend on booksellers' shelves of remainders, in the wake of comet Halley. Comets have been kind to publishers since the earliest days of printing, but Edmond Halley changed the whole nature of the game. Before him, comets took the world by surprise, and were above all an astrological property; after him, they were material objects subject to Newton's laws of motion, and those who traded on apprehension and fear were obliged to shift their stance. After Halley it became a question of calculating whether a comet would collide with Earth. Would it, perhaps, introduce a poisonous gas into our atmosphere? (Cyanogen was the 1910 favourite.)

Books suggesting that such things might happen are currently with us again – though none is on the list below – and are ephemeral by design, rather than by lack of merit. The same has to be said of books whose chief purpose is to describe the sort of spectacle we may expect. Most books of this sort are written in apparent ignorance of the story of the boy who cried wolf. The comet will "blaze across the sky", as several publishers have been assuring us vaguely for months, judiciously avoiding all the best evidence, not to mention the opinions expressed by their authors. In any case, it takes more to astonish us than it did: Westland helicopters notwithstanding, press, radio and television are full of Halley's comet, and if there is ever a publishers' inquest into the demise of the goose that lays the golden egg, the verdict is likely to be "death from boredom".

Both Halley and the comet deserve better. It was he who first recognized a series of cometary records from various epochs as evidence for a single entity. Although much credit goes to him for his discovery, one should forget neither the long tradition of recording cometary positions, on which it rested, nor the mathematical theory published by Isaac Newton (1687) that Halley used in the early 1700s to determine, with "a prodigious deal of calculation", the elements of the orbits of

---

* For details of books reviewed here, see p. 283.

more than a score of different comets. Newton had used the comet of 1680 as his example. Halley was struck by the similarities between the calculated orbits of the comets seen in 1531, 1607, and 1682. He postulated a single cause, and suggested that it should manifest itself again in 1758. (He later revised this estimate to late 1758 or early 1759.)

By the time he came to this conclusion, his comet had been recorded, as we now know, at fairly regular intervals over a period of more than two millennia. What must surely be counted as the most fortunate discovery of recent years in this connection is that of Babylonian cuneiform tablets, at present on exhibition in the British Museum. As told in an "exclusive" in Brian Harpur's *Official Halley's Comet Book*, a gap in the records was filled when Richard Stephenson of Durham asked Hermann Hunger of Vienna whether he knew of Babylonian observations comparable with the Chinese. Hunger searched the papers of the late Abe Sachs, and found records for the appearance of Halley's comet in 164 BC in no fewer than three different tablets. To Harpur, the discovery of the vital link within a mile or so of where Halley was born "seems to stretch the arm of coincidence out of its socket". His love of coincidence encourages him to draw up a series of significant episodes in world history that have coincided – give or take a few years – with the comet's return. It ends with the birth of Richard Nixon in 1913, only three years after the 1910 return. There, presumably, goes the other arm.

The comet's first appearance after Halley's prediction was indeed in 1759. Its return was a prime test, not only of the original prediction, but of a more elaborate mathematical analysis that had been offered by A. C. Clairaut a year or so before. At last, the comet had really come into its own as a phenomenon of scientific concern. Since then, it has been back three times: in 1835, in 1910, and now. On each return, the scientific world has risen to the occasion. One can have an International Year of the Squirrel at the drop of a hat, but it makes sense to concentrate the mind on an important comet when it is in the neighbourhood. No profession has a stronger urge than the astronomical to confer at precisely chosen intervals, and cometary visitations make a change from centenaries. In fact one of the endearing things about Halley's comet is that it returns roughly once in a lifetime. Mark Twain noted that he had arrived with the 1835 apparition. With his inimitable sense of timing, he left with that of 1910.

The proceedings of a Rome conference, *Dynamics of Comets*, edited by Andrea Carusi and Giovanni B. Valsecchi, give a comprehensive account of the present state of knowledge as to cometary origins, meteor streams, gravitational and other forces that determine their motions. They give, too, something that astronomers in 1910 could leave to the imagination of such as H. G. Wells – an idea of the planning of the several space missions aimed at comet Halley. A fuller and more recent account of them, very well illustrated, will be found in Fred L. Whipple's *The*

Mystery of Comets.

The orbits of comets around the Sun are a small but important part in Albert van Helden's well-written general history of changing ideas on cosmic dimensions, *Measuring the Universe*. Although he ends his account with Halley, he points out that there were people alive in 1700 who had first learned all that they knew on this score from the Ptolemaic scheme. Halley's contemporaries had learned that the distance from Earth to Sun was greater than Ptolemy had thought, by a factor of about sixteen, and distances within the solar system were tolerably well-known by the time he announced that the path for the comet of 1682 had amounted to forty times the distance the old astronomers had assigned to the fixed stars, and four times the distance even then assigned to Saturn at its furthest.

Comets are still known to us because they orbit the Sun, but they come in from distances far greater than Halley ever dreamed, and from all directions. These facts suggested to the Dutch astronomer Jan Oort the existence of a vast cloud of comets, with perhaps as many in it as there are stars in our Galaxy. (Nearly a hundred named examples are separately discussed in *Dynamics of Comets*, and as though to show that astronomers on the whole disapprove of stars Hollywood-style, Encke's comet is mentioned as often as Halley's.) Since our Sun, however, has passed a number of times round the Galaxy, with its own considerably larger giant clouds of gas and dust, would such an "Oort cloud" not have been stripped away, had it ever existed? Do comets in fact come from the nebula out of which the Sun was formed, or are they drawn from that vast reservoir, the supposedly cometary cloud, extending almost to the distance of the nearest stars? The Rome conference proceedings show that there is plenty of room for lively controversy on these points.

Two very clear and concise elementary introductions to such matters will be found in both Whipple's and the equally aptly illustrated book by Jack Meadows. Meadows does not take his title, *Space Garbage*, from the book trade: it refers rather to the meteorites, asteroids, comets, meteors and interplanetary dust that remained after the planets and their satellites were formed. He likens himself to an archeologist, sifting through rubbish to piece together the early history of the solar system and beyond. By contrast, most of the popular books brought into being by the comet in the last year or two have been rag-bags of information, compounded rather than sifted – a snippet or two of Halley biography, a few scraps of social history, which is to say the odd newspaper-cutting and a picture postcard or two, a smidgen of science, and information on the Giotto mission, courtesy of the Dynamics Group, British Aerospace. If *The Official Halley's Comet Book* falls into this category, it scores over much of the opposition simply because it has been compiled with enormous zest. It must have been written as the presses were turning. Harpur registered Halley's Comet Limited as long ago as 1975, with charitable intent. His evident love for the comet is such that he would bottle and sell it if he could.

What seems to me the best of the new books for the general reader, however, is Whipple's *The Mystery of Comets*, written with the help of Daniel Green. Whipple is a retired Harvard Professor of Astronomy whose qualifications include the discovery of no fewer than six comets in twelve years. For some peculiar reason, perhaps out of deference to Halley, his publisher has refrained from mentioning that fact on the dust-jacket. The book includes a dash of history (apart from what now seems to be a permanent feature of Cambridge University Press books, namely a comprehensive history of the Press on the title-page). Not the least interesting bit is Whipple's sketch of how he arrived at his own highly influential account of the origin and nature of comets based on the so-called "dirty-snowball" theory. Among the various phenomena he managed to explain on the basis of the model is the way comets seem reluctant to conform with Newton's law of gravity. The reason, he found, is that the "snowball" at the head of the comet, when in sunlight, boils off gas on the side towards the Sun, and acts as a sort of jet engine. A. D. Dubiago in the Soviet Union had been trying out similar ideas when he put forward his theory. It is a curious fact that that was all in the late 1940s, the Golden Age of the jet engine.

Like Halley's comet itself, the books it has spawned in such numbers convey – at least to the historian – a distinct sense of *déjà vu*. Not in every respect, of course. No theatre producer in 1910 had an audience so captive, if not captivated, as the television audience in 1986. For more than a year it has been possible to buy cometary graphics for your home computer (there are at least three programs on the market), another privilege unknown in 1910. There must surely have been comet pop-up books then. They are happily still around, and although they do not exhale the Spirit of the Age, are likely to be less frustrating than things that do. Time will tell, but it is doubtful whether the comet's social impact in 1986 will quite compare with that of 1910. As readers of Bruce Morton's annotated bibliography *Halley's Comet 1755–1984* will be able to discover, more than 800 published items – varying from newspaper articles to weighty monographs – appeared on the theme between 1905 and 1914. Since his book is more or less restricted to the English language, and has a heavy American bias, that number clearly represents only the tip of the iceball. While the United States in 1910 seems to have been second to none in the way of alarmism – Comet Cocktail and Cyanogen Flip were at the other extreme of social response – one has to turn perhaps to France for the comet's most stylish literary and artistic reception, and to Germany for works of the deadliest seriousness. The whole subject will sooner or later be discovered for what it is, an unlimited source of post-deconstructivist literary theses.

The trade in artefacts has far to go before catching up with that of 1910.

Has the cutting of cometary diamonds yet begun? In Haiti in 1910 one could buy voodoo comet pills – one to be taken each hour, until such time as the comet began to recede from the Earth. In New York in the same year, dealers in telescopes sold more in three months than they had until then since the Civil War. Come March, will the Japanese optical industry crack up under the strain? At a humbler level, the postcard, alas, is not what it was, as we are reminded by Roberta Etter and Stuart Schneider's lavishly illustrated *Halley's Comet: Memories of 1910*. My impression is that poster futures are firm, but that 1986 is unlikely to match, for sheer economic consequence, the printing of 277,000 posters by the Christian Literature Society of China in 1910. No one should doubt that comets do affect human destiny.

Before Halley's time, comets were like V2s, alarming only as they arrived. Halley put our anxieties on a firmer footing. As early as 1755, three years or so before the returning comet was actually seen, John Wesley pondered the consequences of a collison with our planet. The Earth, he thought, would be set on fire and would burn like a coal. Among many letters to the *Gentlemen's Magazine* that took a different view was a censorious one which spoke of the "cruelty of terrifying weak minds with groundless pains". Even so it was a strong mind, facing the prospect of being enveloped in the comet's tail, that took comfort in the thought that the heavens declare the glory of God. Of course nothing happened, and the 1835 return – by contrast with those before and after it – seems to have been greeted by a less anxious public. Not that it was unimpressed. Devotees of the Brontës will not need to be reminded of the poem on the subject that appeared from the pen of the Rev Patrick Brontë in the second number of *The Bradfordian*, twenty-six years after the event. There was also Tennyson's "Harold" of 1877, but this should not be allowed to count, since he introduced a deliberate anachronism, just to make a point about how credulous were those who lived in the time of Edward the Confessor. He should have lived to experience the 1910 affair, heralded by H. G. Well's *In the Days of the Comet* (1906).

The novel illustrates Wells's favourite blend of lower-middle-class social allegory and scientific romance. *In the Days of the Comet* is not one of his better-known works, but it bears comparison with the more successful *Tono-Bungay* (1909) for its portrayal of the imagined collapse of the rich, entrepreneurial, English society from which the author felt himself excluded. The story is of an episode in the youth of the narrator, a grey-haired old man whose perspective is that, more or less, of our own part of the century. The foreground of the setting is a baleful Black Country town. The wise, but once so gauche, Leadford tells of his rejection by the unreliable Nettie, who prefers the wealth, not to say the style, of Verrall, the industrialist's son. One part of the background is filled by the threat of "the obvious waste and evil that would result from a war between England and Germany". The war begins, and is blamed by the narrator on those "damned little buttoned-up professors". Fortunately for humanity it was not long before there came about what is known throughout the novel as The Change, and this in the trail of the comet.

The comet itself is wonderfully unobtrusive and, after Leadford's mother, quite the most sympathetic character in the novel. When Wells comes to describe it, with its bright, ghostly, phosphorescent glow that changed the starless sky to an extraordinary deep blue, he really lets rip. A paragraph later, and we have been dropped into the cobbled streets of Monkshampton; and then in the next we have a news-boy announcing the loss with all hands of a British battleship in the North Sea. The second half of the novel tends to be almost a Wellsian self-parody. The story-teller, with the advantage of living now under a world state, is able to expatiate on the evils of those periods of spasmodic violence that occasionally disturbed the "odd slackness" of the "old epoch", before they were all philosophically ironed out. The English class system has disappeared, and with it its grimy industry and indeed all that is not truly Fabian.

Intent on murder, the insanely jealous young Leadford sets out to track down Nettie and her lover, but they are all saved at the last moment by the pacific effects of the gases injected into the atmosphere by the comet. The timely catalytic change in all human behaviour brings to an end the war with Germany, the price of eggs falls and the rest is history – or would have been, had Halley's comet been on the same course as Wells. His most perceptive writing is not predictive, and certainly not scientific. Although too often he shows a maudlin sentimentality, few have conveyed quite so well what it felt to be a talented youth on the wrong side of a certain class barrier.

Pipe-dreams of peace are one thing, but the idea that comets might affect the Earth by filling the atmosphere with dust is no laughing matter. Two decades ago, Harold Urey proposed that an asteroid or a comet was responsible for the massive extinction of many living species marking the end of the Cretaceous period, in which dinosaurs had been a dominant form of life. In 1980, Luis Alvarez, with his son and others of the University of California, put forward the same idea, blaming dust from the catastrophe for a serious interruption of sunlight, and a rapid drop of temperature that they supposed led to the massive extinction of species. Very recently, Edwards Anders and colleagues at the University of Chicago have announced the discovery of evidence of a global firestorm that swept the world (the evidence comes from Denmark, Spain and New Zealand) at the very time of the supposed impact, 65 million years ago. The Bering Sea is the assumed place of impact, and its force is estimated at a hundred million megatons. The general idea is that species of smaller creatures could have slipped into caves and hibernated, until sunlight could once again penetrate to the surface of the Earth.

Unlike the dinosaurs, we no longer have any real cause for alarm, since we now have the technology to detect and track objects on a collision course, and to deflect or destroy them. (How long before President Reagan finally admits that his Star Wars programme is really aimed at protecting us all from comets?) On the other hand, given the right sort of publicity for the fate of the dinosaurs, publishers' sales graphs might yet turn skywards, and before the end of the year we might well be reading headlines comparable with those of 1910, such as "Excited over comet, a Montreal girl suddenly expired", and "In fear of comet, confesses murder".

HARPUR, B. [1985], *The Official Halley's Comet Book*, London: Hodder and Stoughton, pp. 184.

CARUSI A., and G. B. VALSECCHI (editors) [1975], *Dynamics of Comets: Their Origin and Evolution*, Dordrecht: Reidel, pp. 439.

WHIPPLE, F. L. [1985], *The Mystery of Comets*, Cambridge: Cambridge University Press, pp. 276.

HELDEN, A. van [1985], *Measuring the Universe: Cosmic Dimensions from Aristarchus to Halley*, Chicago: University of Chicago Press, pp. 203.

MEADOWS, J. [1985], *Space Garbage: Comets, Meteors, and other Solar-System Debris*, London: George Philip, pp. 160.

MORTON, B. (editor) [1985], *Halley's Comet, 1755–1984: A Bibliography*, Westport, Connecticut: Greenwood Press, pp. 280.

ETTER, R., and S. SCHNEIDER [1985], *Halley's Comet: Memories of 1910*, London: Pandemic Press, pp. 96.

WELLS, H. G. [1985], *In the Days of the Comet*, London: Hogarth, pp. 249.

13

SCIENCE AND ANALOGY

I have chosen to speak about the use of analogy in scientific argument, and I ought to begin by clearing away a number of potential misconceptions: I cannot completely ignore the historical origins of the word 'analogy', but the time at my disposal is much too valuable to be spent in tracing its Aristotelian pedigree in detail. On the other hand, I have no wish to do what so many contemporary philosophers do, that is to say, treat the word as synonymous with the word 'model.' What I shall say has much to do with the notion of scientific model, but my perspective will be rather different from that of a historian searching for the various uses of models in science. I repeat that my concern is with analogical *argument*. As you will see in due course, the sorts of argument that count as analogical are usually reckoned to be rather weak, and even rather dangerous. There is no point in my trying to persuade you that things are otherwise. Mine will not be a history of 'positive science', in Comte's sense, but a history of tentative science, and of certain methods of *conjecture*. Analogy is the basis for much scientific conjecture, but even conjecture is an art, which can be done well, done rationally, that is, even though it might prove in the end to have yielded a false conclusion.

This last remark might well seem highly paradoxical, but it is one that I wish to emphasize, both because I believe it to be true, and because one's attitude to it affects one's whole approach to the history of science. Many historians — more in the past than in the present, I should say — are interested only in scientific success, and in the gradual progress of mankind towards the truth. Other historians, aware of the distorted image created by history practised in that style, boast that they are equally concerned with scientific error, falsehood, and misunderstanding. Some historians — even historians of ideas — profess not to be interested in the *quality* of the arguments they chronicle — 'That's not a historian's job', they say — but anyone who is so interested *must* take into account that no-man's-land between truth and falsehood, i.e. those arguments that were *reasonably* based in the light of the knowledge of the time, but in our own time are judged to have been mistaken. The past, after all, like the present, framed irrational as well as rational truths, rational as well as irrational falsehoods. Not all nonsense is equally foolish. To show what I mean, I shall first take a number of examples from the work of Isaac Newton — a respectable enough scientist, you will agree. I shall then look at the way in which the concept of scientific analogy developed in the middle of the nineteenth century, in the hands of Thomson and Maxwell. I shall not have time to say much about the history of what one might call the *logic of analogy*, but I will include a brief sketch of Mill's ideas on the subject, if only to show that this was not abreast of the best developed scientific *uses* of analogy at the time.

In June 1672, Newton wrote a long letter [1] to Henry Oldenburg, Secretary of the Royal Society, concerning some objections raised by Robert Hooke [2] against Newton's work on colours and the refraction of light. I don't want to be sidetracked into a discussion as to which of the two men was first to maintain that light is a *periodic* phenomenon: following E. T. Whittaker, [3] it is commonly said that Newton's letter of 1672 contains the first statement "that homogeneous light is essentially *periodic* in its nature, and that differences of period correspond to differences of colour"; but this is somewhat too generous. On the evidence, one could as easily ascribe the discovery of periodicity to Hooke — as does Richard Westfall [4] — although for my own part I think neither claim is particularly illuminating. I am content to frame the discussion in the words used at the time. As Newton said, *Hooke's hypothesis* was that the parts of bodies, when briskly agitated, excite vibrations in the aether, and that these in due course, acting on the eye, cause us to have the sensation of light, in much the same way as vibrations in the air cause a sensation of sound, by acting on the organs of hearing. [5] The analogy is all I want to consider, namely the analogy between sound and light. Newton does not actually say he is arguing by analogy when he goes on to say that the *largest* vibrations in the aether give rise to a sensation of red, and the *shortest* a sensation of deep violet; but it is clear that he was indeed using an analogical mode of reasoning. He goes on to draw the parallel: variation in the *size* of the vibration of the air, he says, is responsible for variation in the tones in the associated sound. There is no doubt that by 'size' ('depth' or 'bigness') he means not our 'amplitude', but something like 'wavelength.' He certainly thought of the vibrations as being *longitudinal*, rather than transverse. He knew from his experiments that aether vibrations of various 'sizes', that is, light of various colours, could be separated and recombined by refraction, and he tried his hand — not very successfully in this letter of 1672 — at explaining the colours of thin plates and bubbles. His greatest concern with the Hooke-Newton hypothesis (if I may call it this) [6] was that waves or vibrations in a fluid would be expected to spread out into the adjacent medium, rather than be confined to straight lines. For his own part, Newton believed he could put forward his 'doctrine' without reference to any hypothesis at all. He could, he thought, 'consider light abstractly'; and when, by his example, he shows us what he means by this, we find him falling back on the analogy with sound.

In his new use of the analogy, however, there is a slight change of style. He finds it easy to *conceive*, he says,[7] that the different parts of a body may emit rays of different colours "and other qualities" in the same way as the several parts of an uneven string [on a musical instrument], or of water in a stream or waterfall, or the pipes of an organ all sounding together, or "all the variety of sounding bodies in the world together", should send a confused mixture of sounds through the air. He can, he says, *conceive* of bodies capable of reflecting one tone and stifling or transmitting another, and just as easily can he conceive a body reflecting only that one of a mixture of colours which it is disposed to reflect.

Consider now his next use of the analogy with sound. Hooke had considered the possibility that there are only two fundamental colours. Newton's discussion of the experimental evidence is much clearer than Hooke's, and quite equal to a total demolition of Hooke's conjecture. Nevertheless, Newton throws in a personal aside: he would as soon admit that reds and yellows (or blues and indigos) are merely different dilutions of the same colour as that two thirds or sixths in music are different degrees of the same sound, rather than different sounds. Not much of an argument, you may think; but I will come back to this question later.

The next use I want to consider, by Newton, of the analogy between light and sound occurs in yet another letter to Oldenburg, written in December 1675.[8] Newton is still not prepared to admit to any hypothesis about the nature of light, but he reserves the right to explore the consequences of different hypotheses; and he now confesses that *if* he were to assume an hypothesis it would be that light is *something* capable of exciting vibrations in the aether.[9] You may well be wondering how this differs from the hypothesis of the earlier letter. It is simply that Newton now seems to be thinking of light as a series of small bodies (corpuscles) emitted from luminous bodies, and of the corpuscles (rather than the bright bodies) as stimulants of the vibrations of the aether. Newton does not wish to say categorically, however, that light really *is* a series of corpuscles. Those who wish, he says ironically, may suppose it to be an aggregate of Aristotelian qualities.[10] How does light move? He is inclined to believe in a *mechanical* principle of motion, but others, he says, may look for a spiritual principle, if they so wish. To avoid dispute, and to make the hypothesis general, he adds, "let every man here take his fancy." Newton believes that he can *abstract* from the alternative theories a hard core of indisputable truth. He insists, in fact, that however we may think of light, we are at least to think of it as a succession of rays, differing from one another in such contingent circumstances as size, shape, and strength – as do almost all things in nature. We are to think of light, moreover, as distinct from the vibrations of the aether, which – as we saw worrying Newton earlier – have the unfortunate property of going round corners. We are to think of light as being alternately reflected and transmitted by thin plates, according to their thickness; and we are to suppose that just as light stimulates the aether, so the interaction is mutual, and the aether refracts the light. (The *greater* the *density* of the aether, the greater the refraction.) Reflection is to be considered the result of secondary vibrations in the surface of the reflecting body, some going into the aether within the body, and some being returned to the aether outside it.[11] I will not elaborate further on the precise mechanism suggested by Newton, or on the difficulties he encounters, or even on the fact that he continues to draw analogies with sound vibrations. I wish merely to point out that the model has *changed* in a rather subtle way. Despite the change, the model is still compatible with Newton's experimental findings, and although the old analogy between light and sound is no longer as clear as it was, he continues to develop it.

Newton wants, he says, "to *explain* colours." The emphasis is mine, but the phrase is Newton's. He supposes, he says, that just as bodies of various sizes, densities and tensions[12] "by percussion or other action" excite sounds of different tone, that is, vibrations of different wavelengths ('bignesses'), so *rays of light*, by impinging on the aether both inside and outside bodies,[13] excite vibrations of different wavelengths in the aether. I leave aside the physiological part of the explanation. The first point I want to make is that the analogy is no longer quite as good as it was, for there is nothing carried through the air, in the case of sound, analogous to the rays of light which stimulate the aether on passing through it. My second observation is that Newton has begun to take his analogy very seriously, talking as he does of the 'Analogy of Nature' in the style of his later 'Third Rule of Philosophizing' in the *Principia*.[14] Newton conjectures that just as harmony and discord of sounds proceed from the ratios between their vibrations in the air, so may the harmony and discord of colours proceed from the ratios of corresponding vibrations in the aether.

Here is one argument by analogy, and the analogy is continued immediately after. He describes how he and a friend independently divided the spectrum of light from a prism into its seven component colours. (I assume that the friend was told to distinguish the *seven* colours named by Newton. This, at least, is what his phraseology suggests.) He and the friend arrived at much the same division. Newton took the mean of each pair of alternative divisions, and found on measuring the length of the final divisions of the image of the spectrum that they were in approximately the same ratios as the divisions of a string capable of sounding the notes in an octave.[15]

Newton was clearly very fond of this analogy, in its newly extended form, for when he published his treatise *Opticks* in 1704 he repeated the material on the octave of colour in much the same style.[16] What is more, when he came to summarize his measurements of the diameters of what we now call 'Newton's rings',[17] he again used the musical scale to do so. Remember that at first he did not produce the rings by monochromatic light, and that the rings were therefore coloured.[18] It was natural enough, under these circumstances, that he should extend his musical comparison. What he does is calculate the thicknesses of the wedge of air between the glasses at those points where the rings are made by his seven spectral colours. He finds that these thicknesses are in the ratio of the lengths of a string yielding the notes of the octave, raised to the power 2/3.[19] These thicknesses he subsequently equated with what he calls "the Intervals of the following Fits of easy reflexion and easy Transmission."[20] The *explanation* of the rings offered by Newton on the basis of his theory of fits of easy reflexion and transmission is remarkable and interesting, but does not concern the analogy with sound I am now discussing. As far as I know, Newton does not develop the analogy any further.

I have mentioned so many details in the course of my account of Newton's analogy between light and sound that the shifting character of the analogy has probably been lost to view. I will summarize the six examples I have now given, three from 1672 and three from 1675:

N(1) Correspondences (some would call them analogies) are set up (or implied) between the following concepts:

> air (a^1); aether (a^2); vibration in a sounding body (b^1); vibration in a luminous body (b^2); the tone of sound (c^1); the colour of light (c^2); the sensation of sound (s^1); the sensation of light (s^2); vibration in the air (v^1); vibration in the aether (v^2);[21]

and also between the following:

> causation of v^1 by b^1, causation of v^2 by b^2; causation of s^1 by v^1, and of s^2 by v^2.

As a premiss, we have it clearly stated that c^1 is a function of the length of vibrations in air.

As the *conclusion*, we have

$$c^2 \text{ is a function of the length of vibrations in aether.}$$

N(2) The correspondences in N(1) are obviously still meant to hold, in addition to a correspondence between the two types of reflection, of sound and of light. Newton now offers, not an *argument*, in the usual sense, but a statement of a conceptual possibility. He can, he says, *conceive* a mixture of v^1, selectively reflected, and he can (therefore) *just as easily conceive* a mixture of v^2, selectively reflected.

Some would label N(2) a 'heuristic analogy.' It seems to me to be better described as a new *correspondence relation*. It has illustrative value, but is not an analogy in any of the senses I shall define later.

N(3) The same correspondences apply as before. Newton now says, in effect,

What I regard as two different sounds ($c^1, s^1,$ or v^1?) I cannot regard as different degrees of one fundamental sound. *Therefore*, what I regard as two different colours ($c^2, s^2,$ or v^2?) I cannot regard as different degrees of one fundamental colour.

Insofar as Newton here gives a *reason* for anything, the reason is a psychological one. In fact the 'argument' could be looked upon as a statement of intent, rather in the style of N(2), as to what theoretical concepts are to be utilized. It is no doubt supposed that by showing them to be translatable, their plausibility is increased. What is well worth noticing here is that Newton spent much of his time in the documents of 1672 and 1675 — and of course elsewhere — denying that he made *any* use of hypotheses. The correspondence relations, as a whole, as well as the statements of what is to be taken as conceivable, are, however, good examples of fallible hypotheses.

N(4) In the 1675 document, the correspondence relations are different from those of 1672. As I have already explained, light rays are brought into the aether side of the analogy, having no obvious counterpart in air. (In fact, there is a difference between the functioning of the systems at this point, for, as already noted, sound does not travel along straight 'rays' whereas light was thought to do so.) The change does not affect the argument offered. There are some new correspondences, namely between harmony of sounds (h^1) and harmony of colours (h^2), and also perhaps between ratios of properties of sounds (r^1) and ratios of corresponding properties of colours (r^2). I think one should spell out these correspondences explicitly, even though Newton did not do so, using as he did words ('harmony' and 'ratios') which were the same on both sides of the analogy.

The first 'argument' is now, in effect:

h^1 is a function of r^1. *Therefore* h^2 is a function of r^2.

N(5) Here Newton attempts to determine what to him must have seemed a significant property of colour. How is one colour separated from the rest? It is almost always assumed by commentators on Newton that he took colour to be subject to infinite gradation, but there is clearly a sense in which he wished to preserve the traditional division of the spectrum into a limited number of named colours.[22] His reasoning is not altogether clear, but the four premises seem to be these:

> There are seven points of division of the octave of musical tones.
> The tones are reproduced by the division of a string's length in ratios $k_1 \ldots k_7$.
> There are seven divisions of the spectrum of colour.
> The divisions are (in ratios of the length of a spectrum) approximately $k_1 \ldots k_7$.

The conclusion is then that the divisions are at points dividing the spectrum in *exactly* those ratios.

N(6) The reasoning follows closely that of N(5). The rings are coloured, and by a suitable stretch of the imagination, take the place of the seven (much purer) colours of the spectrum. Newton seems to have thought that in the musical scale he had found an item of conceptual apparatus suitable for investigating the phenomena of light and colour.

It is not, I suppose, really surprising that commentators on Newton's analogies have been unsympathetic towards them. In the words of W. S. Jevons, "even the loftiest intellects have occasionally yielded, as when Newton was misled by the analogy between the seven tones of music and the seven colours of his spectrum."[23] Jevons was discussing analogies where the resemblance is only a numerical one — and in particular, one involving the number 7. As he went on to say, "Even the genius of Huygens did not prevent him from inferring that but one satellite could belong to Saturn, because, with those of Jupiter and the earth, it completed the perfect number of six."[24] I am not in a position to comment on the importance of numerology to Huygens — who certainly, in his study of Saturn's ring, made far more profound uses of analogy than Jevons might lead one to suppose.[25] In Newton's case, however, there was clearly much more than numerology in his arguments — and indeed, it would be difficult to show that he was *here* influenced in any way by the mystical associations of the number 7 — even though his religious writings show distinctly numerological tendencies. This is not the place for a discussion of Newton's mysticism, although I would like to emphasize the very great importance of *theological* debate in the history of analogical thought. I do not think it too strong a thesis, that "The entire vocabulary of religion is based upon the perception of analogies between the material and the spiritual worlds."[26] Both the vocabulary and the arguments of natural religion are heavily dependent on analogy, and the *justification* of the analogy between human nature and the nature of God has always been at the centre of Christian theology.

I mention these things because any historical study of analogy — even as it has been used within science — will be deficient if the philosophico-theological discussions of such writers as Aquinas, Cajetan and Suarez are ignored. There are differences, of course, between scientific and theological analogy. (In particular, in theological argument only one side of the analogy is known.) Philosophical usage of the words 'univocal', 'analogical' and 'equivocal' was nevertheless fixed by the scholastics, as were the closely related expressions 'literal', ('analogical'), and 'metaphorical'. I think there is something to be learned from these writers about the functioning of analogy and metaphor, even in science; but there is still another reason why they are of relevance to scientific history.

As is well known, there was a great efflorescence of natural religion at the end of the seventeenth century and throughout the eighteenth, and the resulting interplay of science and religion is an important part of the intellectual history of the time. Historians of philosophy are obliged to take this intellectual movement into account, if only because (to take a rather parochial English perspective) it was an important influence on John Locke and his philosophical contemporaries. Two theologians who, like Locke, made contributions to the understanding of analogy were Archbishop King, of Dublin, and Bishop Browne, of Cork (and earlier of Trinity College, Dublin), who in turn entered into controversy with the well-known philosopher Berkeley — who was also to become an Irish bishop. There are strong links between the theological discussion of the early eighteenth century and the logical writings of Richard Whately (another Archbishop of Dublin!) and thence John Stuart Mill in the nineteenth century.[27]

Now one of Mill's principal aims in writing his *A System of Logic* [28] was to put inductive reasoning on a secure footing, or, as he explained in his preface, "the task ... of generalising the modes of investigating truth and estimating evidence, by which so many important and recondite laws of nature have, in the various sciences, been aggregated to the stock of human knowledge". Mill was to the nineteenth century what Bacon had been to the seventeenth, namely empiricist philosopher and self-appointed arbiter of scientific method. Neither worked in a historical vacuum, and in fact Mill in his Preface acknowledges a debt to Whewell's *History of the Inductive Sciences*, which he uses, for example, as a source of 'false analogies'! Both Bacon and Mill paid attention to analogy, in their account of induction, and Mill's passage on the subject was especially influential in philosophical circles.[29] I am sure, nevertheless, that neither Mill nor Bacon had very much influence on the analogical *techniques* of scientists. They offered insights into the *character* of scientific argument, but even then the examples they gave were very remote indeed from the best science of their times. Mill, for example, had probably never even heard of a brilliant physical analogy newly developed by the young William Thomson, who had still not reached the age of twenty when Mill's *Logic* was published. Over the following decade Thomson developed numerous large-scale analogical arguments which greatly affected the course of the history of physics. He wrote little about the technique of analogy, but an even younger scientific contemporary, James Clerk Maxwell, made numerous historically interesting asides on the subject, quite apart from developing many important examples of his own. It will be instructive to set Mill and Maxwell side by side, to contrast the logical aspects of their utterances on analogy, and at the same time to compare Maxwell's chosen examples of analogical argument with those I have already spelled out at length from Newton, and with others I shall outline from Thomson.[30] I shall then try to decide whether there was any greater subtlety in the use of analogy in the nineteenth century than in Newton's time. I will begin with Thomson.

To Thomson belongs the credit for drawing the attention of physicists to the *power* of analogy. He did so, not by writing a logic of analogy, but by developing a notable *example*. He showed, in fact, that the equipotential surfaces in a space occupied by electrostatically charged conductors may be made to correspond to isothermal surfaces in an infinite solid in which heat is flowing. An electric charge corresponds to a heat source, and so on. Maxwell later listed the details of the analogy as follows:

Electrostatics.	*Heat.*
The electric field.	An unequally heated body.
A dielectric medium.	A body which conducts heat.
The electric potential at different points of the field.	The temperature at different points in the body.
The electromotive force which tends to move positively electrified bodies from places of higher to places of lower potential.	The flow of heat by conduction from places of higher to places of lower temperature.
A conducting body.	A perfect conductor of heat.
The positively electrified surface of a conductor.	A surface through which heat flows into the body.
The negatively electrified surface of a conductor.	A surface through which heat escapes from the body.
A positively electrified body.	A source of heat.
A negatively electrified body.	A sink of heat, that is, a place at which heat disappears from the body.
An equipotential surface.	An isothermal surface.
A line or tube of induction.	A line or tube of flow of heat.

Thomson's analogy with Fourier's theory of heat was only the first of several he developed. They may be summarized in brief as follows:

T(1) 1841. The foregoing analogy with heat (this being the same as M(4) below). Faraday does not seem to have known of the analogy until 1845. By 1850 at the latest, he had begun to make conceptual use of it in his formulation of the notion of a field, with lines of force in empty space independent of conductors, dielectrics, or magnets.

T(2) 1845.[31] Analogy between Coulomb's theory of electrical action at a distance and Faraday's theory of action by contiguous particles in a continuous medium. The common formal element was a mathematical framework in which Green's potential function played an important role.

T(3) 1845 (British Association meeting, Cambridge). Sketch of possible analogies between optics, electricity and magnetism. This strongly influenced the direction of Faraday's research. One consequence of Faraday's exchanges with Thomson was the discovery of the rotation of the plane of polarization of light by magnetism.[32] Within a year or so Faraday had been led to formulate a number of new concepts, including that of diamagnetism, and that of continuous and polarized lines of force capable of vibration, and thus of transmitting optical 'forces'. He no longer needed his aether particles.[33]

T(4) 1846.[34] Analogy between elastic solids and magnetic and electrical phenomena. (The magnetic induction was made to correspond to the curl of a vector-potential representing the elastic displacement. An electric force corresponded to an absolute displacement of a particle of the solid.) Thomson, perhaps influenced by Faraday's rejection of his old view of the aether as a particulate elastic solid, soon lost interest in his new analogy, although he took it up again in 1889.

T(5) 1849.[35] Analogy between magnetism and a system of 'imaginary matter', capable of attraction and repulsion. (Cf. Maxwell's 'imaginary properties', p. 129 and n. 54 below.) Thomson made 'solenoidal' magnetism equivalent, for example, to equal concentrations of opposite polarity at its two ends. He noted the analogy between the familiar 'equation of continuity' (for an incompressible fluid) and the mathematical condition for a solenoidal distribution of magnetism. He subsequently made use of the 'close analogy which exists between solenoidal and lamellar distributions of magnetism' to lead him to several new formulae.

Thomson's analogy led him to conclude that magnetic action, determined as it is by the imaginary matter, is not located in ordinary matter. One might look upon this conclusion as the thin end of some ontological wedge, but Faraday was converted, and henceforth argued that lines of magnetic force belong to space rather than to matter.[36]

T(6) 1856.[37] Multiple analogy involving electro-magnetism, a luminiferous aether with elastic properties, microscopical vortices within it, and the dynamical theory of heat. (The spiral structures are not to be interpreted as vortex atoms, at this stage.) The aim was to explain magneto-optic rotation, and Thomson seems to have believed that he had found the only possible explanation for it. Knudsen (see Note 37) argues that Maxwell inherited this conviction, and that he retained the vortex theory through all his revisions of the electromagnetic theory. Heimann suggests that Maxwell began by taking lines of force as the fundamental physical entities, but later (1861?) sought to explain them on the basis of molecular vortices.[38]

In 1855, fourteen years after the first of Thomson's analogies, Maxwell expressed himself very strongly in favour of what he called 'physical analogies'.[39] Maxwell was then only 25. He was at Cambridge, but had already spent three years at the University of Edinburgh, where he came under the influence of the physicist J. D. Forbes and the philosopher Sir William Hamilton. To them he undoubtedly owed much of the style of the 1855 paper — 'On Faraday's lines of force.' That he owed methodological as well as scientific ideas to Thomson is also clear, and in fact several letters from Maxwell to Thomson are extant, in which the evolution of Maxwell's ideas is well illustrated.[40] Thus in a letter of 15 May 1855, Maxwell writes:

I am trying to construct two theories, mathematically identical, in one of which the elementary conceptions shall be about fluid particles attracting at a distance while in the other nothing (mathematical) is considered but various states of polarization tension etc. existing at various parts of space. The result will resemble your analogy of the steady motion of heat. Have you patented that notion with all its applications? for I intend to borrow it for a season, without mentioning anything about heat (except of course historically) but applying it in a somewhat different way to a more general case to which the laws of heat will not apply.[41]

On 13 September 1855 he refers to Thomson's 'allegories', showing that he knew of all Thomson's work to date:

In searching for these notions I have come upon some ready made, which I have appropriated. Of these are Faraday's theory of polarity . . . also his general notions about 'lines of force' with the 'conducting power' of different media for them. Then comes your allegorical representation of the case of electrified bodies by means of conductors of heat . . . Then Ampère's theory of *closed* galvanic circuits, then part of your allegory about incompressible elastic solids and lastly the *method* of the last demonstration in your R.S. paper on Magnetism. I have also been working at Weber's theory of Electro Magnetism as a mathematical speculation which I do not believe but which ought to be compared with others and certainly gives many true results at the expense of several startling assumptions.

Now I have been planning and partly executing a system of propositions about lines of force etc. which may be *afterwards* applied to Electricity, Heat or Magnetism or Galvanism, but which is in itself a collection of purely geometrical truths embodied in geometrical conceptions of lines, surfaces etc.

The first part of my design is to prove by popular, that is not professedly symbolic reasoning, the most important propositions about V and about the solution of the equation in the last page . . . and to trace the lines of force and surfaces of equal V.[42]

On 14 February 1856, three days after Maxwell had read the second part of his paper 'On Faraday's lines of force' to the Cambridge Philosophical Society,[43] he notes that he left the paper with Thomson, whom he asks to return it, because he wishes to write up the second part 'On Faraday's electrotonic state.' "I think I left an abstract too", he adds.[44] On an unspecified date in the same month, Maxwell read an essay on analogy to the Apostle's Club in Cambridge.[45] This light-hearted essay, in a flippant style characteristic of university societies, adds nothing to the arguments offered in the scientific paper, although it might well be used to settle a number of disputes over Maxwell's early *Weltanschauung*. I am less concerned with this than with his rather specific claims on behalf of analogical arguments. I will begin with some remarks made at the beginning of the paper 'On Faraday's lines of force.'

In simplifying previous investigations one may choose to express the results, Maxwell said, either in a purely mathematical formula or in a physical hypothesis. In the first case, he thought, we lose sight of the phenomenon to be explained, and fail to obtain an extended view of the connections of the subject. In the second case "we see the phenomena only through a medium, and are liable to that blindness to facts and rashness in assumption which a partial explanation encourages". The middle way, the way of analogy, allows us to grasp "a clear physical conception, without being committed to any theory founded on the physical science from which that conception is borrowed . . .". When he goes on to speak of our thus avoiding being "carried beyond the truth by a favourite hypothesis", he reminds us very much, not only of Mill, but of a very powerful 'Newtonian' tradition of methodological comment, particularly strong in Scottish philosophy, and in time influencing Maxwell's mentor, William Thomson.[46]

Maxwell now says what he means by 'physical analogy', namely

that *partial similarity between the laws of one science and those of another* [the emphasis is mine] which makes each of them illustrate the other. Thus all the mathematical sciences are founded on relations between physical laws and laws of numbers, so that the aim of exact science is to reduce the problems of nature to the determination of quantities by operations with numbers. Passing from the most universal of all analogies to a very partial one, we find the same resemblance in mathematical form between two different phenomena giving rise to a physical theory of light.[47]

Maxwell goes on to outline the following analogies (I will refer to that between the laws of science and the laws of number as M(1)):

M(2) That between light undergoing refraction and a particle moving in an intense force-field.

M(3) That between light, the vibrations of an elastic medium (elasticity being a sort of midwife); and electricity.

M(4) That between attraction at a distance (according to an inverse square law) and the conduction of heat in uniform media. This is Thomson's first analogy.[48]

M(5) That (which it is the purpose of the paper to explore) between a system of electrical and magnetic poles, acting under an inverse square law, and a field of incompressible fluid, moving within tubes directed along Faraday's lines of force. The lines are analogous to the streamlines in the fluid.

Maxwell's comments on these analogies are of some interest. It is said that M(2) was "long believed to be the true explanation of the refraction of light", and that "we still find it useful in the solution of certain problems, in which we employ it without danger, as an artificial method". I will anticipate a passage in which Maxwell says, in effect, how analogical argument does not oblige us to accept the prior theory. He is now saying that *we may continue to use an analogy of highly restricted validity*. We may use it for 'certain problems'; but how the argument is to be kept under control, Maxwell does not say.

Of analogy M(3), Maxwell says that it extends further, and yet "is founded only on a resemblance *in form* between the laws of light and those of vibrations". (He here adds a sentence that I find confusing, which can be ignored.)[49] The drift of his meaning is plain as soon as he discusses analogy M(4), which we might refer to as 'Thomson's first analogy':

The laws of the conduction of heat in uniform media appear at first sight among the most different in their physical relations from those relating to attractions. The quantities which enter into them are *temperature, flow of heat, conductivity*. The word *force* is foreign to the subject. Yet we find that the mathematical laws of the uniform motion of heat in homogeneous media are identical in form with those of attractions varying inversely as the square of the distance. We have only to substitute *source of heat* for *centre of attraction, flow of heat* for *accelerating effect of attraction* at any point, and *temperature* for *potential*, and the solution of a problem in attractions is transformed into that of a problem in heat.[50]

When he said of M(3) that the resemblance was 'only one of *form*, Maxwell was, even if only half consciously, making a distinction between this sort of analogy and analogies in which the '*objects*' in the two related domains so closely resemble each other that the same word may even be used for both. (An example is N(1), where vibrations (in air) correspond to vibrations (in aether).) One may well ask about the danger that two domains will be confused in a carefully prescribed scientific analogy. This is a genuine problem, but it does not seem to have been what was most worrying to Maxwell, who went on to say, in connection with M(4), that

the conduction of heat is supposed to proceed by an action between contiguous parts of a medium, while the force of attraction is a relation between distant bodies, and yet, if we knew nothing more than is expressed in the mathematical formulae, there would be nothing to distinguish between the one set of phenomena and the other.[51]

In other words, the phenomena *are* very different, and the formulae fail to *reveal* the difference; and yet part of the value of the best analogical argument is that it allows the mind "clear physical conceptions".[52] The *stimulation of mathematical ideas* is only a part of what one should hope for:

It is true, that if we introduce other considerations and observe additional facts, the two subjects will assume very different aspects, but the mathematical resemblance of some of their laws will remain, and may still be made useful in exciting appropriate mathematical ideas.[53]

In introducing his own analogy, here numbered M(5), Maxwell made no more than the modest claim that he would show, by the applications of Faraday's methods, the mathematical ideas underlying the phenomena of electricity. If this were all he had done, it would be stretching a point to say that Maxwell had offered a truly analogical argument. What of his hopes for a 'clear physical conception'? At its most physical, his 'conception' is one grounded in what one may call a 'geometry of fluids'.[54] He establishes what he calls a "geometrical model of the physical phenomena", by which he hopes "to attain generality and precision, and to avoid the dangers arising from a premature theory professing to explain the cause of the phenomena".[55] Thomson, I shall later suggest, gave Maxwell a healthy reluctance to lay claim to causal explanations. His geometrical model is one side of an analogy, to be sure, but it is *artificially created* for the purpose of 'arranging and interpreting' experimental findings. The principles governing the motion of the incompressible fluid in his so-called 'geometrical model' are the principles of classical fluid mechanics, but the model itself is a *particular system*, one out of many possible systems.[56] The function of even the most 'formal' of analogies is obviously not to transfer mathematics as a whole from one domain to another. Maxwell's analogy M(1), between the laws of science and the laws of number is of such generality that it can hardly count as a particular *example*. The transfer of mathematics in an analogical argument is one of *particular* mathematical results, derived in the prior system after the imposition of *specific mathematical conditions*. The conditions may be of a very general sort — as, for example, the laws of mechanics in a Langrangean form. And, historically speaking, these conditions might have been imposed long before — as in Newton's musical-harmony analogue — or they might be newly imposed in a very contrived way — as in Maxwell's case M(5) — by the man who is creating a new conceptual analogue, or model. If an analogy is an explanation of the unfamiliar by the more or less completely familiar, then this is not a case of analogy. Perhaps we should distinguish between *established analogues* and *newly contrived analogues*.

Putting Maxwell's own views aside, for the moment, we can now see that analogies between two different domains are likely to be of value in *argument* only if the prior domain (and therefore also the other) is structured by restricting conditions, that is by laws or rules of some sort. This is what such traditional logicians as Whately meant when they said analogy was a 'resemblance of relations' (a statement repeated in effect by Maxwell), and it is a point obscured by Mill when he rejected the traditional view.[57] Of course, if one forms analogies between simple and familiar situations, as did Mill for his examples, the rules governing terms in the prior domain will be a matter of common sense and – one hopes – of common agreement. But the rules are nevertheless there, to be taken into account in any formal rendering of a logic of analogy.

In discussing Newton, I spelt out in some detail six analogical 'arguments', whereas in Maxwell's case I have only hinted at the *basis* for such arguments. Analogy M(1), as I have said, lends itself to no argument in particular. Analogies M(2) and M(3), providing the familiar corpuscle and wave explanations of light phenomena, are no doubt so familiar that I need not describe them in detail. I would like to point out that both M(2) and M(3) were, in a sense I have explained, based on *newly contrived* analogues.[58] This applies also to M(4) and M(5), the systems which are considerably more sophisticated than M(2) and M(3). (Even so, I think it could be argued that M(3), namely Huygens' wave analogy, contained what to fellow physicists was the least familiar analogue of the four, at the time of framing.)

I should like to make a distinction here between two sorts of analogy. Some of Maxwell's examples were, in a sense I have explained, based on *analogues* specially contrived for the occasion. Huygens' wave analogy is an example. How different are analogies with an *artificial* basis (models, in one sense of that word) from those with a *pre-established* analogue – as in Newton's case, where he did not have to invent a theory of sound? A preliminary and obvious answer is that the first may be modified again and again until it satisfies its creator, whereas with the second – something conceived of as *given* – we are obliged to distinguish between so-called 'positive' and 'negative' analogies, i.e. respects in which the analogues agree and disagree. But matters are rarely so simple. We are reminded of the analogies under headings N(2), N(3), and N(6). There we found Newton establishing conceptual possibilities, rejecting and refining a concept, and confirming the value of a concept. Newton was there, in fact, establishing in this way restrictive conditions of a sort which I mention later as having been obscured by Mill in his analysis of the subject, conditions limiting the functioning of key concepts (selective reflection, dilution of sound, octave division, and so on).

There is another side to this question of the difference between pre-established and artificial analogues. I refer to the ontological problem. When Maxwell said of his incompressible fluid that it was "merely a collection of imaginary properties",[59] he was not saying anything likely to colour our views of the real nature of the space occupied by electrical charges. This is less true of Thomson's analogy (viz. M(4)), although the influence of this was oblique. Here is what Maxwell wrote in his *Elementary Treatise on Electricity*:

We must bear in mind that at the time when Sir W. Thomson pointed out the analogy between electrostatic and thermal phenomena men of science were as firmly convinced that electric attraction was a direct action between distant bodies as that the conduction of heat was the continuous flow of a material fluid through a solid body. The dissimilarity, therefore, between the things themselves appeared far greater to the men of that time than to the readers of this book, who, unless they have been previously instructed, have not yet learned either that heat is a fluid or that electricity acts at a distance.[60]

No one reading his work was likely to come away with the idea that heat is really electricity, or that electricity is really heat;[61] and yet Thomson's paper did persuade many that Faraday had been right to suggest that electrical action was effected through a continuous medium. This was the outcome of an interesting clash of paradigms. Throughout the eighteenth century (and up to about 1820) Newtonian dynamics had been considered as an almost essential mode for physical science, and the successes of molecular physics, as practised by such men as Laplace, Navier, Cauchy and Poisson, confirmed a majority of the scientific community in their belief — a belief which lingered on in England rather longer than it did on the continent of Europe. There it had been challenged indirectly by Fourier's theory of heat. In commenting in his second paper (first printed 1845)[62] on the analogy between Fourier's theory and Faraday's theory of electrical action in a medium, Thomson hints at the uneasy compatibility of action at a distance and contiguous action. Since his style was mirrored in some degree by that of Maxwell's paper of 1855–6,[63] I will quote a more extended passage than is necessary to illustrate the new ontological situation:

Now the laws of motion for heat which Fourier lays down in his *Théorie analytique de la chaleur*, are of that simple elementary kind which constitute a mathematical theory properly so-called; and therefore, when we find corresponding laws to be true for the phenomena presented by electrified bodies, we may make them the foundation of the mathematical theory of electricity: and this may be done if we consider them merely as actual truths, without adopting any physical hypothesis, although the idea they naturally suggest is that of the propagation of some effect by means of the mutual action of contiguous particles; just as Coulomb, although his laws naturally suggest the idea of material particles attracting or repelling one another at a distance, most carefully avoids making this a *physical hypothesis*, and confines himself to the consideration of the mechanical effects which he observes and their necessary consequences.

All the views which Faraday has brought forward, and illustrated or demonstrated by experiment, lead to this method of establishing the mathematical theory, and as far as the analysis is concerned, it would, in most *general* propositions, be even more simple, if possible, than that of Coulomb. (Of course, the analysis of *particular* problems would be identical in the two methods.) It is thus that Faraday arrives at a knowledge of some of the most important of the general theorems, which, from their nature, seemed destined never to be perceived except as mathematical truths.[64]

If Fourier had not displaced mechanics from its position as unchallenged king of the physical sciences, the analogies of Thomson and Maxwell had certainly done so by the late 1850's.

I want now to leave Thomson and Maxwell for a time, to consider some passages in John Stuart Mill's famous *System of Logic*, first published in 1843, that is, at much the same time as Thomson's first analogy. I shall later make some comparisons between Mill's ideas and those of the philosophically unsophisticated Maxwell.

Mill begins his account of analogy by trying to put the subject on a very general footing. He rejects tradition in rejecting Whately's equation of analogy with a *resemblance of relations*, and in its place explains analogical reasoning by the following formula:

> Two things resemble each other in one or more respects; a certain proposition is true of the one, therefore it is true of the other.[65]

This formula covers, as he points out, induction as well as analogy. I will not say anything about Mill's views on induction — a very thorny subject — except that for him, in what he calls a 'complete induction', the properties shared by the two things are *invariably* conjoined. In an analogy, on the other hand, "no such conjunction has been made out."

In order to simplify Mill's rather lengthy explanations of his meaning, I shall introduce a notation which extends the one used in connexion with Newton. The two 'things' resembling each other in the properties f will be denoted by a^1 and a^2, and the proposition by $p(a)$ ($a = a^1, a^2$, etc.). Mill's formula then becomes simply:

$$f(a^1), f(a^2), \qquad p(a^1) \vdash p(a^2).$$

He insists that the fact[66] stated in proposition p must be dependent on some property of a^1 (and presumably a^2), but we simply do not know on which. If a^1 and a^2 resembled each other in *all* their ultimate properties, he claims, the truth of p would be guaranteed, and the greater the number of resemblances, the greater the probability of the truth of p.[67] (The statement is highly controversial but I cannot discuss it here.)[68] Mill's remarks on ultimate properties and derivative properties are of dubious value. Let us denote the ultimate properties by F, G, etc., and derivative by $f_1, g_1, \ldots, f_2, g_2 \ldots$ etc. By definition, then,

$$F(a) \rightarrow f_1(a),$$
$$F(a) \rightarrow f_2(a), \quad \text{etc.}$$

Mill claims that if "resemblance be in a derivative property, there is reason to expect resemblance in the ultimate property on which it depends, and in the other derivative properties dependent on the same ultimate property".[69] In my notation,

$$f_1(a^1) \& f_1(a^2) \qquad \vdash F(a^1) \& F(a^2) \quad \text{'is reasonable'},$$

whence, by definition,

$$\vdash f_2(a^1) \& f_2(a^2),$$

and so on for f_3, f_4, and all other derivative properties.

It would be a mistake to suppose that Mill set great store by arguments of this sort. Analogy, for him, was "a mere guidepost, pointing out the direction in which more rigorous investigations should be prosecuted". It must be said that his attempt to make analogical reasoning look respectable has a modestly convincing appearance, until he provides an example. Here my f_2 may be taken as 'has inhabitants', while a^1 and a^2 are the Earth and the Moon. We may take f_1 to be a conjunction of such properties as follows: spherical form, solid, opaque, volcanic, receiving light from the Sun. Mill considers two counter-arguments: first, there may be factors, dissimilarities, working against the inference to life on the Moon. It is in connexion with these that his vague references to probability are introduced. Second, different ultimate properties may give rise to the same derivative properties (and so the inference from f to F will be invalid). This 'giving rise to properties' was meant to be a matter of causation, and to pass from f to F would thus be to commit the 'fallacy of many causes', so called.[70] But the greatest of all the shortcomings of the analysis outlined by Mill is that he entirely evades the enormous problem of what it means to be an 'ultimate property'. Even in his example, he talks vaguely of the property of having inhabitants as "depending . . . on some of its properties as a portion of the universe, but on which of those properties we know not". If his view of the overall structure of science had been clearer, he might have been able to explain his meaning better; but his account of scientific procedure was very rudimentary, and even in his discussion of analogy it is noticeable that the foreground is constantly occupied by what was for him the greatest problem of all, namely the problem of induction.[71] He began, bravely enough, speaking in very general terms of resemblance in certain 'respects', rather than merely resemblance of relations; but still he spoke of resemblance of 'things'. He gave no clear sign that he had considered one of the most important uses of analogy in scientific argument, namely *analogy between entirely different scientific domains*. In short, he threw no light whatever on the sort of arguments I gave from Newton, Thomson, and Maxwell.

Perhaps Mill would have claimed that his analogies extended to the case of different domains, but I think not. At all events, this case is only found in Mill, so far as I can see, under the rubric 'false analogy'.[72] The examples given are not such as to have gained much sympathy from his readers, involving as they do numerology and Pythagorean harmony.

I have now considered excerpts from the writings of four men — Newton, Thomson, Maxwell, and Mill — three of whom made important use of analogical argument in a scientific context, and two of whom wrote about the theory of analogy. I have tried to avoid imposing my own logical views on the historical material, and I hope the result was not too loosely shaped. The subject of analogy is a large and difficult one, extending as it does into every region of human activity. Analogies have two sides to their nature: they are instruments of *argument*, prediction, and validation, and they are instruments of *cognitive meaning*, understanding, formalization and classification. The problem of meaning and categories is not easily disentangled from the problem of argument and law. Are the planets the *same sort* of thing as bodies in free space? Newton said Yes, Descartes thought No. Are terrestrial motions governed by the *same laws* as celestial? Aristotle said No, Galileo said Yes. But what a thing *is* is obviously to a large extent decided by *how* a thing *behaves*; and this is what scientific laws inform us about.

This contrast between problems of meaning and of argument is closely related to one of these dogmas of the logic of analogy which has been so often repeated that it is frequently taken for granted. I am referring to the idea that analogies can be easily divided into two classes — namely of so-called *substantial* and *formal* sorts. A substantial analogy is supposedly one where there is a correspondence of simple *properties*, while a formal analogy is taken to involve a correspondence between *relations*, or, in a more sophisticated version of the idea, between *relations among constituents* somehow stripped of all their properties. Now it is very difficult to comment on this view, unless we are clear as to the epistemology of the person who is proposing it, but the view usually goes with the doctrine that linguistic conventions (and the formulae of scientific theories) are somehow models of complex facts. The 'formal' relations are, it seems, regarded as though they were fixed for all eternity — for otherwise, how can we be sure that phenomena which are now (to take a crude case) explained predominantly in terms of properties will not in future be better explained in terms mainly of formal relations? If I am being much too vague, it is in response to what seems to me to be a very vague thesis, namely that analogies are either substantial or formal. On close inspection, those two sorts of analogy turn out to be only special cases of what I earlier called *correspondence rules*, rules linking air with aether, sound with light, and so on, in my earlier examples. This is only the beginning of an argument by analogy. But wait! There are some who would like to get a valid argument out of the correspondences by the following procedure: if there is a *total* substantial analogy between two things (they say), then *all* properties are shared. From this it follows that the things are identical; and therefore the constituent parts, stripped of their properties, will be identical, and thus there will be a *formal* analogy.

This argument is taken from the work of a reputable modern philosopher (Mario Bunge), who follows very closely a lead given by an excellent historian of science (Hélène Metzger).[73] One is also reminded of what Mill said about 'ultimate and derivative properties'. It seems that those who write on the structure of analogical argument cannot relinquish the idea that there is a formula to be found which will guarantee the outcome of at least some arguments from analogy. Most would want to deny the charge. ("Arguments from analogy may be fertile but they are all invalid" – Bunge.) But why, then, arguments of the sort I have just given?

The sharp division of analogies into formal and substantial is often associated with another misconception. There is a *comparative* concept of substantial analogy, it is sometimes said. The *degree* of similarity is supposed to be determined by counting attributes (cf. Mill, Note 68 above). This seems to me to be very naïve. The properties of an object are not countable. What is countable are the properties human beings agree to count as properties; and *they* are far from being absolute, or stable enough for an argument to be based on a count of them.

The views I have been referring to here are, it seems to me, a by-product of a traditional phase of analysis of the logic of analogy which should have ended at the time of Thomson and Maxwell, if not before. Roughly speaking, one may say that this phase was marked by undue attention to analogies between *things* (or between corresponding *terms*, in more cautious accounts), and the subsequent categorization of those things. Perhaps the biggest impetus ever received by the theory of analogy of this sort was at the hands of Aquinas, who wanted a theory in order to justify the meaning of predicates which were to be applied to things in different Aristotelian categories. There might well be parallels between the epistemological needs of Aquinas and those of the modern contemplative scientist. But these common problems have little or nothing to do with the problem set by those who, in the natural sciences, have *advanced* their knowledge by analogical techniques that they have seldom tried to justify. One often reads the platitude that analogical arguments are inevitably limited in their scope, because what is radically new is precisely that which cannot be accounted for in familiar terms. The whole purpose of analogies, however, is to *explore*, and to explore in the hope that what *seems* to be radically new will have unsuspected elements in common with what is familiar.

Perhaps the sociologist will categorize this as reactionary thinking. Perhaps the logician will dismiss it as invalid or illogical. It is the philosopher's job, nevertheless, to offer an analysis of this very common mode of thought, and if he can offer a satisfactory analysis of real historical examples, so much the better. I hope that I have shown something of the way in which the use of analogy matured between the seventeenth century and the nineteenth, and how in Thomson and Maxwell there is a conscious awareness of the function of a mathematical calculus as an intermediary between analogues — between, that is to say, a theory and a model for that theory. And theirs was more than an idle philosophical observation, for it suggested ways of applying a powerful tool for conjecture and for the unification of the physical sciences.

NOTES

[1] Newton to Oldenburg, 11 June 1672 in Isaac Newton, *Correspondence*, ed. by H. W. Turnbull. 7 vols. Cambridge, Cambridge University Press, 1959–1977. Vol. I: *1661–1675*, 1959, pp. 171–193.

[2] *Ibid.*, pp. 110–16.

[3] *Opticks*, repr. New York, Dover, 1952, Introduction, p. lxviii. (The edition is an enlarged reprint of that published by G. Bell, 1931, and is based on the 4th ed. of 1730.)

[4] "Uneasily Fitful Reflections of Easy Transmission", *The Texas Quarterly* 10 (1967), 86–102. See especially p. 88 and n. 31. Westfall speaks of 'periodicity' as being 'strongly implied' by the pattern of (interference) rings observed by Hooke, but it is clear that the strength of the implication depends on the ambiguity of the word 'periodic'.

[5] Newton to Oldenburg, *op. cit.*, p. 174.

[6] *Ibid.*, p. 175. Newton speaks of this hypothesis, which was one of several espoused by Hooke, as being in conformity with his own theories. He does not seem to have wanted his name associated with it at the time, as he showed by the sentence: "But how [Hooke] will defend it from other difficulties [viz. those I mention below] I know not . . .". Even so, where he now admits (p. 175) that the "fundamental [part of Hooke's] supposition" seems impossible, he was earlier (p. 174) prepared to say that "the fundamental part of it is not against me".

[7] *Ibid.*, p. 177.

[8] *Ibid.*, pp. 362–89.

[9] "Were I to assume an hypothesis it should be this if propounded more generally, so as not to determine what Light is, farther than that it is something or other capable of exciting vibrations in the aether . . .'' (*Ibid.*, p. 363).

[10] *Ibid.*, p. 370.

[11] *Ibid.*, p. 374. Cf. Huygens' secondary wavelets.

[12] He is thinking of strings.

[13] That this is what is meant by "refracting superficies of bodies" is evident from this same letter at p. 371, *op. cit.*

[14] The stimulation of sensations of colour is effected, he says in his letter, "much after the manner, that in the sense of Hearing Nature makes use of aerial vibrations of several bignesses to generate Sounds of divers tones, for the Analogy of Nature is to be observed''. *Ibid.*, p. 376.

[15] *Ibid.*, p. 377.

[16] *Opticks*, ed. cit., pp. 126–8. (I. ii. Prop. III. Prob. I.)

[17] Hooke was the first to publish experiments on the rings (*Micrographia*, Observation 9).

[18] He did realize that by illuminating his glasses with light of fewer colours than were found 'in the open air' he could cut down the coloration and greatly increase both the number of the rings and their sharpness.

[19] Newton, *Opticks*, p. 212 (II, i. Obs. 14). He uses the Dorian mode. The thicknesses are then proportional to the following numbers:

$$1, \left(\frac{8}{9}\right)^{\frac{2}{3}}, \left(\frac{5}{6}\right)^{\frac{2}{3}}, \left(\frac{3}{4}\right)^{\frac{2}{3}}, \left(\frac{2}{3}\right)^{\frac{2}{3}}, \left(\frac{3}{5}\right)^{\frac{2}{3}}, \left(\frac{9}{16}\right)^{\frac{2}{3}}, \left(\frac{1}{2}\right)^{\frac{2}{3}}.$$

[20] *Ibid.*, p. 284 (III. iii, Prop. XVI).

[21] I could add: sound, light. But I take it that these were for Newton synonymous with ν^1 and ν^2, respectively, at least under the hypothesis he was exploring in 1672.

[22] I know of no study of the various conventions of division of the spectrum into six, seven, or other numbers of colours.

[23] Jevons, W. S., *The Principles of Science*. 2 vols. London, Macmillan, 1874.

[24] *Ibid.*

[25] *De Saturni Luna observatio nova*, The Hague, 1656 (*Oeuvres*, Vol. 15). Huygens made much use of the analogy between the innermost satellite of Jupiter and the Moon (of the Earth), each of which has a period much longer than the period of rotation of the parent body. The same is true of the period of Mercury in relation to the Sun's period of rotation.

Cf. Galileo's argument – satellites: Jupiter:: Moon: Earth.

[26] Joyce, G. C., art. 'Analogy', Hastings *Enc.*, p. 416.

[27] The connections are several. Note, for example, the Copleston-Grinfield controversy of 1821, which revived a certain interest in the earlier writers. Whately was a friend and follower of Copleston, and reprinted King's *Discourse on Predestination*, with additional notes. Mill quotes Hooker, Copleston, and Whately, on analogy. (*System of Logic*, V.v.6.)

[28] The full title speaks for itself: *A System of Logic, Ratiocinative and Inductive, being a Connective View of the Principles of Evidence and the Methods of Scientific Investigation*, London, Parker, 1843 (etc.).

[29] Most modern discussions of analogy show signs of its influence.

[30] Maxwell, J. C., *An Elementary Treatise on Electricity*, Oxford, Clarendon Press, 1881 (posthumous ed. W. Garnett), p. 51. The heat flux through a spherical surface due to a point source at its centre is inversely proportional to the area, and hence varies inversely as the square of the radius. The analogy with Coulomb's inverse square law should be obvious.

Thomson's paper, written in 1841, when he was only 17, was first printed in *Cambridge Mathematical Journal* 3 (1843), 71–84. The paper was reprinted in the *Philosophical Magazine* for 1854, and (as paper I) in *Reprint of Papers on Electrostatics and Magnetism* by William Thomson, London, Macmillan 1872. It contains no methodological asides of the sort given by Maxwell. E. T. Whittaker was apparently following Maxwell in mentioning an analogy somewhat similar to Thomson's, arrived at by Ohm in 1827. (See *A History of the Theories of Aether and Electricity*. Rev. and enl. ed., 2 vols, London, Nelson, 1951. Vol. 1, p. 241.)

[31] See Thomson, *Reprint of Papers*, p. 29.

[32] For the circumstances of the new experiments, see Williams, L. P., *Michael Faraday*, New York, Basic Books, 1965, pp. 383–94.

[33] 'Thoughts on Ray-Vibrations', *Phil. Mag.* 28 (1846).

[34] Published 1847; repr. in *Math. and Phys. Papers*, 1, pp. 76–80.

[35] 'Mathematical Theory of Magnetism', *Phil. Trans. Roy. Soc.*, June 1849 and June 1850; paper XXIV, *Reprint of Papers* ... pp. 340–405 (see especially pp. 381–5; 398–401).

In addition to this early work, note the important series of articles (dated 1870 and 1872) in the same volume, at Sections 573–583 and 733–763, on the hydro-kinetic analogy for magnetism, an analogy Thomson holds to have been first appreciated by Euler.

[36] By 1852 he would argue that magnetism in matter is wholly dependent on the surrounding medium (which, he said, was 'perhaps the aether').

[37] 'Dynamical Illustrations of the Magnetic and Helicoidal Rotatory Effects of Transparent Bodies on Polarized Light', *Proc. Roy. Soc.* 8 (1856), 150–8; repr. *Phil. Mag.* 25 (1857), 198–204; reprinted in *Baltimore Lectures on Molecular Dynamics and the Wave Theory of Light*, London, 1904, pp. 569–77. For a close study of this work, see Knudsen, O., 'The Faraday Effect and Physical Theory, 1845–73', *Archive for Hist. of Exact Sciences* 15 (1976), 235–81.

[38] Heimann, P. M., 'Maxwell and the Modes of Consistent Representation', *Archive for the History of the Exact Sciences* 6 (1970), 171–213. See esp. p. 189.

[39] 'On Faraday's Lines of Force', repr. in Niven, W. D. (ed.), *The Scientific Papers of James Clerk Maxwell*, Cambridge, Cambridge University Press, 1890, 1, pp. 155–229. See esp. pp. 155–9. The report of the original, as printed in *Proc. Cambridge Phil. Soc.* 1 (1843–65), 160–6 (read 10 December 1855 and 11 February 1856) does not contain the preamble on analogy, but this is found in the version in *Trans. Camb. Phil. Soc.* 10 (1856), 27–83.

[40] Published by Sir Joseph Larmor, in 'The Origins of Clerk Maxwell's Electric Ideas, as Described in Familiar Letters to W. Thomson', *Proc. Camb. Phil. Soc.* 32 (1936), 695–750.

[41] *Ibid.*, p. 705.

[42] *Ibid.*, p. 711.

43 See Note 39 above.

44 Maxwell to Thomson, *op. cit.*, p. 714.

45 Printed in full by his biographers, Campbell, L. and Garnett, W., *The Life of James Clerk Maxwell*, London, 1882, pp. 235–44. Some of the main points made are: that analogies do not exist without a mind to recognize them; that "causes . . . are reasons, analogically referred to objects instead of thoughts"; that, from a scientific point of view, *relations* are of paramount importance; that there is a remarkable analogy between the *intention* of a man making a machine which will work, and the *principle* according to which it is made. Number, space, and time are discussed, but nothing more scientific.

46 A recent work, which I have not yet seen, but which presumably deals with this subject, is Richard Olson's *Scottish Philosophy and British Physics*, Princeton University Press, 1975.

47 Maxwell, 'Faraday's Lines of Force', p. 156. The point had been made by others before Maxwell, of mathematics in general rather than simply of numbers. Thus Joseph Fourier: "Mathematical analysis . . . brings together phenomena the most diverse, and discovers the hidden analogies which unite them". *The Analytical Theory of Heat*, tr. by A. Freeman, Cambridge, Cambridge University Press, 1878 (from the 1824–6 edition), pp. 7–8. Thomson often remarked on the strong influence Fourier had on his own ideas (see S. P. Thompson's *Life*, 1910, 1, pp. 13–20), and it is significant that Thomson's demonstration that the formulae derived for electricity from Coulomb's law are identical with those for heat flow (i.e. T(1), based on an apparently different conceptual basis, viz. contiguity rather than action at a distance) was a demonstration of one of Fourier's 'hidden analogies'.

48 See the notes to T(1) above.

49 "By stripping [the analogy] of its physical dress and reducing it to a theory of 'transverse alternations', we might obtain a system of truth strictly founded on observation, but probably deficient both in the vividness of its conceptions and the fertility of its method". Maxwell, 'Faraday's Lines of Force', p. 156.

50 *Ibid.*, p. 157.

51 *Ibid.*

52 See above, p. 297.

53 Maxwell, 'Faraday's Lines of Force', p. 157.

54 "The substance here treated . . . is not even a hypothetical fluid which is introduced to explain actual phenomena. It is merely a collection of imaginary properties which may be employed for establishing certain theories in pure mathematics in a way more intelligible to many minds and more applicable to physical problems than that in which algebraic symbols alone are used." *Ibid.*, p. 160.

55 *Ibid.*, pp. 158–9.

56 The beginning of the explanation as to how the particular system is selected is explained at pp. 158–9.

57 *Op. cit.*, p. 9.

58 Unlike sound (analogue for light) in Newton's analogy. There was a pre-existing theory of sound, however weak, not artificially set up for the purpose of the analogy.

59 See Note 47 above.

60 Maxwell, *Elementary Treatise on Electricity*, p. 52.

61 Maxwell went very far in the direction of caution, in his *Treatise on Electricity and Magnetism*, 1st ed., Oxford, Clarendon Press, 1873, para. 72. After drawing parallels

between fluid pressure, electrical potential, and temperature, he went on: "A fluid is certainly a substance, heat is as certainly not a substance, so that though we may find assistance from analogies of this kind in forming clear ideas of formal relations of electrical quantities, we must be careful not to let the one or the other analogy suggest to us that electricity is either a substance *like* water, or a state of agitation *like* heat". (My italics.)

[62] Paper II in Thomson, *Reprint of Papers* . . . ; paper first appeared in *Cambridge and Dublin Math. Journal*, November 1845, and then in an extended version in *Phil. Mag.* 1854.

[63] That is, the paper from which I have quoted much already ('On Faraday's Lines of Force'; see Note 39). It was read in two parts, in December 1855 and February 1856.

[64] Thomson, *Reprint of Papers* . . . , p. 29.

[65] Mill, *System of Logic*, III. xx. 2 (see Note 28 above).

[66] He calls this 'the fact *m*'.

[67] *Loc. cit.*

[68] Cf. *Ibid.*, III. xx. 3: "If, after much observation of *B*, we find that it agrees with *A* in *p* out of 10 of its known properties, we may conclude with a probability of 9 to 1 that it will possess any given derivative property of *A*". Dissimilarities are said to furnish counter-probabilities. Mill makes no attempt to decide whether some properties may not be more fundamental than others; or whether there is any limit to the known properties of a thing; or to their triviality.

[69] *Ibid.*, II. xx. 2. Further references to Mill are to this section, unless said to be otherwise.

[70] Cf. *Ibid.*, V. v. 6: "It has to be shown that in the two cases asserted to be analogous, the same law is really operating; that between the known resemblance and the inferred one there is some connection by means of causation".

[71] Mill wished to settle for nothing less than absolute truth in science. He was deeply suspicious of hypotheses, which he admitted might be fruitful; but for this very reason they might − if fruitful yet false − be an "impediment to the progress of real knowledge by leading inquirers to restrict themselves arbitrarily to the particular hypothesis which is most accredited at the time". The wave and emission theories of light are instanced as "unsusceptible of being ultimately brought to the test of actual induction", even though they are not "worthy of entire disregard"!

[72] *Ibid.*, V. v. 6.

[73] Bunge, M., *Scientific Research*, 2 vols., New York, Springer, 1967, Vol. 2: *The Search for Truth*, esp. ch. 15; Metzger, H., *Les concepts scientifiques*, Paris, Alcan, 1926, passim.

14

HIGH VICTORIAN SCIENCE: THE RECIPES FOR SURVIVAL
OF THREE SCIENTIFIC CINDERELLAS

In 1910 the Executive Committee of the British Science Guild, probably acting under Lockyer's guidance, set a prize essay on the subject "The best way of carrying on the struggle for existence and securing the survival of the fittest in national affairs." The chosen subject not only reflects the influence of Huxley and social Darwinism but is almost an epitaph for an era of Self Help, an era which was fading away in the face of educational reform. Poverty was not the key to the need to struggle for scientific survival. Of those three scientific Cinderellas, Michael Faraday, Thomas Huxley and Norman Lockyer, only Faraday knew real poverty. Although Huxley left school at ten, his father did at least own books: and Lockyer's more or less conventional middle-class education had scarcely ended when he joined the War Office. The essence of the struggle was the powerful cultural stream against which it was necessary for the would-be scientist to swim, especially if he had no university education.

As for survival, each of the three men had his own recipe. Lockyer went through his somewhat off-beat scientific life campaigning for this and that, quarrelling, waxing indignant, and preaching like the prototype Victorian paterfamilias. Huxley was a more incisive campaigner, if not a more effective one. Both men worked with Victorian intensity, but there was a relaxed air about the War Office in Lockyer's day more redolent of All Souls than of the medical practice in the Rotherhithe dockland to which Huxley was apprenticed at the age of fifteen. Huxley's secret was sixteen hours' work a day. "If you cannot do that you may be caught out any time." Fame, of course, was the spur, although even this meant different things to the three, and Lockyer's acceptance of a knighthood would not have pleased Huxley:

> Newton and Cuvier lowered themselves when the one accepted an idle knighthood, and the other became a baron of the empire. The great men who went to their graves as Michael Faraday and George Grote seem to me to have understood the dignity of knowledge better when they declined all such meretricious trappings.

FARADAY, M. [1972], *The Selected Correspondence. Vol. 1: 1812–1848: vol. 2: 1849–1866*, ed. by L. P. Williams, Cambridge: Cambridge University Press, pp. 1078.

BIBBY, C. [1972], *Scientist Extraordinary. The Life and Scientific Work of Thomas Henry Huxley, 1825–1895*, Oxford: Pergamon Press, pp. 208.

MEADOWS, A. J. [1972], *Science and Controversy. A Biography of Sir Norman Lockyer*, London: Macmillan, pp. 331.

As he informed Salisbury, men of science should find their proper recognition in the judgment of their scientific peers. Huxley's life was a successful fight for this recognition, both on his own and on Darwin's behalf. (When he was made a Privy Councillor, he said that the only ambition left to him was the Archbishopric of Canterbury.) If Lockyer was bitter, and avid for public recognition, perhaps that is because the scientific establishment was sceptical, tending to find his ideas speculative and marked by dilettantism.

Faraday needed no such compensation as Lockyer's, nor did he ever try to bully the world into agreement with him. The selection of 814 letters to and from him which has been edited by L. P. Williams is remarkable for the almost total absence of polemic. Perhaps social gratitude is proportional to the social distance climbed, or perhaps Faraday was simply made that way. The first letter was written when he was twenty-one, a self-educated man with reasons enough for bitterness, who could remember delivering newspapers to eke out the family finances, and who could recall a week which he survived on a single loaf of bread. When he died in 1867 he was perhaps the most renowned and certainly the most generally liked of the leading English men of science. He had helped to turn science into a profession; he had replaced many of the prevailing orthodoxies with others of his own making; and he had managed to persuade technologists of the value of his work. These things we can learn without an edition of his correspondence, but, to adapt a saying of Dr. Johnson's, one can get closer to Faraday's heart by reading one of his letters than by reading the whole of his *Experimental Researches in Electricity*. Throughout his correspondence he retains the equable temper of a well-adjusted Christian (a Sandemanian, in fact), whose modest ambition was merely to reveal a due proportion of hidden truth – and he almost succeeds in persuading us that his success was fortuitous.

Faraday designed his own career as surely as he planned his experiments. As a boy, he was greatly influenced by Isaac Watts's *The Improvement of the Mind*, and when in the first letter (1812) he opens a correspondence with his young friend Benjamin Abbott, he is not without a word or two of explanation:

> . . . however let me notice, before I cease from praising and recommending epistolary correspondence, that the great Dr. Isaac Watts (great in all the methods respecting the attainment of Learning;) recommends it, as a very effectual method of improving the mind of the person who writes, & the person who receives . . .

Only two more pages of philosophizing in this vein, however, before he is well and truly on his way: "I have lately made a few simple galvanic experiments merely to illustate to myself the first principles of the science." From this point, his gentle interrogation of Nature is disciplined, restricted, and systematic. Not for him the scheme drawn up by Huxley in his youth thirty years later, a scheme which encompassed almost all of human knowledge. One could turn to Shelley for a contrast with a man near to himself in age, and especially to Shelley's Oxford years. Both Faraday and Shelley shared a character of effervescent scientific curiosity, but while the one was drawing sparks from expensive machines in his expensive Oxford rooms, and doing so, one suspects, largely in the interests of Art, the other was being moved by less costly experiments to seek out new and acceptable explanations, albeit of what he could only describe in excited, awkward and untutored prose.

Within a year or so, Faraday's letters reveal all the marks of true scientific genius in embryo – great manual skills, an abounding natural curiosity, and an imagination tempered with self-control and precision in the formulation of hypothesis. He shows an early concern for the art of lecturing, in which he was later to excel, and which – perhaps because he escaped a conventional university education – he never assumed to be his birthright. "A Lecturer falls deeply beneath the dignity of his character when he descends so low as to angle for claps & asks for commendation . . ." Stern comment, but perhaps also tinged with that puritan ethos which scientists have so often found appealing.

Faraday had his share of luck. His first education in the sciences was had from books he was employed to bind. He was fortunate in being able to attend at the City Philosophical Society, and fortunate even in his choice of an elementary book on chemistry, by Jane Marcet, who had been fired by Davy's lectures at the Royal Institution. In 1812 Davy was temporarily blinded by an explosion, and the unknown Faraday was appointed his amanuensis. The former bookseller's assistant was conscious enough of the privilege of travelling in France and Italy with Davy, and there meeting leading European scientists. Soon he had convinced the learned world of his own mettle. The letters barely touch upon his achievement of 1820, when he first synthesized compounds of chlorine and carbon, but the early work on electromagnetism is well illustrated. Here, admittedly, a note of controversy creeps in, for it was rumoured that Faraday took from Wollaston the idea of the possibility of a motor effect. In a letter to Wollaston he vehemently protests his innocence. After a reply suggesting that the matter was a storm in a teacup, Faraday is more circumspect. As he steadily accumulates fundamental discoveries – of diamagnetism, of the laws of electrolysis, of the effects of magnetism on optical rotation, and so forth – his desperate desire to please gradually takes on a secondary importance. Towards the end, he knows that his mind has weakened, and what he writes makes a sad contrast with the self-confident letters of half a century earlier.

Between these limits he shows himself the peer of such correspondents as Ampère, Berzelius, Liebig and Oersted. In the presence of the young Maxwell he clearly feels inspired, but in a sense outclassed, and he could see that he was one of the last of a generation of physicists capable of making great theoretical progress with a minimum of mathematics. He asked Maxwell: "When a mathematician engaged in investigating physical actions and results has arrived at his own conclusions, may they not be expressed in common language as fully, clearly, and definitely as in mathematical formulae?" Faraday here has many modern sympathizers, and not merely among those incapable of understanding Maxwell. Faraday's letters, admirably edited and indexed, quite adequately annotated, and excellently printed, are never more interesting than when we catch him in the act of manipulating alternative concepts to produce a coherent theory of electrical phenomena. Many of the letters he seems to have written only to sort out his own thought, and often he seems to be talking to himself as much as to his correspondent. The resulting atmosphere is one of disengagement from the affairs of man as a political being, and no doubt his other-worldliness explains something of his own magnetism.

As a young medical student, Huxley had approached the great man outside the Royal Institution with a scheme for producing perpetual motion. Faraday gently knocked the idea on the head, but he confirmed Huxley in his ambition to live by science. Huxley could see the difficulty, however, as he confided to his fiancée in the year of the Great Exhibition of 1851:

> To attempt to live by any scientific pursuit is a farce. Nothing but what is absolutely practical will go down in England. A man of science may earn great distinction, but not bread. He will get invitations to all sorts of dinners and conversaziones, but not enough income to pay his cab fare.

By 1851, Huxley had already taken part in a voyage of exploration in the South Seas, as an assistant naval surgeon abroad HMS Rattlesnake. Researches begun on this voyage established, for example, the group of coelenterata (two-layered jelly-fish), and entailed a fundamental revision of the classification of the mollusca and ascidians (which Kowalewsky was subsequently able to show to have an affinity with the vertebrates). This was the work of a fine intellect, if not in the highest class. As Cyril Bibby explains, Huxley was no great experimenter, although a skilled dissector and draughtsman. But above all, his work was essentially critical and synthetic, ranging easily over his own as well as over previous discoveries. While still on the voyage, he had sent back his researches at intervals in the form of substantial papers to learned journals, and almost immediately on his return he was elected FRS. This, of course, brought him no bread. He was given three years' leave of absence, but was at length struck off the Navy List, and went to teach at the School of Mines (later the Royal College of Science), where he stayed for most of his life. He declined a chair at Oxford, on the grounds that a man used to the freedom of London might be unsuitable for it, and later he declined the position of Master of University College, Oxford. To his son Leonard he explained, "I do not think I am cut out for a Don nor your mother for a Doness", and it is hard to reconcile the idea of an Oxford college accepting Huxley as its head, with his notorious lack of belief in revealed religion. (His choice of metaphor is well exemplified in his comment on the ceremony of baptism as "a kind of spiritual vaccination without which the youngsters might catch sin in worse forms as they grow up".)

Dr. Bibby opens his book – his fourth to deal with Huxley – on a wildly extravagant note, describing his subject, for instance, as "perhaps the most brilliant and certainly the most influential scientist of the century", and suggesting that in him England lost "the most powerful scientist she has ever known". This is uncharacteristic of the biography as a whole, which is presented with an absolute minimum of value-judgment, and which does not cease to be entertaining as long as Huxley is there standing in the background, and more especially as long as Huxley is allowed to speak for himself.

A first-class biologist Huxley certainly was, but his reputation as a great controversialist, a sort of Socratic gadfly, was well deserved. "Controversy is as abhorrent to me as gin to a reclaimed drunkard", Huxley told John Morley, Gladstone's biographer, in 1878, a few years before he had occasion to cut in shreds Gladstone's interpretation of the Book of Genesis. Not for nothing was Huxley known as Darwin's bulldog. His triumph over Bishop Wilberforce is well known, but chiefly because his ready wit coined a memorable idea (the phrases he is supposed to have used are as varied as the accounts are numerous), namely that descent from an ape is preferable to descent from an intelligent and influential prelate who chooses to abuse his gifts. Less familiar is Huxley's triumph over some ideas of a scientifically worthier opponent, Sir Richard Owen. Owen held that a small structure known as hippocampus minor occurred only in the human brain, and when Huxley proved its existence in the brain of the gorilla, he refused to give way with good grace. He remained our leading comparative anatomist, but he was wrong, and Huxley's reputation was notched up by several points. The timid Darwin, sixteen years older than he, was grateful for an alliance.

Huxley turned his talent for controversy in many directions. He campaigned for working-class education, and at the very end of his life joined the movement for turning London into a professional university. He had no wish to found an "Established Church Scientific" with a "professorial Episcopate", which some of his fellow campaigners seem to have had in mind. Some of them must have wondered whether they had been wise in allowing him to lead their cause when he expatiated in *The Times* (1892) on the danger of government by professors only, when the professorate was bound to include men "ignorant of the commonest conventions of official relations, and content with nothing if they cannot get everything their own way". As always, he made his pronouncement only after having first anatomized a few chosen specimens.

If Dr Bibby does not anatomize quite as much as the case merits, his biography is highly readable, and with its eighteen portraits of Huxley it provides us with a veritable flick-over picture of a man's life. A. J. Meadows writes his biography of Norman Lockyer, whom he chooses to compare with Huxley, in a much lower key. Lockyer was a man of great vigour and he left behind him a mountain of relevant paper. In his capacity as first editor of Macmillan's journal *Nature* he corresponded with an enormous number of scientists. He wrote a great deal himself, much of it at what might now be somewhat unkindly regarded as a semi-professional level. He left his mark on the British Association, the Royal Astronomical Society, the Royal Society, and the South Kensington Museum, to name but a few of the dioceses within Huxley's "established Church Scientific". Confronted by this mass of material for a biography, Professor Meadows has meticulously put in order, duly added his references, and generally left behind one of those granite Victorian monuments which will not be easily eroded.

His portrait is honest. Lockyer was, like Huxley, a born controversialist, but was as often wrong as he was right. If he frequently speculated on scientific matters, it was not that he did so without reason, but that his reasons were too often insubstantial. He did valiant work in spectroscopy, using the spectroscope to study sunspots and other solar effects, and making important discoveries in the process; and yet for many years he was as active in maintaining that the Sun's corona, visible during an eclipse, was an optical effect produced by the earth's atmosphere. He explained the evolution of the stars in terms which were no less speculative and undisciplined because it is now possible to say, in commenting on his ideas, that they "received support" in such and such a later discovery. There is a much better case for Lockyer's "dissociation hypothesis", which is easily related to the discovery of cathode rays; but always Lockyer manages to find some small way of letting down his would-be apologists by going beyond the evidence or by making plain mistakes (as when, ignoring Henry Roscoe's advice to intone the words "get thee behind me Sodium", he claimed to have decomposed the pure metal by heating, and to have obtained hydrogen in the process).

But Lockyer, right or wrong, calm, explosive, or in some intermediate state of irritability, was a force to be reckoned with, and a detailed biography is long overdue. He was, for example, one of the first to make a serious study of the possibility that ancient Egyptian buildings had been given some astronomically significant orientation, which would, among other things, allow them to be astronomically dated. In this he had a number of Egyptologists on his side. If he was a thorn in the flesh of others, he did nothing to improve relations between astronomy and archaeology when he argued that Stonehenge was the work of astronomer–priests from Egypt whose lineal descendants were the druids. As Professor Meadows points out, in his preface, there was a great breadth to Lockyer's interests, as there was to Huxley's.

But he also suggests that this gives us a measure of Lockyer's stature, which it certainly does not. Had these two controversialists been enemies, rather than friends, Huxley's trenchant intellect would have made short work of the ebullient opposition. Above all else, Huxley was a master of language. "Cogito ergo sum", he once wrote, "is to my mind a ridiculous piece of bad logic, all I can say at any time being 'Cogito'." Lockyer would probably have opted for the "sum".

15

THE TIME COORDINATE IN EINSTEIN'S
RESTRICTED THEORY OF RELATIVITY

It was predictable enough that, after the first flush of enthusiastic hero-worship, historians would begin to question Einstein's authorship of the several principles on which the restricted theory of relativity is founded. Given a little more time, and no doubt we shall see their ancestry traced back to William the Conqueror. There are few, even so, who are prepared to deny that Einstein alone derived from his few basic principles, borrowed or not, the theory as it is now generally accepted. One might imagine, therefore, that the interpretation of the theory is no longer a matter for dispute, but this is not so, and even in recent years a good deal of controversy has been concerned with interpretations of Einstein's time coordinate, and of those equations which suggest the idea of "time-dilatation". It is on time-dilatation and the so-called "clock paradox" that I want to dwell. In particular I shall consider versions of the paradox which, by introducing three or more clocks, avoid accelerations. Little of what I have to say is in substance new, and I should make it plain at the outset that I shall not argue either for or against the overall empirical viability of Einstein's restricted theory under any interpretation. Amongst other things, however, I wish to show that there are certain charges of inconsistency from which the theory may be saved, although not necessarily by following Einstein himself to the letter.

What subsequently became known as the "clock paradox" was given a prominent enough place in Einstein's 1905 paper, and yet at first it provoked less discussion than did Einstein's denial of absolute simultaneity, and his finding that there are events – those outside the observers' light cones, in Minkowski's parlance – whose order in time is different for different observers in relative motion. Once these ideas on simultaneity were widely accepted, it became impossible to assume uncritically an objective and universally acceptable time order. But if the old absolutes had gone, there was a new absolute soon to take their place in the form of an absolute of events. No-one did more to fuse space and time into such an absolute than Minkowski, with the much quoted opening words of his famous paper of 1908, *Raum und Zeit*.[1] Before long, Weyl and Eddington took up the cry with great eloquence, speaking of the human consciousness as something which ranged over a universe pre-existing in space and time, and which singled out its own unique section of experience.[2] The objectivity not so much of time as of time *in isolation* had been called in question. Local, personal or proper time, the time appropriate to any place as taken from a clock at that place, continued to be held meaningful; but all else was inevitably a construction from such local times and from spatial coordinates. When we come to discuss time dilatation and the clock paradox, it will be useful to make a strict division – strict, but founded on an epistemologically unpretentious thesis – between the time of local and distant events. But this is a distinction which cuts across certain others which are more fundamental in the sense that they do not concern a particular scientific theory or group of theories. These more fundamental distinctions I wish to consider first.

1 Eleven important papers on relativity, including *Zur Elektrodynamik bewegter Körper* (Einstein, 1905), *Raum und Zeit* (Minkowski, 1908), and two earlier papers by Lorentz (1895 and 1904) are readily available in English translation in "The Principle of Relativity", London, Methuen, 1923, reprinted by Dover subsequently. The translation was made from the German edition "Das Relativitätsprinzip", 4th edition, Teubner, 1922. References below are to the English edition, abbreviated as P. R.

2 Hermann Weyl's *Raum, Zeit, Materie* is available in English translation, unfortunately from the fourth German edition of 1921, rather than the fifth, with the title *Space – Time – Matter,* London, Methuen, 1921, subsequently reprinted by Dover. For A. S. Eddington, see especially *The Mathematical Theory of Relativity*, 2nd edition, Cambridge University Press, 1924, p. 26: "the contraction [of a moving rod] and retardation [of a moving clock] do not imply any absolute change in the rod and clock. The 'configuration of events' constituting the four-dimensional structure which we call a rod is unaltered; all that happens is that the observer's space and time partitions cross it in a different direction".

Some Preliminary Divisions

Grandiose metaphysical schemes are a thing of the past, and few would wish it otherwise. One thing to be said in their favour, however, was that their authors were as often as not careful to say precisely what they meant in speaking of the reality (or unreality) of time. Today, "time is real" may mean such disparate things as that "clocks are real", and that "the relative rate of two given clocks is single-valued". We can do worse, under the circumstances, than turn to Kant, who, to say the least, drew his distinctions more finely than we are prone to do.

Time, for Kant, is the form of the intuitions of the self and of our internal state, rather than of outward phenomena.[3] It does not subsist of itself or inhere in things as something objective.[4] It is in fact a presupposition made prior to our appreciation of the coexistence and succession of things in time, and may be thought of independently of phenomena, whereas the converse is not the case.[5] Time was for Kant a necessary condition of all our internal experience, and the reality of the objects of experience was not in doubt. We notice that in all this he confined his attention to time at the observer – proper time. Had there been a generally accepted theory of time relations analogous to the Euclidean theory of space, there is little doubt but that at this point Kant would have argued that time necessarily has a corresponding structure, exactly as he held space to be necessarily Euclidean. He came near to such an argument when he wrote:

> And precisely because the internal intuition presents to us no shape or form, we endeavour to supply this want by analogies, and represent the course of time by a line progressing to infinity, the content of which constitutes a series which is only of one dimension ...[6].

It seems that he has taken time to be a form without structure except insofar as it is pulled into shape by certain apodeictic (clearly demonstrated) principles, or "axioms of time in general", examples of which are, according to him, "Time has only one dimension", and "Different times do not coexist, but are in succession".[7] Such axioms are valid as rules through which experience is possible; and they are not conceived of as coming from experience, for in that case they would give "neither strict universality nor apodeictic certainty". This is a weak point of Kant's argument. As G. J. Whitrow among others has demonstrated[8], there are axioms governing temporal experience of a value which can only be judged by experience, and then only in conjunction with other principles, if

---

3 *Kritik der reinen Vernunft*, 2nd edition, 1787, I. ii. 7 (b).
4 *Ibid.*, I. ii. 7 (a).
5 *Ibid.*, I. ii. 5 (1–2). This did not mean that it had absolute reality in the sense of inhering in things as a condition or property. *Ibid.*, I. ii. 7 (c).
6 *Ibid.*, I. ii. 7 (b).
7 *Ibid.*, I. ii. 5 (3).
8 Whitrow, G. J.: *The Natural Philosophy of Time*. London: Nelson, 1961, chapter 4.

anything of scientific moment is to be derived. The old problem of the viable alternative, so familiar in relation to Euclidean geometry, presents itself again here, to the detriment, as I believe, of Kant's position. On the other hand, we might consider Kant to have made, as it were, a distillation of human experience of space and time, a distillation so pure that *ordinary* experience is always likely to include it as an ingredient.

Kant's philosophy of space and time is often regarded as concealing an essentially psychological thesis on the nature of the human mind. Almost anything one chooses to say about time – indeed about almost anything – may be translated into a statement of psychology, and I have no intention of pursuing the idea here. Time, proper time, treated as an *a priori* necessity, we may call "Kantian time". Time for which axioms provide a structure, and which "instruct us in regard to experience", to use Kant's phrase[9], we might call "physical time". (The two sorts of time might overlap, but need not necessarily do so.) The difference between the modern physicist and Kant is that the physicist tends not to regard his axioms as apodeictic. Einstein's claim that such axioms must involve space as well as time, however, is of little consequence in any reformulation of Kant's philosophical position.

Kant did not consider the possibility that the truth of an interpreted set of axioms may be preserved by varying the interpretation placed on them, but neither did he consider the axioms themselves, in isolation from their interpretation. The uninterpreted axioms embrace what we might call "mathematical time", by analogy with the term "space" as used by the pure mathematician without any thought as to its physical interpretation. (William Rowan Hamilton perceived this, when he argued that as geometry is the mathematics of space, so algebra must be the mathematics of time.)[10] Any particular species of mathematical time may be said to "exist" as long as the axioms describing it are consistent.

Although few mathematicians have arrogated to themselves the word "time" as they have done the word "space", many a physicist defending the truth of Einstein's restricted theory has considered his task done when he has underlined ("proved" would be too strong a word) the consistency of the theory. I think it was Eddington who once said that the Astronomer Royal's time was in the position of a vested interest, and was therefore deemed to require no philosophical support. *Plus ça change, plus c'est la même chose.* Today it is Einstein's time which

---

9 See note 7.

10 *Theory of conjugate functions, or algebraic couples; with a preliminary and elementary essay on algebra as the science of pure time* (read in 1833 and 1835), Transactions of the Royal Irish Academy, Vol. 17 (1837) pp. 293–422. Cf. p. 297: "The notion of intuition of ORDER IN TIME is not less but more deep-seated in the human mind, than the notion or intuition of ORDER IN SPACE; and a mathematical Science may be founded on the former, as pure and demonstrative as the Science founded on the latter". Since Hamilton hoped to derive algebra from a prior intuition of time, this was not strictly mathematical in the sense used here.

is the vested interest. Although, as I have said, I shall myself accept it in order to find some of its implications, I hope to make it clear by adding "mathematical time" to my list that Einsteinian time needs to be much more than a term in a consistent mathematical system.

When a physicist speaks of "time observed" or "time experienced", he is usually referring in the last analysis to clock readings, rather than to that special sort of physical time (in our earlier sense) which we might reasonably call "psychological time". The mind plays tricks with time – which is another way of saying that under a psychological interpretation we cannot expect our physical postulates to prove acceptable. This problem is not without relevance to a strict analysis of the "twins" version of the clock paradox, but there are problems enough when we suppose time to be registered locally by a well-behaved and impersonal clock, without going into this further complication. Initially it is necessary to take this and a good deal more for granted, both of a physical and a philosophical sort. Is it not to beg a great many questions to suppose that mechanical clocks must needs keep the same time under all circumstances as clocks based on electromagnetic principles? And what meaning are we to give to the statement that a clock beats at a constant rate? Although not perhaps without solutions, it is as well to remember that such problems are not trivial. Henceforth, nevertheless, the idea of local or proper time will be accepted as a datum, and it will be assumed that no other immediate interpretation of the time coordinate in terms of experience is possible.

Ideas of Time Dilatation before Einstein

Formulae for the transformation of the time coordinate had appeared on several occasions in scientific writings of the decade ending in 1905. From some time before 1895, Lorentz had been investigating transformation equations for electromagnetic systems moving within the aether. He had made use of the idea of a "local time" (t') appropriate to a moving observer, this being defined in terms of the true and absolute time (t) by the equation $t = t' + vx'/c^2$. (Here v is the velocity of the moving observer, and x' the spatial coordinate with respect to an observer on the Earth, both being reckoned in the direction of the motion.) At first Lorentz ignored quantities of higher order than v/c, and not until 1904 did he find the precise form of the transformation under which the Maxwellian equations of the aether are invariant.[11] Following Poincaré, we call the full set of equations by Lorentz's name. (They are quoted in full subsequently.) In 1900 Joseph Larmor had extended Lorentz's original analysis so as to include quantities of the second order in v/c. In the course of doing so he gave extensive consideration to the transformation $t'' = t' - \varepsilon vx'/c^2$, where

---

11 For an ample bibliography of relativity, see the references given in E. T. Whittaker's *History of the Theories of Aether and Electricity*, Vol. 2. London: Nelson 1953, chapter 2. Lorentz's paper of 1904 (not 1903, as given in error by Whittaker) is printed in P. R.

$1/\varepsilon = (1 - v^2/c^2)^{\frac{1}{2}}$, and where the symbols t'' and t' are retained in deference to Larmor's notation.[12] On one or two occasions after 1905, Lorentz admitted that he had never considered his "local time" as having anything to do with real and absolute time. He considered his time transformation, he said, "only as a heuristic working hypothesis", a convenient grouping of physically meaningful terms, as we might say.[13] Larmor similarly made nothing of the paradox which his transformation equation might have been thought to entail, although he fully appreciated that "The change of the time variable in the comparison of radiations in the fixed and moving systems, involves the Doppler effect on the wavelength".[14] Time dilatation as an idea associated with important physical effects had to wait for Einstein.

The complete Lorentz equations were derived on the basis of the idea of a fixed aether, an extended form of Maxwell's theory of electromagnetism, and an "electric theory of matter". As is well known, Einstein's basic assumptions were somewhat different. Accepting the principle of relativity itself, that "the same laws of electrodynamics and optics will be valid for all frames of reference for which the equations of mechanics hold good", and the postulate that "light is propagated in empty space with a definite velocity c which is independent of the state of motion of the emitting body", he placed no reliance on Lorentz's electrical theory of matter. He nevertheless accepted Maxwell's theory for stationary bodies, and he made use of the concept of a rigid body. His derivation of the Lorentz equations was inevitably different from Lorentz's, while others subsequently offered derivations different from his, though usually based on comparable principles. By common consent, however, a great conceptual gulf lies between Einstein on the one hand, and Poincaré and Lorentz on the other.

The different approaches of the three men cannot be adequately dealt with in a few sentences. It is sufficient here to point out that Lorentz to the end of his life denied that the Lorentz-Fitzgerald contraction is only an appearance. He is difficult to follow on this point, however, since he was capable of adding these words to his denial: "On the contrary, the contraction could actually be observed by A as well as by B, nay, it could be photographed by either observer ...".[15] This seems an odd way of denying that the contraction is only an appearance. Similarly he emphasized that time dilatation is real. When A says that B's clock goes slower, this, he believed, "expresses a real phenomenon". Since B may make a similar statement, is it not perverse of Lorentz to say that the phenomenon is real? He is in fact saved from inconsistency – at least in regard to

12 *Aether and Matter,* Cambridge University Press, 1900, chapter 11, especially pp. 167–77.

13 See in particular Astrophysical Journal, Vol. 68 (1928), pp. 385–8, and *Lectures on Theoretical Physics*, Vol. 3, London: McMillan 1931, pp. 181 to the end. These lectures on relativity date from 1910–12.

14 *Op. cit.* (note 11), p. 177.

15 *Op. cit.* (1931, note 12), p. 303.

time, which is all that concerns us here – by the realization that the statements of *A* and *B* "express entirely different things, although both observers use the same words". I shall shortly have more to say of these "different things" and their nature.

As for a more fundamental distinction between the writings of Einstein and the other two, suffice it to say that Lorentz and Poincaré seem to have thought of the impossibility of detecting absolute motion as an empirical matter, whereas for Einstein the principle of relativity was, if not exactly accorded the status of *a priori* certainty, at least not one he would have willingly sacrificed in the case of a clash between the theory as a whole and observation. In his 1905 paper, indeed, Einstein spoke of raising the conjectured principle of relativity to a *postulate*, and in due course it became customary for him and exponents of his theory to say that it was actually *meaningless* to assert that a certain body has a certain absolute motion. However philosophically enlightening this kind of remark may be, it cannot, as some people seem to think, save the restricted theory of relativity as a whole, should sufficiently damning experimental evidence turn up. What it can do, however, is attract the attentions of armchair scientists, and in particular those who would like to prove the theory as a whole incompatible with it.

Einstein's Clock Paradox

By far the best known of the inconsistencies – if that is the right word – are those which grew around section 4 of Einstein's 1905 paper.[16] I shall take it that the clock paradox, as it was to be called, or paradox of the twins, following Langevin's anthropomorphic version, is in outline well known, for it has drawn to itself an immense literature. Two synchronized clocks separate and re-unite, and one is held to be retarded in relation to the other, although motion is apparently inherent in neither, but is relative, involving both equally.[17] Here, they argued, is a contradiction, for which "paradox" is nothing more than a polite word.

How did Einstein reconcile himself to the idea in the first place? He phrased the problem rather strangely, subjecting the "moving" clock, not (as is now usual) to a single deceleration between outward and inward journeys undertaken at constant velocity, but to a constant acceleration, the whole journey being in a curved path (covered at constant speed). This device he must later have realized to be illegitimate, for he did not use it again. Later, presenting the paradox in its now usual form, he ascribed the predicted discrepancy between clock readings to a real asymmetry between clocks: a force, he said, was required to move

---

16 P. R., pp. 48–50.
17 That the criticism did not pass unnoticed is suggested by Bishop Barne's *Scientific Theory and Religion,* Cambridge University Press, 1933, p. 114, for example.

one of the clocks, that is, to change its inertial frame.[18] This explanation of the asymmetry has since become fairly standard, although there are several variants of the simple idea. It may be said, for instance, that the moving clock may be identified in absolute terms because it changes its disposition with respect to a very special inertial frame, namely that at rest with regard to the universe as a whole. But such an absolute, it has been held, is not in the spirit of Einstein's restricted theory. Without pursuing the question as to whether in 1905 Einstein showed a hidden preference for one set of space and time measurements rather than another, I will simply point out that in envisaging two clocks, one at the Earth's equator and the other at one of the poles, it was that at the *equator* which he took to go more slowly. The implications are obvious.

The kinematic part of the restricted theory of relativity deals, of course, with a limited class of physical situations, referring them to coordinate frames not in relative acceleration. If it is to be proved capable of leading to inconsistency, then clearly this cannot be by a consideration of problems outside that restricted class. On the other hand, we plainly say nothing for the restricted theory if we choose to solve the problem of the space-travelling twin by recourse to other theories, such as Einstein's general theory of relativity. The paradox, if it is indeed a paradox, is one only with regard to the expectations of the restricted theory. It has been said that if the restricted theory contains a genuine inconsistency, then the very foundations of Einstein's general theory crumble, and hence the fact that the general theory offers a solution involving accelerations is of no consequence. This is not unduly pessimistic. Certainly there is no inconsistency in kicking over the ladder by which one has climbed; but the fact that the general theory reduces to the special in the absence of gravitating matter suggests that inconsistencies of the special theory are inevitable in the general. If they exist, then they are not to be taken lightly.

The Lorentz Transformation Equations, and Some of Their Consequences

Although it is impossible to give an entirely verbal analysis of the problem, I shall reduce the mathematics to a minimum, and leave the details to my footnotes. Our first problem is one of notation, which if unchecked may itself easily become a source of error. At the expense of simplicity, I shall represent by the symbol $t_A^{E_B}$ the time assigned by an observer A to an event E_B which occurs at a place B. Where there is no cause for ambiguity, I shall denote this more simply by the symbol t_A^B. The merit of this notation is that it removes the need for such circumlocutions as are commonly found introducing the formulae of the theory, where for instance we read of events "viewed from the system S", "observed from the perspective of an imaginary observer O" and so forth. Equally valuable is the fact that the notation allows us to distinguish at a glance

---

18 Naturwiss., Vol. 6 (1918), p. 697.

between local time, written in the form t^A_A, and the time assigned to events at a distance, written in the form t^B_A, where B and A are different.

Using this notation the inherent form of the "Lorentz transformation equations" and their inverses is plainly evident, restricting attention to one spatial dimension only, as is not unusual. I shall take the two x-axes of coordinates in opposite directions, in order to make the symmetry of the two sets of equations fully apparent:

$$x^E_B = \beta(vt^E_A - x^E_A) \tag{1}$$

$$t^E_B = \beta(t^E_A - \frac{v}{c^2} \cdot x^E_A) \tag{2}$$

$$x^E_A = \beta(vt^E_B - x^E_B) \tag{3}$$

$$t^E_A = \beta(t^E_B - \frac{v}{c^2} \cdot x^E_B) . \tag{4}$$

Here E is an arbitrary event, v is the relative velocity of each of the two coordinate frames with respect to the other, c is the velocity of light, and $\beta = (1 - v^2/c^2)^{-\frac{1}{2}}$. The notation for x-coordinates follows that for t-coordinates, and clearly for any A, $x^A_A = 0$, whence follow the Eqs. (1′)–(4′):

$$x^A_B = \beta v t^A_A \tag{1′}$$

$$x^B_A = v t^B_A \tag{1″}$$

$$t^A_B = \beta t^A_A \tag{2′}$$

$$t^B_B = \beta(t^B_A - \frac{v}{c^2} \cdot x^B_A) \tag{2″}$$

$$x^B_A = \beta v t^B_B \tag{3′}$$

$$x^A_B = v t^A_B \tag{3″}$$

$$t^B_A = \beta t^B_B \tag{4′}$$

$$t^A_A = \beta(t^A_B - \frac{v}{c^2} \cdot x^A_B) . \tag{4″}$$

The Eqs. (3), (4), and their derivatives need not have been given at all, of course, but those unfamiliar with Einstein's theory will perhaps take comfort in the reciprocal nature of the two sets of equations, which in this respect at least conform with the principle of relativity.

In order to illustrate the use of the Lorentz equations with this notation, consider the usual derivation of the so-called "Lorentz-Fitzgerald contraction" of moving rods. In general, if P is an event at one end of a rod, and Q an event at the other end, a double application of Eq. (1) gives, after subtraction,

$$(x^P_B - x^Q_B) = -\beta(x^P_A - x^Q_A) + \beta v(t^P_A - t^Q_A) .$$

The last bracket is usually suppressed from the outset, on such grounds as that "we may consider the case where $t = 0$". If we do in fact equate the bracket to zero, however, we arrive at an equation which is flatly inconsistent with the similar equation which we have an equal right to derive from Eq. (3): referring to the absolute value of the difference of x-coordinates by the symbol $\varDelta$, from (1) we have $\varDelta_B = \beta \varDelta_A$, whilst from (3) we have $\varDelta_A = \beta \varDelta_B (\beta \neq 1)$. The mistake was in taking the events at the two ends of the rod as simultaneous *according to both* A *and* B. There is of course nothing new in the idea that the notion of simultaneity enters into the concept of a rigid rod, and it is a simple matter to show that if events at the two ends of the rod are taken as simultaneous with respect to either A or B, but not both, then the two derived equations are no longer inconsistent, but are indeed one and the same equation. All this foreshadows a problem to come, namely that of laying down a definition of a journey in which two observers separate.

Problems of a less tractable sort begin with Eq. (2′) or (4′). Written in the conventional way, $t = \beta t'$, it appears that (2′) makes an absolute distinction between "observers" A and B. But "from another point of view", goes the typical argument, "we can write $t' = \beta t$ (i.e. (4′)), and therefore relativity is upheld". A glance at (2′), however, shows that the principle of relativity is never in jeopardy, so long as we are prepared to distinguish between local time (t_A^A) and time as a coordinate of distant events (t_B^A). It is well known that Einstein introduced such time-at-a-distance into his theory by means of a definition in the first section of his 1905 paper. He imagined a system of clocks made synchronous with any given clock by means of an ideal experiment involving the reflection of light from any event. The time of the event was by definition to be the arithmetic mean of the times of despatch and return of the synchronizing signal. It is for time defined in this way that the symbol t_B^A is here reserved.

At least five years ago H. Dingle introduced a somewhat unusual paradox of his own making with the purpose of proving that although the special theory of relativity is mathematically self-consistent, it "requires clocks to behave in an impossible manner, for one set of synchronized clocks must concomitantly go both faster and slower than another set, and this is impossible".[19] Dingle observed that when Einstein deduced the rate-ratio of two clocks he chose to compare times assigned to a very special pair of events, namely a pair at the origin of one of the two systems. Taking the first event as that of coincidence of A and B (in our previous notation), and the second as any event at B, then it seems that the clock at A runs faster than that at B. But, as Dingle pointed out, why not take the second event on the other clock? When we do so, we shall decide that the clock at B runs faster than that at A. This is what Dingle meant by saying that clocks are required to behave in an impossible manner.

---

19 British Journal for the Philosophy of Science, Vol. 15 (1964), p. 46.

If we denote the interval between the two events by "dt", then Dingle's two supposedly conflicting propositions, when expressed in the notation of (2') and (4'), are as follows:

$$\frac{dt_A^B}{dt_B^B} = \beta$$ — A judges the interval to be greater than does B, and therefore "clock B runs faster than clock A"; (i)

$$\frac{dt_B^A}{dt_A^A} = \beta$$ — "clock A runs faster than clock B". (ii)

The propositions might well seem flatly contradictory, and yet it is obvious that the verbal forms contain only a part of the meaning of the equations which precede them. I am afraid that I simply cannot see how clocks have been required to behave "in an impossible manner", although Dingle has certainly drawn attention to the dangers of expressing Einstein's theory too casually. Einstein's own interpretation was confused, or how, without further explanation could he have supposed an equatorial clock to go faster than a polar one? And as Dingle remarks, almost all subsequent textbooks have interpreted the rate-ratio in the same way as Einstein, who took the second calibration event to be at the "moving" clock.

The moral of all this is that we should not speak of relative clock rates at all without specifying the events with respect to which they are measured. Einstein was inconsistent to the extent that he spoke as though the ratio were *objective*, but he was under no compulsion to do so. Dingle's purpose was to discredit Einstein's interpretation, and not to indicate ways of saving it. He maintained, however, that "unless [a certain equation leading to (ii) above] is held to rule out [an equation leading to (i)], it may be proved – the proof is withheld for want of space – that the special relativity theory cannot fulfil its purpose of justifying the classical electromagnetic equations".[20] I am unable to envisage this proof, since (i) and (ii) are absolutely identical in form. In considering the kinematic interpretation of relativity, however, one should certainly keep an open mind as to possible inconsistencies in their electromagnetic consequences.

Turning to the clock paradox in its traditional guise, there are certain necessary distinctions which make it desirable that we pay even more attention than before to matters of notation. Suppose that identical clocks A and B, synchronized and reading zero at the moment they coincide (the event E_0), separate at constant velocity v (in the estimation of both) to some distance. Obviously

---

20 *Loc. cit.* (note 18).

we must distinguish between the event at B (E_b, say) when the journey is reckoned by an observer at B to cease, and the event at A (E_a) when he reckons the journey to cease. It is a simple matter to show [21] that

$$t_A^{E_b} = \beta t_B^{E_b}$$

and

$$t_B^{E_a} = \beta t_A^{E_a} .$$

If we were to make the mistake of supposing that there was a unique, that is universally valid, moment at which the journey ended, and of identifying the time of either E_a or E_b as estimated by the two observers, then each equation is self-inconsistent, even in isolation, since $\beta \neq 1$ *ex hypothesi*. It goes without saying that to suppose the events simultaneous as judged from a single place (that is, to write $t_A^{E_a} = t_A^{E_b}$ and $t_B^{E_a} = t_B^{E_b}$) would give rise to a pair of inconsistent equations. These difficulties are easily avoided by saying that in each equation the left hand side is conventional or coordinate time, rather than a proper time, and therefore the equations do not compare real with real. To do so, and obtain the clock paradox proper, it is necessary to bring the clocks back into coincidence, and this we cannot do without introducing accelerations into the problem. As H. E. Ives [22] and others have indicated, however, the problem may be re-cast by the introduction of another moving clock. Before adopting this approach it is necessary to look more closely at the meanings which can be assigned to the phrase "time of the journey".

Suppose that A and B are in agreement when at relative rest as to their scales of length and time. The event E_b, marking the end of B's journey, may then be regarded as the event when B reaches a point (say the end of a rod) at distance $\varDelta$ (measured by A) from A. Earlier I said that E_a was an event at A when he reckoned this same journey to cease. This might mean at least two different things. Henceforth I reserve E_a to denote the event at A when, in their relative motion, A reaches a distance $\varDelta$ (measured by B) from B. By $E_{a'}$, on the other hand, I denote an event at A which occurs at the instant when A judges E_b to happen at B. In short, $t_A^{E_{a'}} = t_A^{E_b}$, by definition. E_b and $E_{b'}$ are defined by analogy. By "time of the (outward)) journey" it seems that we may mean at least four sorts of time in the estimation of each observer, conceptually if not all numerically different. The relations between them are

---

21 It is a consequence of (4′) that

$$\int_{E_0}^{E_b} dt_A^E B = \beta \int_{E_0}^{E_b} dt_B^E B ,$$

whence (since $t_E^{A_0} = t_E^{E_0} = 0$, by definition) the two equations of the text follow.

summarized in the following table, the entries in which are justified more fully elsewhere:[23]

time	$t_A^{E_a}$	$t_A^{E_{b'}}$	$t_A^{E_b}$	$t_A^{E_{a'}}$	$t_B^{E_a}$	$t_B^{E_{b'}}$	$t_B^{E_b}$	$t_B^{E_{a'}}$
scale factor	t/β	βt	t	t	t	t	t/β	βt

It is of some interest to consider the status of the different sorts of "journey time" in this table. We may call $t_A^{E_a}$ "real", for it is the local time of some event which could in principle be directly and locally observed – that of passing the end of a rod fixed in relation to B, for instance. At first sight, $t_A^{E_{a'}}$ is just as real, but on closer examination it emerges that the event $E_{a'}$, although local to the clock at A, was simply postulated to be such that $t_A^{E_{a'}} = t_A^{E_b}$. It is no less conventional, therefore, than the quantity to which it is equated, which ultimately takes its meaning from Einstein's definition of the time at distant events. The fourth quantity, $t_A^{E_{b'}}$, is doubly conventional, for it is the time of a distant event $E_{b'}$, itself postulated in the same way as was $E_{a'}$. The last four sorts of journey time to be listed in the table all have a counterpart in the four already discussed.

Four Clocks and No Paradox

Consider now the following symmetrical movement of four clocks. Clocks A and B separate at constant velocity v under the same circumstances as before. A clock C later passes A (the event being E_a) and a clock D passes B (the event being E_b), the clocks C and D approaching along the same line as A and B formerly separated, and with the same relative velocity. For simplicity let it be supposed that C and D meet when they both read zero, and that as before each measures x-distances in the direction of the relative velocity of the other. They will of course assign negative t-coordinates to the moments at which they pass A and B respectively. If we allow ourselves to speak of the "joint experience" of A and C (or B and D), and suppose it to be given

---

23 As a matter of definition (see the text),

$$t_A^{E_{a'}} = t_A^{E_b} \quad \text{and} \quad t_B^{E_{b'}} = t_B^{E_a} .$$

By Eqs. (4') and (2'), respectively,

$$t_A^{E_b} = \beta t_B^{E_b} \quad \text{and} \quad t_B^{E_a} = \beta t_A^{E_a} .$$

These equations may be repeated with $E_{b'}$ and $E_{a'}$, giving two more relations.

$$t_B^{E_a} = t_A^{E_b} .$$

This follows from the fact that, by hypothesis, scales of length and time have been agreed upon, and each side of the last equation represents the time B and A respectively assigned to a journey of distance Δ at velocity v. (This is merely to affirm Lorentz transformation equations (3') or (3'') of the text.)

by the total proper time of the round trip, then we may write these totals of proper time as

$$t_A^{E_a} - t_C^{E_a} \quad \text{and} \quad t_B^{E_b} - t_D^{E_b}.$$

Reference to our tables above shows that the positive terms in the two expressions are equal. (Each of A and B covers a distance – the length of the other's rod – which he judges to be equal to the estimate made by the other of his own rod; and the velocities are the same.) But what of the remaining terms, representing times as measured from C and D? By what argument could we prove them to be equal or otherwise? We could say that at the event E_a (at C) there is associated with D a rod which just reaches from D to C, and at the event E_b (at D) there is a rod from C to D, but we cannot assume without proof that these two rods are equal in length, for we cannot, as in classical kinematics, say that the two events E_a and E_b are simultaneous. It is tempting to produce an argument by "running time backwards", by appealing to symmetry, or by a *reductio ad absurdum*, which amounts to the same thing; but this would be to pre-suppose the very self-consistency of the theory which is in doubt. On the hypothesis that Einstein's restricted theory of relativity is consistent, however, it seems to me that, with the present formulation of the problem, the aggregate proper time predicted for one pair of clocks is equal to that for the other pair. The clock paradox has refused to appear, but up to a point it has been ruled out by *fiat*. We notice, however, that no hypothesis was made as to the relative motion of C and A (or B and D), and there is no reason why, for example, A and C should not be taken as relatively stationary, in which case we are left with a three-clock problem. We might equally well stipulate that the return velocity of C with respect to A is equal to that of D with respect to B.

The four-clock problem was introduced here partly in an attempt to engender contradiction – and the attempt has, so far, failed – and partly to find a symmetrical analogue to the traditional problem of *two* clocks (or twins), but without introducing accelerations. It falls down as a satisfactory analogue, however, in the absence of absolute simultaneity. With only two clocks, there are no problems of definition in regard to the *limits* of the journey of separation. A way of overcoming this problem with four clocks will be explained shortly.

The version involving three clocks may initially be characterized very briefly. According to B, he requires a time of $\Delta/\beta v$ to reach the end of A's rod, while according to D, the return journey along A's rod takes the same length of time. The total of proper time kept by B and D together $(2\Delta/\beta v)$ is less than the time assigned by A to the round trip.[24] With a little elaboration, the argument

---

24 The total "time assigned" to the journey by A and C is $t_A^{E_b} - t_C^{E_b}$. (It is to be remembered that clocks C and D are zeroed at the end of their journey, making $t_C^{E_b}$ etc. negative.) Applying Eq. (4') of the text, this is β times greater than the time kept locally by B and D jointly.

soon gives us the asymmetrical aging of twins, and so forth, in which for a variety of reasons the great majority today apparently believe. But one way of avoiding this solution is to say that there is an error in equating the "time assigned" by A (and C) to the aggregate time kept on the two clocks A and C. The time assigned is certainly not the time kept on a clock or clocks, but a coordinate time deriving from Einstein's definition of time at a distance. We are under no obligation, therefore, to accept this "asymmetrical aging" as real without further proof, but such proof is forthcoming, as I shall show.

We return to a fully symmetrical statement of the problem in terms of four clocks. If we stipulate that A and B separate with a relative velocity of v, that C and D approach with the same relative velocity, and that each has the same velocity relative to the clock it first passes, then it is easily shown that this last velocity turns out also to be v. In other words, C and B are relatively stationary, as are D and A. It is this fact which makes for a simple description of the double journey. A holds, as before, that he takes time $\Delta/\beta v$ to reach to a distance which B claims to be Δ. (B maintains that A takes time Δ/v and C, being at rest with respect to B, agrees.) In reaching D, which we may imagine to be at the end of A's rod, again of length Δ, C reckons time $\Delta/\beta v$.[25] The time jointly experienced during the round trip by A and C will be simply $2\ \Delta/\beta v$, and it is usual to add that B and D will assess that time as $2\ \Delta/v$. But repeating the argument for the other pair, B and D will likewise jointly experience a time $2\ \Delta/\beta v$, which it may be said A and C will assess as $2\ \Delta/v$. Overlooking the assessments of times by distant observers, there is no paradox, for the aggregates of time experienced by the two pairs are the same. It is nevertheless imperative, if Einstein's restricted theory is to be saved from contradiction, that the assessments of times $2\ \Delta/v$ be explained away. There is no longer the obvious asymmetry of the three-clock version for us to fall back on. One alternative would be to examine more critically the nature of the times assessed by distant observers, which as already emphasized are not proper times kept on clocks, but coordinate times stemming from Einstein's definition of time at a distance. In this particular example these "mere coordinate times" are a somewhat incongruous mixture. The first, for instance, is a time at A assessed by B or C, added to a time at C assessed by A or D. What claims can a time of this sort have to physical significance, even when C is reckoned to be A's *alter ego*, as it were, undergoing the return journey on his behalf? Clearly what is needed is that such times assigned to distant events be in some way correlated with local events, which may in turn be compared with a local clock. Unfortunately, when this is done, the "mere coordinate times" have an uncomfortably real look about them, for they simply turn into proper times. (I considered the (proper) times of receipt of continuous

---

25 The v here is C's velocity with respect to A, not with respect to D. Given that the latter were different from v so would the former be different. C and B (and A and D) would no longer be relatively stationary, and this simple statement of the problem would not be possible.

light signals.) There is, after all, no easy way out of the dilemma for those who would like to make a category distinction between "proper time" and "assigned time", and then say that there can be no conflict between the two expressions ($2 \, \Delta/\beta v$ and $2 \, \Delta/v$) for the time. I shall come back to this point at the end. First, however, the four clock problem will be re-considered in terms which make it unnecessary to stipulate equal velocities of outward and return journeys.

The argument from four clocks in the general form given at the outset had certain weaknesses. Even overlooking C and D, in what sense do A and B perform the *same* journey? The assumption that they do was tacitly made when we compared B's estimate (rather than D's or that of some other observer) of A's journey with his own, and so on. The reason was, of course, that the "clock paradox" is generally explained in terms of only two observers, and then only a single division into outward and return journeys is required. It is a mistake to suppose that in the four clock version we must necessarily be describing a journey uniquely when we say, in effect, that A goes to the end of B's rod while B goes to the end of A's. There are two unique events involved, and the Lorentz equations do not themselves contain any means whereby the two may be iden-tified with anything which could be termed the end of the journey of separation of A and B. To claim to identify them is to claim to have some absolute stand-ard of simultaneity, and this, we know, Einstein's theory does not have. (The whole problem is reminiscent of that of saying what we mean by "the contraction of a moving rod".) The most we can say is that A and D can agree that the events are simultaneous, as can B and C. This does not mean that the theory cannot be re-interpreted in terms of a universal time – and the most natural way of doing so would seem to be to begin from the methods of "kinematic relativity" of the late E. A. Milne and our President Dr. Whitrow. Since, however, the object of the present exercise is to remain more or less within Einstein's inter-pretation of his theory, the idea of absolute simultaneity must here be rejected. It should, none the less, be possible to introduce a standard of simultaneity achieving much the same purpose, by simply supposing a neutral observer (O, say), positioned between A and B in such a way that all three are simultaneously concurrent, and such that O subsequently judges each of the others to have the same x-coordinate and velocity (w) with respect to him. Some of the conse-quences of this supposition will be outlined.

An immediate consequence is that the time-relations of the Lorentz trans-formation equations are symmetrical about O:

$$t_A^O = \gamma t_O^O = t_B^O \, ,$$

$$\gamma t_A^A = t_O^A \, , \quad t_O^B = \gamma t_B^B \, ,$$

and $\qquad t_B^A = \beta t_A^A \, , \quad \beta t_B^B = t_A^B \, ,$

where $\qquad 1/\beta = (1 - v^2/c^2)^{\frac{1}{2}} \quad$ and $\quad 1/\gamma = (1 - w^2/c^2)^{\frac{1}{2}} \, .$

It is a simple matter to show that once v is given, w is uniquely determined, and that w and γ are simple functions of v and β.[26] O is then determined uniquely in velocity and position with respect to both A and B for all times. From Eq. (1″), since by hypothesis $x_O^A = -x_O^B$ (taking O's x-coordinate to be positive in the directions of A), $wt_O^A = wt_O^B$. Since w is not zero, $t_O^A = t_O^B$, and therefore from the last group of equations we find also $t_A^A = t_B^B$, and thus $t_B^A = t_A^B$. In other words, there is a way of defining events at A as simultaneous with events at B, and this defined simultaneity involves not merely the equality of coordinate times, but also of proper times (t_A^A and t_B^B). (The actual existence of the auxiliary "observer" O is, of course, totally irrelevant to the validity of this definition.) To this extent, the definition supplements Einstein's definition of simultaneity at different places. We notice, furthermore, that we may also equate x_B^A and x_A^B, and we may therefore take as the two "simultaneous" events the E_a (at A) and E_b (at B) of our first example, that involving four space-travelling clocks, A, B, C and D. It should by now be clear that in this example, where C and D meet after passing A and B respectively at events E_a and E_b, the aggregate proper time recorded by A and C will be equal to that recorded by B and D. For this to be true, C and D need not approach with the same velocity as that at which A and B separated, although for O to be the same on both outward and return journeys, C and D must have the same speeds with respect to A and B respectively. There is no obvious sign of any clock paradox here.

No doubt it will be objected that since there is an element of *definition* in all this, nothing conclusive has been proved. This misses the point, however, that the definition laid down is a possible, i.e. consistent, definition of a journey.

Three Clocks – Asymmetry without Contradiction

Without changing notation, we consider the problem of three Clocks: A and B separate at constant velocity, and D subsequently passes first B and then A along the same line, also at constant velocity. The time assigned by A to the two parts of the journey will certainly be greater than the sum of those recorded locally by the "travellers", but is the sum of the times assigned by A necessarily equal to the total proper time *recorded* by A between B's departure and D's return? We can easily see how doubt could be cast on their equality. Let us simply

---

26 By Einstein's formula for combining relative velocities, $v = 2w/1 + w^2/c^2$. Solving this quadratic for w in terms of v, we find – after rejecting one root as involving a velocity in excess of the velocity of light –

$$w = \left(\frac{\beta-1}{\beta}\right) \cdot \frac{c^2}{v}, \quad \text{where} \quad \beta = (1 + v^2/c^2)^{-\frac{1}{2}}.$$

By definition, γ is the same function of w as β is of v. By substitution of the above value for w in this function, it is found that $\gamma = \sqrt{\left(\frac{\beta+1}{2}\right)}$.

denote by E_0, E_1 and E_2 the events when B leaves A, when B meets D, and when D meets A. When writing down the equation comparing A's assessment of the time of E_1 with B's, we are reckoning forward from the event E_0, and yet when comparing A's assessment with D's, for the second part of the journey, the time is reckoned backwards from the event E_2. The events E_0 and E_2 are the only events of the journey of B and D of which A can be sure. Considering A's proper time as a line stretching from E_0 to E_2, how do we know that we are entitled to consider the "time of E_1," as represented by a unique point on the line? One way to be sure is to consider B and D as sources of light. If A observes B continually from its departure to E_1, he may then immediately observe D, also at E_1, since according to Einstein the velocity of light is independent of the velocity of its source. A may then observe D until the two meet. It is a simple matter to integrate the time kept locally by A during his observation of B and D, and to show that this is precisely the time he assigns to the outward and return journeys of B and D, and equal to β times the aggregate of their local times during the double journey.[27] (This is so even with a different return velocity.) There can be no doubt but that Einstein's restricted theory of relativity applied to three clocks A, B, and D as explained, does predict that the proper

----

27 E is the element of emission of photons from a source at B, and R is the event of their receipt at A. Then

$$t_A^R = t_A^E + \frac{x_A^E}{c},$$

and therefore

$$dt_A^R = dt_A^E\left(1+\frac{v}{c}\right).$$

(By Eq. (4'), $dt_A^E = \beta\,dt_B^E$; and equating photons emitted and received, $v_B^B\,dt_B^E = v_A^B\,dt_A^R$. Combining, we obtain the Einsteinian Doppler formula for a receding source:

$$v_B^B/v_A^B = \beta\left(1+\frac{v}{c}\right).)$$

For the return journey of D, at velocity w, the equation will be

$$dt_A^R = \left(1-\frac{w}{c}\right)\cdot dt_A^E.$$

Suppose that B goes out to a distance $\varDelta$ as measured by A, and D returns along the same distance. The total time recorded by A during his observations of B and D will be $\int dt_A^R$,

$$\text{namely} \quad (B) \quad \left(1+\frac{v}{c}\right)\int dt_A^E = \left(1+\frac{v}{c}\right)\cdot\frac{\varDelta}{v},$$

$$\text{and} \quad (D) \quad \left(1-\frac{w}{v}\right)\int dt_A^E = \left(1-\frac{w}{c}\right)\frac{\varDelta}{w}.$$

The sum of these quantities is $\varDelta\left(\dfrac{1}{v}+\dfrac{1}{w}\right)$, which is the time A would arrive at on the basis of the coordinate time he assigns to the distant event E_1.

time recorded by A is (for equal velocities out and back) β times the aggregate proper time recorded by B and D.[28]

There is an asymmetry here as between recorded local times; but where is the paradox? One might expect that the factor β would appear on the other side of the equation if light from A were received first at B and then at D. Here certainly, would be a paradox, a true contradiction. But a simple calculation shows that treated in this way the problem yields the same answer as before: the proper time at A is β times as great as that at B and D. There is no contradiction here, strange as the asymmetry of the result might at one time have appeared to be. It seems at first equally strange that there was no comparable asymmetry in the journey involving four clocks, but the two results are not mutually contradictory. Why need we say more than that the structure of special relativity is very different from that of the corresponding sector of everyday belief?

On the Cause of the Asymmetry

The word "perspective" is one of the most useful in the vocabulary of the apologist for Einstein's restricted theory, for we are all of us familiar with the alteration and distortion of spatial experience as a result of changes in our point of view. But what would it mean to a young astronaut to be told that his twin brother's long grey hair and general decrepitude were the result of his peculiar time-perspective? Would he be satisfied with the Lorentz equations, or with a Minkowski diagram, and agree not to ask for such an antiquated thing as a causal explanation? One would like to have subjected those brilliant rationalistic twins, Weyl and Eddington, to a real test of this sort. Although it is no argument for the need to provide causal explanations, it is clear enough that most physicists who discuss the paradox of the twins do have a desire to go beyond the restricted theory to a physics where "physical" causes operate. This is as much a psychological fact as a symptom of concern for a legitimate methodology, and I doubt whether many of those who are prone to distinguish between physical and kinematic explanations would find it easy to justify the distinction. Bishop Barnes, writing in the late 1920s, was untypical only insofar as he was honest enough to say openly that he did not believe that the discrepancy between the time of consciousness of the two men had been satisfactorily explained, although on the previous page he had given the standard Einsteinian explanation.[29] Einstein

---

28 It might be asked why no use is made here of the idea of introducing the quasi-universal time of an auxiliary observer O, as in the last section. Would this not lead to a situation in which A and B agreed about the outward journey, and A and D about the return? In fact it would be necessary in this three-clock version to introduce *two* auxiliary observers, O_1 and O_2, one for the outward and one for the return journey. We should then be committed to combining times recorded by A on two incompatible scales.

29 *Scientific Theory and Religion,* (The Gifford Lectures, 1927–1929), Cambridge University Press, 1933, p. 113.

himself, and most orthodox relativists since, have put the effect down to the acceleration of one and only one observer with respect to an inertial frame. Accelerations of this sort have qualified as causal at least from the time of Newton, a fact which might perhaps have weighed in their favour as explanations of the discrepancies between clocks or twins. But even without accelerations we have seen that similar discrepancies are possible. Are we to ascribe them to the *motion* of the observers B and D in the three-clock example? Clearly not, for that motion is not inherent in them any more than it is inherent in A, with respect to whom it is measured. Is *relative* motion the cause? Even allowing the naive desire to find a simple cause, it can hardly be supposed that we have found it in the relative motions alone, for two bodies in relative motion share the motion equally. And was there not relative motion enough in the four-clock examples, where there was no sign of asymmetrical aging? On the other hand, since the relative velocity v is present in the factor β, the cause-and-effect philosopher might be inclined to speak of a multiple cause of asymmetrical aging involving relative velocity and an asymmetry in the distribution of the moving clocks. (We recall that in the four-clock problem we compared the aggregate time of one pair of relatively moving clocks with that of another pair, whereas in the case of three clocks we compared the aggregate time of such a pair with the time of a *single* clock.)

In passing, we notice that no violence is done to any principle of relativity. In fact it is only when we pass beyond the restricted theory, and treat B and D in that example as one and the same clock, with its motion reversed by deceleration, that even the unrestricted principle of relativity is called in question; for then, the acceleration being relative to A, it should therefore be just as capable of making his clock register the lesser time. There is one clear way of avoiding this unacceptable (because contradictory) alternative, and that is to bring back something akin to the much despised aether – under some other name such as "universe" of course. Those who are honest enough to adopt this course openly are not without their causal explanation of the asymmetrical readings of relatively moving clocks. But those who wish to stay within Einstein's restricted theory must be satisfied with the frail causal ingredients mentioned earlier – relative velocity and an asymmetrical situation. Where does the searcher for causes go from there? In all probability he will devise a model displaying all the characteristics of those situations which are deducible from the Lorentz equations. If he is Minkowski's equal, the chances are that his model will differ little from that which Minkowski put forward more than sixty years ago. And in the unlikely event of his feeling that he now has reason for greater confidence in the reality of the situations with which Einstein's theory deals, he is simply deluded.

The syntactic structure of the theory is totally independent of confirmatory experience, and is yet again independent of the rules of correspondence between them. Each of these three things may be modified to a greater or lesser extent.

Since in discussing some of the problems associated with time dilatation I have modified none of them, have kept within the theory and interpretation offered by Einstein, and have ignored all experiment bearing on its plausibility, I realize that I have scarcely done more than scratch the surface of the problem. I hope at least that I have shown special relativity, one of the great corner-stones of twentieth-century physics, to be free from some of the kinematic inconsistencies with which it has been charged. Although I have not found an acceptable answer to the problem of the space twins, that was never my intention. As for the apparently incompatible results obtained with three and four moving clocks, however surprising at first, they seem to me no more difficult to accept than the theorems that two sides of a triangle are together greater than the third, and that two pairs of sides of a rhombus are equal. The unfamiliar is not on that account illogical.

Addendum

In his *The Structure of Time*, London, 1980, at pp. 194-5, W.H. Newton-Smith writes as follows:

> North, in a careful study of the 'clock paradoxes', considers a three-clock version akin to the one I have discussed. However, he fails to realize that there is, in fact, reciprocal retardation. This leads him to raise the question of the causal explanation of the retardation of C . . . However, if my account of the matter is correct, there is no non-reciprocal retardation to be explained . . .

As far as I understand his argument (which is marred by many mistakes and printing errors) it is identical to mine except in one fundamental respect, namely that he works with a single 'composite clock C' on occasions where I believe he must make a distinction between two clocks. In his own account (especially pp. 192-4) he speaks of C's as a single frame, where it is really two. What I admit I cannot understand is the usefulness of his concept of 'reciprocal retardation'. Here I can only plead once again for a more careful use of notations powerful enough to avoid the confusions into which ordinary language tends to lead us.

16

A HUNDRED YEARS OF THE CAVENDISH

In 1764 Richard Watson became Professor of Chemistry at Cambridge. "At the time this honour was conferred upon me", he wrote, "I knew nothing at all of Chemistry, had never read a syllable on the subject, nor seen a single experiment in it; but I was tired with mathematics and natural philosophy, and the *vehementissima gloriae cupido* stimulated me to try my strength in a new pursuit, and the kindness of the university (it was always kind to me) animated me to very extraordinary exertions." Seven years later the Regius Chair of Divinity became vacant, and within nine days, "by hard travelling and some adroitness", Watson was created DD. A day later he was elected to the Chair. "On being raised to this distinguished office", he now wrote, "I immediately applied myself with great eagerness to the study of Divinity." Although unusual even in eighteenth-century Cambridge, and not wholly unparalleled in our own times, these transmutations of office tell us something of the light-heartedness with which the experimental sciences were regarded in Watson's day. In fact he did all the right things and turned out to be a good choice: he appointed a good "operator" from Paris, became a very popular lecturer, and was a prototype for his successors when he advised the government on improvements in gunpowder. He even resisted the temptation to meddle with history. (The nearest he got was when he accepted membership of the Massachusetts Historical Society.) In short, he was in the best tradition of Oxbridge experimentalists.

It goes without saying that until well into the nineteenth century the government and teaching of Oxford and Cambridge were predominantly clerical, and that the empirical sciences had very little part to play in the curriculum at either place. From an undergraduate point of view, chemistry was little more than an amusing pastime, and physics mostly a source of problems for mathematics – and especially so for the Cambridge Mathematical Tripos. The Tripos was regarded with the utmost suspicion by most of those who claimed to purvey a liberal education; and as J. G. Crowther reminds us in his introduction to *The*

CROWTHER, J. G. [1974], *The Cavendish Laboratory 1874–1974*, London: Macmillan, pp. 464.

HEILBRON, J. L. [1974], *H. G. J. Moseley: The Life and Letters of an English Physicist, 1887–1915*, Berkeley: University of California Press, pp. 312.

Cavendish Laboratory 1874-1974, even so scientifically informed an educator as William Whewell, Master of Trinity, thought that "such departments of science as Chemistry are not proper subjects of academical instruction", and that a century should pass before scientific discoveries were certain enough for a course of instruction. "Are the students of Cambridge to hear nothing of electricity?" asked Peel. When the leading Cambridge physicist of the nineteenth century, James Clerk Maxwell, asked the great mathematical coach Issac Todhunter whether he would like to see an experimental demonstration of conical refraction, Todhunter declined. He had taught it all his life, and did not want his ideas upset. But such coaches as Todhunter, serving the needs of perhaps the most taxing of undergraduate mathematical courses then devised, were producing a small but growing number of mathematicians with different tastes from their own. There were the men who combined an interest in experimental physics with a command of mathematical techniques, and who were soon to make so much of the work of their eighteenth-century predecessors look utterly trivial. Cambridge was their principal habitat, although Glasgow, King's College (London), and Oxford were marginally earlier in offering systematic practical physics tuition. And at Cambridge in 1871 Maxwell was persuaded to take the newly created Chair of Experimental Physics, a position whose creation had in fact been recommended by the Royal Commission of 1850.

The university committee which finally brought the Chair into being also proposed the appointment of a Demonstrator, and, of course, the establishment of a laboratory with apparatus. The laboratory took its name from Henry Cavendish, the illustrious scientific kinsman of the Seventh Duke of Devonshire, University Chancellor, Second Wrangler (1829), and fount of the money which paid for it. The building was completed a hundred years ago last year, and Mr Crowther's book commemorates the fact. As he says with perfect justice in his preface, reckoning not unreasonably by Nobel prizes, no other physics laboratory in the world has a comparable record. The most natural English comparison is with Oxford, where the Clarendon Laboratory, built by a Cambridge man (R. B. Clifton), was a few years older; but as a teaching laboratory, let alone as a research institution, it was a far less distinguished place until long after the First World War. A brief moment of Oxford glory before the war was when H. G. J. Moseley worked freelance under J. S. E. Townsend. This is a part of J. L. Heilbron's story, and by taking his book together with Mr Crowther's one can form a very good idea of the divergencies and similarities between experimental physics as developed in two of ostensibly the most conservative centres of learning in the modern world.

The new Cambridge Chair was created, but Sir William Thomson did not wish to leave Glasgow for Cambridge: Hermann von Helmholtz would not leave Berlin for either Cambridge or Oxford; and Maxwell, rather reluctantly, came out of retirement at Glenlair. He was thirty-nine when he returned to Cambridge, and died of cancer only eight years later. Judging by the numbers attending his lectures, he was not the most popular of lecturers. (J. A. Fleming, inventor of the thermionic valve, was for a whole term in an audience of two.) Physics remained unpopular until a syllabus change in 1832. Mr Crowther has little to say of Maxwell's philosophy of experimental physics, but, from an aside, we gather that he wanted to attach to the laboratory small bands of graduates, each of whom would undertake a definite piece of work "after having received his training in measurement". This has a ring of Kelvin about it. Of course, here and throughout Mr Crowther's long book there is a strong temptation to forget that Cambridge physics was not coextensive with the Cavendish, even though after Maxwell there was a progression of more pragmatic professors.

The first in the line was Lord Rayleigh, a past Senior Wrangler with a genius for good organization. In 1882 Rayleigh opened his classes to women on equal terms with men. Marriage to Evelyn Balfour had confirmed his suspicion that John Stuart Mill was right on the question of equal intellectual opportunity for the sexes. (A more cynical publisher would have exploited the fact on the dust jacket.) Rayleigh resigned at the first opportunity (in 1884), and retired to his private laboratory. J. J. Thomson, who was not yet thirty, came next in succession. Appointed over the heads of several older men, Thomson was a Mancunian whom Mr Crowther, for twenty years the scientific correspondent of *The Guardian*, treats very sympathetically. Had "Manchester" been included in the index, one could indeed have more easily quantified a very real impression that the fortunes of the Cavendish and the affairs of Manchester have long been closely connected.

By his discovery of the electron, Thomson soon convinced the electors that they had acted wisely, and also set the tone for Cavendish research for years to come. A change in the statutes had brought an influx of scientists from other universities – including Ernest Rutherford from New Zealand – and this at length transformed the laboratory into an important international centre for research. Much historic work was done under Thomson. There was C. T. R. Wilson's work with the cloud chamber, for example, described by Rutherford as "the most original apparatus in the whole history of physics", in due course used to photograph the tracks of ions, but owing its invention to the awesome sight of Wilson's own shadow on a pond on Ben Nevis. There were the remarkable researches of such as C. G. Barkla, W. L. Bragg, E. V. Appleton and F. W. Aston, and there was a veritable chain reaction within the world of the physics of the fundamental particle, with Rutherford leaving for Montreal in 1898, Townsend for Oxford in 1900, and R. J. Strutt (Rayleigh's son) for Imperial College in 1902. Thomson failed to hold on to Niels Bohr, who in 1912 chose to follow Rutherford –

by now in Manchester. Thomson was perhaps the most eminent experimental physicist in Britain; but, as Mr Crowther remarks, "the visible spearhead of British physical research had clearly gone to Manchester".

Rutherford took on the mantle of Elijah in 1919 (not 1918, as Mr Crowther states), at an age when Maxwell's life was over. The war had disrupted not only the Cavendish, but the lives of its occupants, many of whom had been reduced to the status of inventor in the war effort. (W. H. Bragg worked on anti-submarine devices, and J. J. Thomson on mines. Thomson and the Cavendish were mainstays of the Board of Invention and Research.) After the war Rutherford could call on most of the collaborators he wanted – although William Kay, with an uncommon sense of values, chose to stay in Manchester, where his wife's friends were. James Chadwick and Charles Ellis went to the Cavendish, Chadwick as Assistant Director of Research. They shared the distinction of having run a small research laboratory with the help of German friends at Ruhleben, where they had been interned throughout the war. In 1921 P. L. Kapitza came from Russia as a research student. It was to a discussion club Kapitza founded that Chadwick first announced the discovery of the neutron. The year was 1932, a vintage year, for then it was that J. D. Cockcroft and E. T. S. Walton first disintegrated nuclei by means of electrically accelerated protons.

Rutherford, as we are told, ran the laboratory like a school of which he was the headmaster, but one with more energy than most of the boys. He died suddenly in 1937, and the Australian-born W. L. Bragg took his place, over the head of E. V. Appleton. Some thought the Cavendish was giving too much attention to nuclear matters, and that a change of direction was needed. Under Bragg, molecular biology and radio astronomy advanced more noticeably. (Under Rutherford, the entire radio research programme was run on £50 a year – a sum which will now buy only two copies of the book from which the statistic is taken.) To Bragg – and to M. F. Perutz – goes the credit for having the good sense to support the research of J. D. Watson and F. H. C. Crick on the double helix, the molecular mechanism by which hereditary characters are transmitted in living organisms. Bragg described theirs as "the most important single scientific discovery of the twentieth century", and so far as biology is concerned there are few who would disagree.

After Bragg, N. F. Mott (1954), and after Mott, A. B. Pippard (1971). As the book nears its end, molecular biology is moved to a separate Institute (1962), and the whole pattern of university research in high energy physics changes, with the establishment of centres such as that at Harwell. With the growing complexity of science, a chronicle like Mr Crowther's must either be highly selective or grow steadily more terse. He opts for the second alternative. The buildings are not ignored, however, whether the small one in the centre of the town, or the vast new complex opened earlier this year. The laboratory assistants get their due. (British industry owes much to them.) But the main approach is biographical, and Mr Crowther draws heavily on relatively few works, most of them in the form of memoirs. He knows his subject well, has a good intuition for the thought-processes of the people about whom he is writing, and describes their scientific achievements accurately, but in easy and general terms. He thus puts to shame a number of contemporary academic historians, who talk of science as a commodity in much the same way as a Sotheby's warehouseman might talk of art. In focusing so closely on the sort of physics done at the Cavendish, he has something of a blind spot for much of the theoretical work being done in Cambridge – as when he rightly lavishes praise on P. A. M. Dirac, but only twice mentions A. S. Eddington, and then casually. Another part of the picture which is slightly blurred is that which concerns the effects of Cavendish-style physics on collegiate self-sufficiency, not to say complacency.

Relations with society at large are not overlooked, although Mr Crowther makes a number of rather forced attempts to prove that the idea of organized research and teaching of experimental physics was taken over from the factory system, that the creation of machine and power industry led to "the substantiation of the atomic theory of matter", and that the Cavendish was, generally speaking, a sort of/factory, a "manufactory for discoveries". (One cannot help recalling a different argument, with similar ingredients, in which Maxwell passed from the uniformity of atoms to the existence of God-the-manufacturer.) No one will cavil at the idea that Cambridge has always – with a time lag of half a century, perhaps – reflected the class-structure of the nation as a whole. There might even be an atom or two of truth in Mr Crowther's claim that American scientists are less given to frivolity than British, but can this really be "a residue of the aristocratic factor" in our scientific tradition? The sociology is dubious, but not drab. The doubtfully humorous Cavendish verse which Mr Crowther sees as a symptom of middle-class tastes is surely a symptom of an international phenomenon, and nothing more than a mark of the simple-minded enthusiasm of overstrung scientists, overjoyed at the discovery that they speak a private language. Or perhaps not. The social axes Mr Crowther has to grind are, admittedly, small and unobtrusive. They leave no offending mark on his invaluable record of a very great institution.

J. L. Heilbron's monograph, *H. C. J. Moseley: The Life and Letters of an English Physicist*, is ostensibly of a different character. Although it is primarily a biography, it is nevertheless the study of the physical sciences at Oxford before the First World War. In the recording of detail it has all the virtues and a few of the vices of scholarly history. And Professor Heilbron's greatest virtue, after his capacity for taking pains, is an ability to present the essential physical principles – which will certainly not be well known to all his readers – in a clear and concise way.

The War brought Harry Moseley's short and brilliant career to an end. He died in the battle for Gallipoli. His earlier researches had occupied little more than three years, and yet in the course of this short time, working on radioactivity and X-ray spectroscopy, Moseley had the great good fortune to hit on a connexion between the ordering of the elements and their atomic structure. (More precisely, he identified the rank of an element in the periodic table with the number of electrons in the constituent atoms.) Moseley's insight into atomic structure led to a number of improvements in the quantum theory of the atom, but Moseley was not to see them.

Professor Heilbron begins the *Life* according to a conventional enough pattern. It seems that Moseley's ancestry was not unscientific: even allowing for some family hyperbole, there was a very colourful Calvinist great-grandfather who earned a prison sentence for practising medicine without any qualification, and a more staid Anglican grandfather who introduced a blend of Christianity and hydrodynamics to London's newly founded King's College. Harry Moseley's father was still more closely involved in the front-page science of his day. He was with Norman Lockyer in Ceylon, later joined the voyage of HMS Challenger as a naturalist, and at length became the Linacre Professor at Oxford. He died before his son was yet four. Not the least part of his legacy was a passion for natural history, shared by his three children.

Moseley's academic career is duly recorded – with an attention to detail which may surprise readers unfamiliar with the Thames Valley culture – from Summer Fields, through Eton, and thence in 1906 to Trinity College, Oxford. Oxford's general hostility to the physical sciences is perhaps best summarized in these words of Frederic Harrison:

> The attempt to foist these special physical researches on Oxford, which still [1910] remains an aristocratic gymnasium and essentially a theological seminary – where not one student in a hundred intends to pursue a scientific profession, where there is little scope for post-graduate study, in a world totally devoted to the "humanities", to Church, to "good society", and sport – this is a sheer waste of labour and money.

It is clear that Moseley, and the best undergraduate physicists were able to rise above the mediocrity of their milieu through junior scientific societies and private reading. His future colleague C. G. Darwin (grandson of Charles Darwin) judged Moseley "without exception or exaggeration" the most brilliant man he had ever met; but with their not unaccustomed perversity, the examiners disagreed, and Moseley had to be reconciled to second-class honours. His salvation was an appointment in 1910 to a post in Manchester, under Rutherford. There Moseley taught and experimented – at first with the beta radiation originating with the decay of radium. Relations between Rutherford and the several members of his research team are perceptively described by Professor

Heilbron, and it is obvious that Moseley was always a trifle uneasy at Rutherford's colonial banter. When, however, in 1912, Moseley and Darwin wished to work with X-rays in order to attempt to reconcile their corpuscular nature (as suggested by W. H. Bragg's work) with their diffraction properties, Rutherford let them have their way, even though the Manchester group was not the best equipped for the work in question. Their findings were highly significant, and in some respects they were running a neck-and-neck race with Bragg at Leeds.

By November 1913 Moseley was back in Oxford, without an official position or financial support, but working in Townsend's Electrical Laboratory. By the end of January 1914 he had done the work for which he is best remembered. It is perhaps noteworthy that in the course of arguing for the best theoretical interpretation of his findings he crossed pens with K. L. F. Lindemann and irritated L. V. de Broglie. At the end of the war, Lindemann was to be given the Oxford Chair vacated by R. B. Clifton. Professor Heilbron speculates that this would have annoyed Moseley immensely, and that, had he lived, the chair might have been his.

Professor Heilbron's biography of Moseley is an unusually deft weave of many styles, varying from neat summaries of Moseley's scientific findings, and of relevant contemporaneous belief, to numerous asides on the Oxford curriculum. (But how many universities *were* teaching relativity to undergraduates before 1910?) There is also a fund of personal detail of a sort which for most young men is preserved only by maiden aunts. Apart from a few with scientific content, the letters (145 in all, including thirty-six from the time of Moseley's recruitment into the army) seldom rise above the commonplace; but time will change that.

Moseley rushed headlong into the firing line with an overwhelming sense of duty that will itself no doubt soon require a footnote of explanation. Writing of his death, Rutherford suggested that using such a man as a subaltern was "economically equivalent to using the Lusitania to carry a pound of butter from Ramsgate to Margate". A less pessimistic note was struck by Niels Bohr, in words used on the title page. Of Rutherford's nuclear atom Bohr argued that it was not taken at all seriously at the time. "There was no mention of it in any place. The great change came from Moseley." Few of Moseley's generation can have had a more enviable epitaph.

17

ON MAKING HISTORY

When these meetings[*] were being planned it was my impression that the aim was to bring together members of the intersection class, so to speak, of the history and philosophy of science. We should, as I thought, be a priesthood advocating – despite our own shortcomings— the marriage of history and philosophy of science, and deploring divorce, unlawful cohabitation, and associations of a kind it used to be polite not to mention. Things are not turning out quite as I expected. Some of our number clearly want to defend the single life, while others put their faith in shotgun weddings. Since in much of what follows it will seem that I am playing the part of spectre at the feast, let me say at once that I do believe in the logical possibility of marriage, as long as it is of the right sort. This morning I shall speak only of the tensions in marriage, and not of the intrinsic worthiness of the partners.

We are all of us concerned with different types of explanation— with their feasibility, their legitimacy, and even with their relative worths. Lest you doubt this last point, let me remind you of W. Stegmüller's characterization of a type of understanding which involves knowing how people react to a situation because one is a person oneself. It is understanding based on "judgements concerning spontaneous human reactions" and is "a typical case where acquiring understanding remains superior to any attempt at historical explanation".[1] Not just *different*, but *superior*. I can sympathise with him at this point. As human beings we are, as he says, competent to evaluate certain reactions as "typically human" without having to resort to generalizations with a hypothetical character. Even historians are human—but they are human enough to know that human beings change from century to century no less than from hour to hour, and that historical explanations can be given of these changes. I shall touch on aspects of this problem again, but I shall do so only in the context of potential liaisons between Moulines' style of philosophy and writings on the history of thermodynamics.

As you know, Dr Moulines writes in the Sneed-Stegmüller tradi-

*The 1978 Pisa Conference on the History and Philosophy of Science.

tion. He looks beyond the statements of the sciences, as such, and beyond the specific theories of the sciences, to the general structures which those theories might share. His ambition is to remove vagueness from discussions of structure, and to do so by axiomatization. He uses the method of defining set-theoretical predicates (which has formerly been used to clarify the structure of individual theories) to clarify meta-theoretical structures. The metaphorical nature of such words as "structure" and "theory-frame" (perhaps better "family") as they are used in general discussion is easily forgotten in a sea of set theory, as is the fact that similar discussions (less strict but more realistic?) have elsewhere had a motivation which was primarily historical. Families, in the historian's world, include infants and old men alike. To forget this when we speak of families of theories is to submerge a vital historical distinction. The discussion of historical change is not new to historians, however loose the language of that discussion. Despite the fact that one of the prime aims of the "rational reconstructions" of recent years has been to show what it is to be a paradigm (in something like one of the Kuhnian senses), these reconstructions have a rather unhistorical look about them, which is not much altered by the inclusion of the parameter t here and there. They usually seem to require, for instance, that the theories which share a structure have shaken down, as it were, so that all who work with them (or a discerning outsider?) are agreed on what it is which allows them to be grouped together (so as to qualify, for instance, as "theories of equilibrium thermodynamics"). This is calculated to make the historian of these theories uneasy. He is likely to say, in effect:

"If you are going to wait for the dust to settle, and look only to those periods of history at which there is a consensus about what constitutes (for instance) thermodynamics, then you will automatically rule out much that the historian finds interesting. You will read history backwards. You will look only for those historical phenomena which fall into your logical scheme, and ignore – even disparage – all else".

We are thinking here, of course, of Moulines' level 3, rather than of such axiomatizations as Joseph Sneed's reconstruction of classical particle mechanics. The historian might also be uneasy with the latter approach, despite its lower level of generality, on the grounds that it picks out elements from quite different historical epochs. (One can hardly "sign" each of the concepts in the axiomatization with a time parameter giving the date of its introduction!) At both levels, the analogy with the empirical sciences is a seductive one, even though most of us have more confidence in the uniformity of the world than in the uniformity of historians, that is, of those very human human beings who provide us (the division of labor being what it is) with our historical data. And at both levels there is the old problem of the *uniqueness* of the historical past, which is what the historian will say he is trying to understand. It is not the same thing, he will say, to understand the evolution of Clausius' thought as to impose upon the overall pattern of his thought a set of axioms (or several sets) which it happens to satisfy. "Don't worry", the axiomatizer might reply, "I am trying, not to catch every aspect of Clausius' thought, but only its general structure, and perhaps also the laws governing its growth. I shall leave to you the task of filling in further historical details". By "historical details" it must be admitted that the philosopher often means whatever he really believes to be trivial—or at best, whatever is relevant to types of explanation other than that he believes he can himself provide. The phrase might refer, for instance, to the precise wording of a quotation, to the color of the ink in which it was first committed to paper, to a scientist's religious life, or to other aspects of his work which are no longer regarded as scientifically reputable. It might even be considered that the logical order of presentation of a theory when it was first announced, or the empirical basis it was then supposed to have, is irrelevant to the deeper historico-philosophical meaning of that theory. Even among philosophers of science there are many shades of autocracy.

I hardly need say that these are topics in the perennial debate between historians and philosophers of history. There are many ways of characterizing the two professional groups, but the philosopher tends to caricature the historian as a man entangled in the minutiae of unarticulated (and hence boring) facts, while the historian sees the structuralist philosopher of history as a naïve and excessively optimistic soul, who has lost his way to the geometry classroom, and whose limited historical experience prevents him from appreciating the infinite complexity of the history of human thought and actions. Of all professional historians, the historian of mathematics and science is likely to be the least hostile to attempts to impose formal structures on history, but even he will be inclined to despair. He will, like all historians, be much concerned with historical change, and he will emphasize (often if only to display his own refined powers of perception) the subtlety and frequency of transitions of thought. Even if he were to grant that a man's thought could be represented axiomatically (with due rejection of historical detail), he might well argue that the effort is not worthwhile—that Einstein's shifting thought, shall we say, might have needed ten axiom systems before breakfast, and that the redundancy they would involve us in, not to mention the sheer pointless effort, cannot possibly be justified.

The historian's deep concern with historical change makes him almost paranoid about anachronism, and the first specific charge he will bring against the structuralist philosopher of science is that of reading history not only backwards, but almost randomly, and of appropriating words so as to impose on them meanings they did not always have. A case in point would be Rankine's "thermodynamic function", later seen by Rankine himself to correspond to Clausius' entropy function. An axiomatization would immediately display (in the absence of some sort of "footnoting") the sameness of the two concepts—even though at the time of its introduction (1854) the role of Rankine's function was very dimly appreciated. No true historian of thermodynamics could be content were this fact lost to view, and most would feel that it is a fact more suitably recorded in prose—or even in verse—than in the profundities of set theory.

The danger of appropriating concepts in an unhistorical way is present at both level 3 and level 2, in Moulines' sense.[3] According to him, for example, all theories qualifying as "thermodynamic" are assumed to satisfy conditions (a) to (g) of his section IV.2. The inspiration for this list of conditions comes from simple equilibrium thermodynamics, for "it is plausible to think that simple equilibrium thermodynamics should be taken as a paradigm for the other, more complex theories in the sense that the essential structure will be the same"[4]. Historically speaking, the word "thermodynamic" seems to have been first used by Thomson in 1840 merely to imply a relation between heat and work. This was before he had given up Carnot's approach and adopted the "dynamical" theory of heat (i.e., when he was using the concept of heat content and not the kinetic theory), and it is ironical that this was all before Thomson was in possession of the concept of entropy. Now Thomson's stance vis-à-vis the kinetic theory is, on Moulines' analysis, of no importance, for Thomson's is a theory which I gather Moulines would place outside thermodynamics (see his (d), which rules that the only thermodynamic concepts are energy, entropy, and the monadic property "is an equilibrium state"). The concept of entropy was not, as I have said, in Thomson's early work; but what if we find that it can be *superimposed* on Thomson's theory (by definitions drawn up in terms of functions which are explicit in his early work) without any crucial emendation of that theory? Is the early Thomson theory to be classified with later "thermodynamic" theories? And if so, has there been any gain in historical insight as a result of this classification? I have a vision of an aeronautics engineer making a rational reconstruction of the Wright brothers' biplane, and finding when he tries it out that it flies at the speed of sound.

I realize that not all of us will wish to give the same answer to my question about historical insight. Those who look for formal qualities in history—that is to say, by and large, those who do not practise history—will argue for formalization, while those who view history as a sequence of unique and unrepeatable events, and who find the richness of the texture of history more important than hidden patterns in it, or than patterns into which (as they will say) it can be forced, will tend to disagree strongly. The split in temperament is as old as history itself; and even in the formalised disciplines there have always been marked differences of opinion as to where the process of generalization should stop. How can we reconcile such fundamental differences of opinion? Not by preaching *in abstracto*, but by being scrupulously severe about matching up our "rational reconstructions" of history to the best historical data. To do less is only one stage removed from the rationalist's *a priori* construction of the physical universe without reference to empirical findings. And this parallel at once underscores the problem of the involvement of theory in data, historical data no less than the data of science. Does this not mean that those responsible for the axiomatization of history must—if they are not to gather their own data—at least become sufficiently familiar with the proclivities of "ordinary" historians to make allowance for the "personal equation", so to say? It seems to me obvious that no historian will even take them seriously unless they do so.

Another problem analogous to one faced by the rationalist metaphysician is that of excessive generality—or, as a critic would say, of vagueness. Dare I draw a parallel between Kant's Principles of the Understanding and Moulines' axioms? As you know, the circumspect Kant spoke of the understanding as both a "power of formulating rules through comparison of appearances" (as in the natural sciences) and as a "lawgiver of nature", conferring upon appearances their conformity to law, and so "making appearances possible" (*KdrV*, A 126–7). Reason was then said to be the judge who compels the witnesses to answer questions which he himself has formulated. For "reason" read "rational reconstructors" and for "witness' read "historian of science". I see no problem here. For Kant, reason has a plan which *must* be followed, not by Nature but by knowledge of Nature. I take it that the axiomatizations which the future might bring at "level 3" will not be deemed to have the same necessity as far as historians of science are concerned. Does Kant not serve as the awful warning that is needed here? His principles had almost no influence on those whose task it has been to extract the (lower order) laws of nature, and in fact there are many who would maintain that Kant's views on space and time have acutally inhibited creativity in corresponding branches of mathematics and physics. In other words, when his philosophy was specific enough to interest the scientist it stood or fell on its scientific merits, and when it was general enough to be safe it was *uninteresting* to the scientist. I must apologise if all this sounds too much like a morality tale, but the analogy with Mouline's axioms seems to me a close one. I simply doubt whether his axioms are ever likely to influence or interest most historians. As he himself admits, "On the whole not many interesting conclusions can be drawn from the axioms of an operational basis because of their generality".[5] If a philosopher can say this, no historian is likely to disagree.

Another objection likely to be raised by the historian concerns the *insensitivity* of the axiomatizations (at all levels), which might even escape the worst forms of anachronism and yet give consistency to a theory which originally lacked it. It would be a bold philosopher who attempted an *irrational* reconstruction of the history of scientific development, but this might be the honest thing to do. Very occasionally the scientific debate itself forces inconsistencies on our attention. (Take, for instance, Poincaré's theorem of quasi-periodicity with which Zermelo hoped to demolish statistical mechanics, whereas Einstein used a very similar procedure to draw a wholly different conclusion. Whether or not there was any strict logical conflict between the two parties to the dispute, I can see no axiomatization on the horizon capable of doing justice to the situation.) For the most

part, historians and philosophers of modern science are the embodi-
ment of kindness and charity towards scientists, and prefer to give
them the benefit of the doubt as regards this question of consistency
(or perhaps I should say "avoidance of inconsistency"). Perhaps this
is only a sign that inconsistency is consistent neither with hero-
worship on the one hand nor with a satisfying axiomatization on the
other.

Another quality most of us will demand in formal representations
of the historical process is conciseness. It is hard to contemplate
those axiomatizations of the future which far exceed in length the raw
data which they are held to encompass; which lead us to conclusions
(with exactness and confidence, it is true!) that had been proposed
long before, albeit in the form of untidy sentences; and which indeed
were framed so as to lead to those conclusions. "But", you can hear
the future historian saying, "I knew all along that psychological
factors play an important part in the genesis of scientific theories".
"No", some future structuralist might reply, "you only *suspected* that
the psi-factor could be introduced into meta-history. I can now show
quite precisely how your suspicions were, broadly speaking, along the
right lines".

This is clearly not enough. The structuralist's scheme must do more
than give a formal guarantee of the coherence of historiographical
principles already in vogue. Like the empirical sciences, it must
reveal the unexpected if it is to be taken seriously. I am not qualified
to comment on the present state of "structuralist" philosophy of
science, but I confess to having been disappointed on this score. In
the explication of so-called "normal" science, for example, we have
an explanation of why it is reasonable to treat theories as safe from
refutation at the hands of contrary experience. But was this point not
being made by Poincaré, and Eddington, and others, more than half a
century ago? Again, according to Stegmüller, normal scientific progress is definable in terms of "cumulative development of Kuhn-
theories", while (as he goes on to say) "in progressive scientific
revolutions 'the displaced theory can be partially and approximately
imbedded into the supplanting theory'".[6] And yet discussions of
"approximative imbedding in the supplanting theory" are not rare in
historical and physical writing. Yet again, when we recall recent
discussions of the question as to whether the decision to reject a
theory is simultaneously a decision to adopt a *new* theory, we tend

not to be astonished, but for a different reason. No longer is it the case that we have heard the answer before—now we can see that there are many possible answers. For, however tightly we define "adhering to a theory", it will never be positively illogical for individuals (or groups of individuals) simply to walk away from the entire subject. There are numerous historical examples of the "emigration syndrome", in fact. Of course, statistically speaking it might prove to be insignificant, and an analysis in terms of community inertia might be much more useful. I do not wish to be sidetracked for the time being into considering the adequacy of current "structuralist" concepts, but merely to draw attention to the present unreadiness of the structuralist analyses for that continuing search for refutation which is usually taken to be a mark of the critical spirit.

The suspicious historian is likely to be delighted at such a state of affairs. We can imagine him drawing an analogy between the astronomer, unable to influence the properties of the distant celestial objects he studies, and the structuralist, unable to affect the unique flow of history. But the analogy with the empirical sciences (an analogy I have used perhaps too often) is a dangerous one when we are concerned with a subject which is largely sociological, and therefore, as it might be crudely expressed, "ontologically poor". There *is* an analogy with the cosmology of the nineteenth century here, as it happens—namely with a cosmology in which no one was quite sure how the nebulae were to be classified. There was no similar disagreement about heat baths, steam boilers, and sawdust. As befits a set of concepts which function at philosophical and sociological levels, "progress", "community", "paradigm", "scientific revolution", even "theory" and "person", are not readily interpreted. Disagreement over the pre-formalised use of such basic concepts, combined with the coarseness of the mesh of the axiomatizations incorporating them, might put the possibility of "historical test" far in the future, but does not rule it out of court. (I suppose some philosophers will say that as soon as structuralism finds itself empirically tested, it will cease to be philosophy. *Per contra*, there will be those historians who use its key phrases because they are useful in structuring historical material, and possibly others who do so because those key phrases have a certain cachet in the academic world.) If I seem skeptical of the possibility of historical test, it is not as a matter of principle. It is rather that it seems to me that much more groundwork remains to be done in liaison with historians of science.

This must all sound both obvious and pretentious. I know better than anyone how pretentious it is, for I have neither a good grasp of the work that has been done already nor a programme for the future. I must be content with a few final disjointed ideas. It seems to me that it is unfortunate that attempts are being made to "rationally reconstruct" two things at once. The new movement seems to have begun with attempts to tighten up physics (as Euclid, in the old story, tightened up the rules of the carpenters), or to put physics on a sound operational footing. For these purposes there is nothing illogical about defining heat (for example) in terms of entropy, which was historically second on the scene. The temptation was then to move to Moulines' level 3, where the entities are whole theories. The step was original and intrinsically interesting. However, complete theories are *not* the fundamental entities of history of science, as one is tempted to believe when hypnotized by the delights of axiomatizing restricted physical theories. Believing that theories are basic leads to a disregard for the early and tentative phases of science. By laying down criteria for a theory to qualify as a theory of thermodynamics, we *force* upon the historian an abrupt transition from the period before there were any "truly thermodynamic" ideas to the period when there were theories which can plausibly be said to have satisfied the criteria. It seems that scientific revolutions and periods of normal science are merely reflections of our desire to pack history into a few neat boxes—and to pretend that what cannot be boxed does not really matter.

The attempt to extend the axiomatization of physical theories to an analysis of history, seen only as a history of theories, has other unfortunate consequences. To theories is assigned the job of uniting individual scientists into the same tradition. (There may be other ways of grouping scientists together, allowing more scope for heresy. This fact is beside the point.) I gather that some writers have been led from this position to one in which the concept of a scientific community is taken as basic. In neither case is the historian of ideas going to be content. Not all history is sociology. Many historians of ideas— I would say almost all—prefer to start with the individual; with the individual responsible for innovation; with the individual whose beliefs are more easily grasped than those of the group; with the individual who holds a pen; with the individual whose ideas may lie fallow for generations, or who may be a century behind his time, and yet still historically significant; with the individual whose personal qualities were such that his beliefs were almost ritually acknowledged as valid. Even at the level of individuals, any attempt to formulate a calculus of belief is likely to run into difficulties, unless it is left in terms so general as to be without useful historical application. Different individuals, especially from different historical periods, are prone to offer different sorts of reason for holding to their beliefs. Such difficulties are not new to sociology, where the thoughts of individuals are customarily smoothed out by approximative methods. Few historians will accept the loss of the fine-structure of history which such methods entail. One might compare the recalcitrant historian with a painter who objects to Euclid on the grounds of his failure to take red lines and blue lines, irregular lines and blurred lines, into account. I cannot see that Euclid would gain anything by affirming and reaffirming that painting is nothing but geometry, or even that it is "essentially" geometry. He might teach the painter something; but in the end, the two must go their separate ways.

My analogy with Euclid and the painter is of course more appropriate to the bad old days when philosophers were chiefly concerned with a "statics" of scientific theories than to our own enlightened times, when the kinetics of theories adds an extra philosophical dimension. The advance is a significant one, but it is not likely to affect the historian very directly. The logic of science, whether a statics or a kinetics, can in turn survive without the historian's help, as long as its practitioners confine their attention to the realm of the possible. But if the two subjects are to interact, it seems to me that the greater influence by far must be from history to philosophy of science. It is easier to act the part of Cassandra than to give precise details of the influences we are to expect, but I can at least give an example of what I have in mind. To take Rankine again: here we have the historical phenomenon of two largely independent traditions—his and the Clausius-Thomson tradition—being developed independently. Broadly speaking, we can say that Clausius and Thomson derived Carnot's law (of the efficiency of heat engines) from the principle that heat cannot, by itself, pass from a cold to a hot body. Rankine, on the other hand, who was for long ignorant of the details of Thomson's proof, and who disliked that given by Clausius, derived the law instead from his hypothesis of molecular vortices (1851, etc.).[7] It was almost twenty years before Rankine abandoned his vortex hypothesis (essentially one which required that absolute temperature is proportional to the square of the vortical velocity of atomic atmospheres around their nuclei) and acknowledged the more conventional kinetic theory of heat. No "sociologist of knowledge" can define away the intellectual and practical influence of Rankine, who must somehow be fitted into the categories of any *historically* acceptable analysis of the kinetics of theories, simultaneously with Clausius and Thomson. Faced with the problem, the structuralist philosopher of science, new style, might incorporate into his schemes a branching (and perhaps a rejoining of branches) from revolutionary nodes, with consequent simultaneous traditions of "normal" (or "extraordinary") science. No doubt there will then be repercussions on the discussion of the concept of progress. No doubt there will be repercussions on the very notion of scientific theory, and scientific community. The idea of divergent traditions seems to fit perfectly well, in fact, with Moulines' idea that a theory of thermodynamics is characterised by (among other things) a parameter *not* presupposing thermodynamics. He would, therefore, presumably distinguish between Rankine and Thomson on the grounds of their different choice of non-thermodynamic parameter. Whatever the possibilities for combining these ideas, they seem to me far stronger than the possibility of a significant feedback of anything more than a few simple philosophical categories and concepts into the historian's world. I hope that history will prove me wrong!

Any such feedback, from the philosopher's total system, is likely to be delayed and indirect. Even so, it is potentially extremely valuable. I am not very sanguine about the axiomatization of painting, but painter and critic alike will not be the poorer for Euclid's structures. Much the same goes for history and the categories of the structuralist philosopher of science. It would be unrealistic, however, to expect too much. The artist is not likely to have much use for the nine-point theorem; and if some zealous philosopher should one day axiomatize a Richter scale of scientific upheaval, he would not only take away most of the fun of courses on Scientific Revolution (attended largely, it seems, by students attracted by the simplest of political analogies), he would almost certainly be ignored by historians. This would not entirely be because historians are simpleminded. It would in part be a consequence of the fact—often overlooked, it seems, in the recent debate over the nature of scientific revolutions—that the historian's categories are often taken from systems of thought within the very societies he is studying. Revolution, renaissance, progressive—the historian and the philosopher might be dissatisfied with, criticize, and revise ways in which these concepts were, at the very time of their invention, applied to contemporaneous events; but it would be wrong to ignore those past applications, which are themselves an ingredient of history. Even in history there is interaction between the observer and the observed.

NOTES

[1] 'Accidental (non-substantial) theory change and theory dislodgement ... [etc.]', *Erkenntnis* **10** (1976), 165.
[2] 'An example of a theory-frame: equilibrium thermodynamics', *op.cit.*, n. 6 below.
[3] *Ibid.*, p. 217.
[4] *Ibid.*, p. 219.
[5] *Ibid.*, p. 233.
[6] *Probabilistic Thinking, Thermodynamics, and the Interaction of the History and Philosophy of Science*, ed. J. Hintikka *et al.*, Dordrecht, 1980.
7 For a comprehensive treatment of Rankine's thermodynamical ideas see the D.Phil. thesis by K.R. Hutchison (Univ. of Oxford, 1975).

18

ONE OF OUR GALLEYS IS MISSING

Diplomatic Fictions in the Editing
of an International Journal

Editors of scholarly journals, like doctors and priests, are custodians of a great many confidences. This is not to suggest that much of what they could tell the world would ever make the pages of *Stern*, *Paris Match*, or even *Private Eye*; but scholars do tend to be sensitive creatures, more afraid of being found out in an unverified reference or a mistaken algebraic sign than of being discovered, shall we say, running off with another man's wife – or another wife's man. Like it or not, an editor is in an unusually good position to discover what Dr A thinks of the work of Professor B, and conversely, but is obliged to maintain a diplomatic silence until such a time as nobody really cares any more. I must make this point at the outset, if only to explain why I shall speak more of dull general principles than of named individuals – except, of course, when I sing the praises of those who merit the song.

Although I shall speak mainly from my own experience, it would be wrong on this occasion to overlook the part played by Aldo Mieli and those others who laid the foundations and established the general character of the *Archives internationales d'histoire des sciences*. The first volume of his *Archivio di Storia della Scienza* spanned the years 1919 and 1920. By 1927, when Mieli used the journal to make an appeal for an international organisation of the history of science – an event from which the Academy traces its origins – the main title had become *Archeion*, and the subtitle was given in no fewer than five languages. The virtues of a short title are no less obvious to writers of footnotes than to librarians, although punctilious librarians might well feel that five subtitles are five too many. At all events, the journal was launched with enough vigour to carry it through the difficult pre-war years, and the international committee of the history of science likewise. Then, as you know, political circumstances forced Mieli to leave Rome, first for Paris and then for Argentina, from which country *Archeion* was published for a short time (1940-1943).

At the fifth International Congress of the History of Science, held in Lausanne in October 1947, the Academy was resuscitated, following a tragic period of war that was starkly symbolized by the death of Hélène Metzger in the Auschwitz Camp. She had been the Academy's administrator and treasurer for thirteen years. The war was now over, if not the period of austerity that followed it, and in November of 1947 there appeared the first issue of the *Archives internationales d'histoire des sciences*, which bore the subtitle *Nouvelle série d'Archeion*. Mieli remained as nominal *Directeur*, and responsibility for editing the journal passed through a succession of hands: Pierre Brunet, Pierre Sergescu, Jean Pelseneer, Maurice Daumas, Pierre Costabel, and Mirko Grmek. This is not the place to recount the somewhat convoluted publishing history of the journal, or the ambiguous relations that then subsisted between the Academy and the International Union [1]. Suffice it to say that the Parisian epoch came to an end when the publishing house of Hermann et Cie decided to tear up our contract unilaterally at the end of 1970.

It was agreed at the Moscow meetings of the Academy and the Union, held in 1971, that the Academy should be solely responsible for the future of the *Archives*, and it has to be said that many of those then present did not honestly believe that the *Archives* had any future at all. A generous grant obtained by Willy Hartner, President from 1971 to 1977, made this possible. The grant was from an old friend of his, the Frankfurt industrialist Ernst Teves, and it was fitting that our publisher should be Franz Steiner Verlag of the neighboring town of Wiesbaden, a firm run by another aquaintance of theirs, Herr Jost.

It is always considerably easier to work with a tightly knit local group than with colleagues a week or more away by mail. We now settled on a triangular arrangement, with Dr Yas Maeyama and Dr Walter Saltzer as editorial intermediaries, able to sort out problems with Frau Inge Dittrich in Wiesbaden — separated from Frankfurt by only a short stretch of the most dangerous *Autobahn* in Germany. In those days, production was the least of our worries. Re-establishing the *Archives* was hardly different from founding a new journal, for commercial momentum had been lost, and it was difficult to persuade most of our colleagues that the phoenix had indeed risen from the ashes. Everywhere there was insistence that our international formula was a guarantee of disaster.

---

[1] For further details, see my edition of the *Directory of the International Academy of the History of Science* (Roma: Istituto della Enciclopedia Italiana, 1985), 24-25.

The Academy is a body with a limited membership, and as such lacks the firm financial basis of journals belonging to the larger national societies, which if they are selective in their membership are likely to have state backing, and if not, are usually sustained by an open – and therefore potentially large – membership. There can be no doubt that the problem of language, too, has always been a difficult one. As those who knew Willy Hartner will remember, his linguistic abilities were exceptional, and when, early in 1973, he wrote to, or telephoned, six or seven of those he called the 'big shots' in our subject, to ask them for articles, he addressed them in their own languages. The result was that most of them promised articles. Of course many of the promises failed to materialize, but one cannot have everything.

We had many pious speeches in favour of a multi-language journal, without much thought of what that implies. It was sad to hear some of them turned into unmistakable murmers of disapproval when I published an article on the translation of Dutch medical works into Japanese, an article liberally quoting both of those beautiful languages. Should any other editor wish to repeat the experiment, may I say that working with the Japanese printer was a pleasure of a sort rare since the days of Gutenberg. I remember his exquisite presswork, which we reproduced of course photographically, and I remember how, on opening the volume with the article in question, the page fell out. It was in a glued binding, one that in English is euphemistically called a 'perfect binding'. It somehow seemed to be symbolic of the translation of Dutch books into Japanese.

This was not a typical experience, of course. We had settled on English, French, German, Italian, Russian, and Spanish, as our main languages, and it was with some guilt that I saw my own language insidiously taking over, as scholars with other native languages set down their thoughts in English, the *lingua franca* of the time. Correcting proofs is usually child's play by comparison with turning this kind of English into a plausible form – and my colleagues who have done the service for French and German have said the same there. Perhaps one day the *lingua franca* will be Japanese or Chinese, but when that day comes I am sure they will not need to send to Europe for printers[2].

Getting our first number into print occupied more than a year and a half – it carried the date of June 1974 on the cover, but cover-dates are another diplomatic convention that only the initiated know how to interpret. Of course the date

---

[2] Since I have mentioned proofs, and Gutenberg, let me comment on the astonishing accuracy of the German printers, and their peculiarly limited view of printing styles. German printwork is wonderfully neat and tidy; but ask for Bembo – ask for it even in Italy – and from the printer's stare you will learn what it feels like to be a museum exhibit.

is just the variable aspect of the outer cover. When it came to designing the cover as a whole, I had to find one that marked us out from other journals. I asked colleagues for their opinion on the dark green colour we finally chose, a colour more redolent of evergreen northern forests than of the Costa del Sol. Their answers fairly accurately separated the puritans from the rest. "Un peu triste" was a characteristically diplomatic Mediterranean reply. The outside is not the only sad thing about the cover of a journal, however. As everyone in the academic community knows, journals almost invariably bristle with the illustrious names of those who have agreed to lend their services; but as every editor knows, the fiction usually ends where it begins. The fault is not necessarily that of the individuals themselves – who protest that they will do anything if only they are asked – but is a question of the difficulties of triangular communication over large distances. It is no doubt chiefly the editor's fault if he treats those names only as cachets of glory, rather like Napoleon's signature on a bottle of cognac.

The same difficulties over the delegation of responsibility arise at an editor's local level. Secretarial help can be had for a price, but it is not easy to find a secretary who knows much about professional competencies in the history of science. An ideal arrangement, it seems to me, would be to have the services of a couple of post-graduate research historians of science, each assisting the editor for a day or so a week, in the same town, and with access to a good institute library. They would find reviewers, pack and post off review copies, and send reminders at regular intervals to recalcitrant reviewers, whom they would often know professionally. They would of course be immaculate typists, and would correct proofs with their inside knowledge of the subject and a working knowledge of five or six languages: and all this they would do in return for a suitably slender stipend.

This was my pipe dream, as with the help of a secretary, or more often my wife, I parcelled up books and walked down to the Kingston Road Post Office with them, knowing that many reviewers would be utterly shameless about ever sending in the review. We sent off about 850 books to reviewers over eleven years, and of those about 200 were sunk without trace. This is an appalling statistic about which there is much to be said. First, it has a little to do with the maxim 'out of sight, out of mind'; and yet people close at hand are often just as bad as those living at the Antipodes. An editor might have to write three or four times to potential reviewers, before finding someone who is prepared to take on a book. At a later stage, reminders to the reviewer who accepted the book will often result in solemn promises four or five times over. An editor ought at this stage to ask for the return of the book, but on the few occasions when I did so, the book rather than the review came back – on one occasion a completely different book came back – and the whole charade had to begin again. Life is short and postage is expensive. I have one reviewer who sends me regular — and unsolicited — messages that he has not forgotten the review of a certain book. Perhaps he does not realise that he has had it for ten years.

None of this will come as a surprise to those for whom time is at a premium, and who are busy, perhaps, with a potentially immortal book or article of their own. Excuses are usually convincing, and some can make an editor feel positively guilty for broaching the subject, as when the reviewer is too busy doing good deeds, like preventing nuclear war, to answer letters or review the accepted book. And then there are those who treat publishers as parasites on the beautiful body of our subject, and who regard review copies as akin to free samples of washing powder. There are people who will ask you to get them a review copy from the publisher, and even then never review it. But where, without publishers, would our profession be? It might be going too far to say that in publishers historians live, and move, and have their being; but without them, certainly, most of us would just be crying in the wilderness.

All in all, a single review — from the arrival of the book to the final printing — might cost the editor and publisher together between six and say twelve mailings, in some cases for a half a dozen miserable lines of prose, dashed off between committee meetings. You might be sent a five-line review for a £ 50 book. You might as well have written it yourself. Why does an editor bother, in view of the fact that running a journal without a review section would be so very much easier? A part of the answer is that reviews represent the most important forum for critical dialogue in our peculiar world of paper and ink. When, at the very outset, I asked a dozen friends what they most liked about the old *Archives*, almost without exception they named the reviews section. For long I tried hard to get lengthy and responsible reviews of only the better books. Here I came up against a sad fact of life: the better the publishers, the less likely you are to get their books for review in an international journal. What need they care about the fact that their products are admired in outer Mongolia, Ethiopia, or Nicaragua? You therefore begin to accept works of unknown publishers in unlikely places, and since their books are less likely to have gone to the large national journals, this is no bad thing.

It is all too easy for those in the large centres of civilization to forget how difficult it is for those who are not so privileged to obtain recognition for their work, let alone engage in any meaningful form of historical discussion[3]. It is

---

[3] To my great regret we have not been in a position to exchange journals with the many organisations in disadvantaged countries who have asked us to do so. Philanthropists more concerned with merit than glory might bear this problem in mind.

also one of the oddities of historical scholarship that debate is often regarded as somewhat ungentlemanly, as though historical truth, properly expounded, is absolute. In philosophy of science, to take the case of a neighbouring subject, there are no such inhibitions. I have often tried to persuade colleagues to air their views on controversial matters in the column 'Questions, réponses et répliques', but when invited to write in that, most of them preferred to nurse a private grudge. There are notable exceptions, worthy of the more vigorous style of the seventeenth and eighteenth centuries, and the danger of letting vigour turn into acrimony is a small price to pay for the proof that the parties to the dispute are committed to what they say, and do not consider themselves to be simply turning out mass-produced truths on the production line of history, every one guaranteed perfect.

Book reviews may have a similar function. No doubt one day university chairs will be established in the theory of book reviewing, that is, if the subject is not first annexed by the polemologists. The sentimentalist school will cite Walter Savage Landor: "He who first praises a good book becomingly is next in merit to the author". The school of realism will echo the words of the Reverend Sydney Smith, who said that he never read a book before reviewing it, for "it prejudices a man so". There is room for both approaches to the art, although I dare say that in our world, where reviewers are so busy writing books of their own, the realists outnumber the rest. The Sydney Smith approach, contrary to popular belief, does not usually result in hypercritical reviews. An editor will be deeply suspicious of a reviewer who heaps too much praise on a book, and one of the most likely reasons for this is that the work has simply not been read, and that the reviewer judges discretion to be the better part of valour. Other reasons are that it is the work of a friend; or of a potential patron [4]. One should not exclude the possibility that it actually is a good book.

There are many ways of judging the care with which a review has been written. One sure sign is when a reviewer, asked for 800 words, writes a long letter explaining why there are 813. Even odder is the reviewer who writes a letter twice as long as the review you hoped for, explaining what is wrong with the book, and why he cannot therefore review it. If a book has not been read carefully, this might be because it is hard reading, and this in turn because it is written in an exotic language, which might of course mean *any* language – for even French is exotic to the Eskimo. It is a sad fact of life that the most expert reviewers are those well on in their careers, and that they do not usually accept

---

[4] I employed one secretary who had an uncanny knack of somehow from time to time inviting authors to review their own books, but as far as I know they all refused.

books for review. Reviewing is time-consuming, and who on a salary of $ 50 000 per annum cares about an epoch-making ten-dollar paperback? Who in his nineties wants to spare a week or two putting the world to rights? The younger the scholar, the more likely the acceptance, but also the more probable it is that fur will fly. If publishers were to start taking out insurance against the accident of the savage review, insurance companies would soon establish tables exactly correlated with tables of motor insurance, with maximum risk from the class of male drivers in their twenties. There is, however, one small social group that defies this rule, namely of raw novices who will send you long lists of books that they are prepared to immortalize. They always write insipid but essentially kind-hearted reviews; but then, they probably drive a Topolino, or a Citroën Deux Chevaux.

There is one ever present hazard to the editor who wants to avoid the flying hub-caps in the fast lane, namely that he will send off a book for review to a scholar with an axe to grind against the author. One is expected to know by heart not only Who's Who, but Who Hates Whom. This makes life exceedingly difficult, for there are important branches of the history of science competently practised by, shall we say, only five people, with perhaps twenty editors trying desperately to place as many review copies. The race is to the fastest, and the prize even then is often a hornet's nest of professional hostility.

One of the consequences of the fact that we inhabit a tower of Babel is that it is very difficult to get books in some major languages – Russian, Chinese, and Japanese, for instance – reviewed outside their native countries. From an international perspective, however, there are more insidious problems than those of language, entirely analogous to the problem of inviting a person's greatest rival to review his or her book. Should a work written within one ideology be reviewed by the adherent of a rival ideology, or should an editor try to find a *sympathetic* critic? One has to be a saint to be entirely disengaged from such potential conflicts – but there you see how deep the problem goes, for in some ideologies saintliness is incompatible with the legitimate workings of the dialectical process. I never found a simple solution to the problem, but I will venture a pragmatic maxim that works in most cases, namely, that in the academic world personal vanity is more powerful than ideology in nine cases out of ten. It is the tenth case that is really dangerous.

The other side of the same coin is the thorny problem of refereeing articles submitted for publication. One has a moral duty to send them to scholars whose interests are close to those of the writer. And if A reviews B's book, or referees B's article, should not B be given a chance to do the same for A's book or article? One must try to read between the lines of any criticism, and there is nothing here peculiar to an international journal, except inasmuch as the criteria used by different national groups vary so much. Take language, for example. Sprinkle a few words of Chinese in an English article on Jesuit astronomy, and in the eyes of members of your own faculty your scholarly reputation will probably rise. A referee in China, however, is not likely to be overawed in the same way. Scholars offer hostages to fortune when editing work in other languages than their own, and this to a far greater extent than they ever did in the past – a principle well illustrated in a number of articles as well as book reviews I have published. At the end of the day, one requires a certain instinct, for no-one is in a position to decide personally on more than a small proportion of relevant linguistic and technical matters. The question is whether criticism is ingenuous, or whether it has some hidden and fraudulent motive. I must admit that my instincts have occasionally let me down, and yet speaking generally, referees are generous to a fault with their time, and we all owe a great deal to those who give their services anonymously.

When a reviewer criticises an author then the author should have the right of reply. In the case of a rejected article, the case is entirely different. There are, alas, some writers who are unable to retire gracefully from the fray. Those who sell cars know better than to malign all who refuse to buy them. They are wise enough to move on – but then, they live in a less confined world. There are some notable instances of scholars convinced that there is an international conspiracy against them. To them, the editor is merely the visible part of the iceberg of criticism, and so must offer endless justification for any rejection, at the risk of being judged a member of the hostile Establishment. As against this situation, there is another almost as tedious, namely the case of the writer who, having been offered devastating criticism in perhaps too kind-hearted a way, adopts one as an intimate and eternal friend, bombarding one with a succession of papers for similar treatment. No doubt morally speaking this is preferable to the charge of personal bias; but many of us have other things to do.

Prejudice and bias are all the more insidious in an international situation, for there they are all the more difficult to appreciate. Oxford might not be Cambridge, or Groningen Amsterdam, but in both cases people know more or less how, in the other place, the cat will jump. In the history of science there are enormous differences between predominant national styles, not to say intellectual standards. The history of science has to a very large extent been the creation of people who were primarily scientists, but whose intellectual awareness extended beyond their immediate circumstances. In many countries, the subject is still in the hands of this group, whose professionalism is unquestioned, but is not that of the professional historian. On the other hand, there are many countries where academic history as a whole is dominated by political, constitutional, and social historians, who shudder at the thought of introducing anything that savours of the history of ideas – which they suppose to be narrow, as though the history of constitutions and governments was all-embracing. In such countries – and Britain is one of them – the great magnet of employment causes many historians of science to perform the most astonishing of intellectual acrobatics, to prove that the history of science is really just one small aspect of that particular 'larger' history. In other places, the history of science is the history of philosophy in a traditional and extremely narrow sense. Then again, in other outposts of civilization a more expansive view is taken, and there the entirely legitimate hope still lingers on that history and philosophy of science belong together. In countries with a prevailing Marxist philosophy there has been a natural tendency to place the main emphasis on technology, on 'praxis' – although it has to be said that the Marxist world is a place extensive enough to be able to show some of the best practitioners in other genres. My time as editor of the *Archives* largely coincided with the great sociology-of-science bandwagon, but this was no respecter of national frontiers. And so I could go on; but I have no intention of providing a historian's guide to the history of science. My point is simply this: no editor of an international journal who tries to impose any one style on his journal is behaving impartially.

One of the consequences of the equilibrium I wish to advocate is that the journal can never reveal that strong philosophy that makes for a truly 'great' journal, in the opinion of many. I would go further, and say that in order to promote the subject in odd corners of the globe, or even to promote odd corners of the subject, an editor may be obliged to lower standards quite consciously. There have been times when I have done the same for no better reason than to prevent a rift with colleagues who have recommended less than perfect material. No doubt this will seem scandalous to those with a monolithic view of truth and excellence; but such a view is a luxury, common only in tightly bound and self-confident communities. (I could name several, but so, I am sure, could you.) From the very beginning, we drew the boundaries of our subject and the style of its representation very broadly, and as far as I know, no-one has objected very vociferously to articles on the history of music, or logic, or to the experiment in which we printed more or less verbatim a lengthy interview (with Eugene Wigner) [5].

The *Archives* publishes obituary notices, and here the problems are not so very different from those already raised, except that there is no longer the right of reply to tacit criticism. This is not a problem peculiar to our subject, of course. As I soon discovered, people do not like writing about their contempo-

---

[5] The *Archives* include the history of medicine only inasmuch as it has a markedly extra-medical scientific element. This is less a point of principle than a consequence of the fact that historians of medicine have a separate corporate existence, with their own academy, for instance.

raries – perhaps since the act of doing so brings with it an intimation of mortality. The moral is that scholars who worry about their image should cultivate the friendship of the young. It is a mistake to think that close colleagues are always the best people to approach. An invaluable piece of advice to editors – of course it is not infallible – is that they approach the bereaved family for ideas as to a suitable writer. It is certainly dangerous to leave the case of your own obituary in the hands of eager young researchers who never knew you. I remember once being asked to write an obituary notice (not for the *Archives*) on someone I knew only slightly. I made an arrangement to meet his widow, to get a few essential biographical details, and was rewarded with a long and fascinating account of his many love affairs [6]. I was about to say that this was not a peculiarly international problem, and yet in his case it was indeed so. The international movement of scholars, which we are in a sense celebrating this week, has a lot to answer for.

The fate of the *Archives* has always rested heavily on that of a relatively small number of individuals. Financial problems concerning our relations with our Wiesbaden publisher were already in sight when in 1981, by a sad blow, we lost both Willy Hartner and Ernst Teves within a space of weeks. Our relations with Steiner Verlag were always good, but we were in a situation known in England as 'contract publishing', where the publisher is not especially concerned to put much effort into advertising the resulting publication to the world. The Academy could not in any case afford to renew its contract, and was obliged once again to look elsewhere. To our inestimable good fortune our member and present host Vincenzo Cappelletti enabled us to bring our *Archives* back to Rome, the city where Aldo Mieli had first set its wheels in motion. The first number to appear from the Istituto della Enciclopedia Italiana carries the date of June 1982. I will not recount the names of all who are concerned with the Roman redaction of the *Archives* – you will of course find them all inside its cover – but I should mention Alberto Postigliola, whose attention to the copy we provide has always been fastidious. I say 'we': at the end of 1984 I handed over the editorship to Robert Halleux, an excellent scholar and organizer – his persuasive powers with the Belgian Ministry of Education leave me breathless. One of the first tasks he undertook was to produce, with the help of his wife, and Alberto Postigliola and others, a most comprehensive index to the *Archives*, running from 1968 to 1984. [7] A change in editorship portends a fresh outlook on both the form and the content of our *Archives*, and perhaps even a new style in apologetics.

[6] It seems that I have here qualified my rule about consulting the family; but it is an excellent rule in general.

[7] See vol. 34 (1984), 329-532.

The compilation of the index I mentioned – it is more than two hundred pages in extent – should remind us of that all too often invisible force in academic collaboration, the partner in marriage, who certainly never wished for, much less contracted for, the generally tedious business of an editor. If my own wife were here she would no doubt quote to Robert Halleux the words of the king of Israel to Ben-hadad, king of Syria: "Let not him that girdeth on his harness boast himself as he that putteth it off"[8].

The new index unfortunately illustrates another kind of force, this time one of those random perturbations that seem to be inevitable in action-at-a-distance. The cards for one whole section of the index were irretrievably lost, when Italian customs officials opened the packet in its transit to Rome. Another version was compiled with much difficulty, and we can only hope that our readers will forgive the delay in publication. (I sometimes feel that the only readers remaining to us are the forgiving ones.) Perhaps, one day, the cards will turn up, and history will repeat itself, in a manner of speaking. I am thinking of an incident in the Steiner epoch, when an entire batch of our old galley proofs was found by the printer. I had corrected them, but he had overlooked them, and for a couple of years they had lain unnoticed. It never occurred to me to verify that all galley proofs had actually been incorporated in the final printed copy; but what was really astonishing to me was the fact that the galleys contained numerous reviews, and not one of the reviewers wrote to ask about the fate of what he or she had written. One reviewer eventually confessed to me that he had noticed the omission, but assumed that I had rejected his review on the grounds that it was not of a high enough quality. I felt much as Francis of Assisi might have felt had someone mistaken him for Jenghiz Khan.

The moral of all this is that editors may not only have more influence than they merit, but much more than they recognize. Just as Molière's Monsieur Jourdain had been speaking prose all his life without recognizing the fact, so an editor is often likely to be conveying a message without quite knowing what the message is. For example, it is inevitable that an editor, who after all can only know a small section of the academic community, will tend to make the journal a focus for his own interests and those of his acquaintances. This might even be a good thing in some instances, but it is emphatically not so in our case, where the journal represents an international community with a wide range of interests and a desperate need for a channel of communication and debate. The case of UNESCO well illustrates how, over the last forty years, naïve optimism on this score can turn to dangerous cynicism. Naïveté and

---

[8] I Kings 20:11.

cynicism are in one respect remarkably similar moods, for both offer a means to self-protection, and insulation against disappointment. From cynicism, though, it is a short step to throwing in the sponge. It is better by far to be naïve, and to go on trying – trying, that is, to keep open the channel conceived by the naïve Mieli and his naïve friends nearly sixty years ago. Our subject can only benefit, and the world is not likely to be any the worse as a result.

INDEX

Notes to the end of chapters are indexed only to the extent that matter is not referred to in the chapter itself. Harriot's text on the calendar is indexed only through its introduction.